Springer Series in Advanced Manufacturing

Series Editor

Professor D. T. Pham
Intelligent Systems Laboratory
WDA Centre of Enterprise in Manufacturing Engineering
University of Wales Cardiff
PO Box 688
Newport Road
Cardiff
CF2 3ET
UK

Other titles in this series

Assembly Line Design
B. Rekiek and A. Delchambre

Advances in Design
H.A. ElMaraghy and W.H. ElMaraghy (Eds.)

*Effective Resource Management in Manufacturing Systems:
Optimization Algorithms in Production Planning*
M. Caramia and P. Dell'Olmo

Condition Monitoring and Control for Intelligent Manufacturing
L. Wang and R.X. Gao (Eds.)

Optimal Production Planning for PCB Assembly
W. Ho and P. Ji

Trends in Supply Chain Design and Management: Technologies and Methodologies
Hosang Jung, F. Frank Chen and Bongju Jeong (Eds.)

Collaborative Product Design and Manufacturing Methodologies and Applications
W.D. Li, S.K. Ong, Andrew Y.C. Nee and Chris McMahon (Eds.)

Lihui Wang and Weiming Shen (Eds.)

Process Planning and Scheduling for Distributed Manufacturing

Springer

Lihui Wang, PhD, PEng
Senior Research Officer
Integrated Manufacturing Technologies
  Institute
National Research Council of Canada
London
Ontario N6G 4X8
Canada

Weiming Shen, PhD, PEng
Senior Research Officer
Integrated Manufacturing Technologies
  Institute
National Research Council of Canada
London
Ontario N6G 4X8
Canada

British Library Cataloguing in Publication Data
Process planning and scheduling for distributed
    manufacturing. - (Springer series in advanced
    manufacturing)
    1. Manufacturing resource planning 2. Production scheduling
    3. Production planning
    I. Wang, Lihui, 1959- II. Shen, Weiming
    658.5'03
ISBN-13: 9781846287510

Library of Congress Control Number: 2007922345

Springer Series in Advanced Manufacturing ISSN 1860-5168
ISBN 978-1-84628-751-0   e-ISBN 978-1-84628-752-7   Printed on acid-free paper

© Springer-Verlag London Limited 2007

Access™, ActiveX®, Internet Explorer®, .NET™, Visual Basic®, Visual C++® and Windows® are trademarks of Microsoft Corporation, One Microsoft Way, Redmond, WA 98052-6399, USA, www.microsoft.com

ACIS® is a registered trademark of Spatial Corporation, 10955 Westmoor Drive, Suite 425, Westminster, Colorado 80021 USA, www.spatial.com

Java™, Java 3D™ and JRE™ are trademarks of Sun Microsystems, Inc., 4150 Network Circle, Santa Clara, CA 95054, USA, www.sun.com

OpenGL® is a registered trademark of SGI, 1200 Crittenden Lane, Mountain View, CA 94043, USA, www.sgi.com

ST-Developer™ is a trademark of STEP Tools, Inc., 14 First Street, Troy, New York 12180, USA, www.steptools.com

Unigraphics® is a registered trademark or trademark of Electronic Data Systems Corporation, 5400 Legacy Drive, Plano, Texas 75024, USA, www.eds.com

9 8 7 6 5 4 3 2 1

Springer Science+Business Media
springer.com

Preface

Manufacturing has been one of the key areas that support and influence a nation's economy since the 18th century. Being the primary driving force in economic growth, manufacturing constantly serves as the foundation of and contributes to other industries with products ranging from heavy-duty machinery to hi-tech home electronics. In the past centuries, manufacturing has contributed significantly to modern civilisation and created momentum that is used to drive today's economy. Despite various revolutionary changes and innovations in the 20th century that contributed to manufacturing advancements, we are still facing new challenges when striving to achieve greater success in winning global competitions.

Today, distributed manufacturing is unforeseeably coming into being due to recent business decentralisation and manufacturing outsourcing. Manufacturers are competing in a dynamic marketplace that demands short response time to changing markets and agility in production. In the 21st century, manufacturing is gradually shifting to a distributed environment with increasing dynamism. In order to win a competition, locally or globally, customer satisfaction is treated with priority. This leads to mass customisation and even more complex manufacturing processes, from shop floors to every level along manufacturing supply chains. At the same time, outsourcing has forged a multi-tier supplier structure with numerous small-to-medium-sized enterprises involved, where highly-mixed products in small batch sizes are handled simultaneously in job-shop operations. Moreover, unpredictable issues like job delay, urgent-order insertion, fixture shortage, missing tool, and even machine breakdown are challenging manufacturing companies, and adding high uncertainty to the fluctuating shop floors. Engineers often find themselves in a situation that demands adaptive planning and scheduling capability in dealing with daily operations in a distributed manufacturing environment.

Targeting the uncertainty issue in manufacturing, over the past years, research efforts have focused on improving flexibility, dynamism, adaptability, agility and productivity of manufacturing, particularly in a distributed environment. Various Web-based and AI (artificial intelligence)-based tools have been developed to deal with issues in process planning, scheduling, and their integration. Many research projects have been devoted to improving the throughput and process efficiency, targeting manufacturing uncertainty. Thanks to recent advancement in AI and network technologies, sensing, monitoring and control, manufacturing research has progressed to a new level in adaptive decision making and trouble shooting, in order to address those problems encountered in today's manufacturing with increasing globalisation and outsourcing. While research and development efforts have been translated into a large volume of publications and impacted the present and future practices in manufacturing, there exists a gap in the literature for a focused collection of works that are dedicated to the process planning and scheduling for distributed manufacturing. To bridge such a gap and present the state-of-the-art to a broad readership, from academic researchers to practicing engineers, is the primary motivation for this book.

Focusing on effective process planning, Chapter 1 presents a function block enabled approach for adaptive decision making. Algorithms for machine-specific operation planning are embedded into function blocks that are called at runtime by a controller, rather than being decided in advance. Owing to its event-driven nature, a function block can also be used for remote process monitoring while in execution at a local machine, providing an interface to dynamic scheduling. It is implemented into a *DPP* (distributed process planning) prototype for concept realisation. The discussion of decentralised process planning is extended in Chapter 2 to include polishing processes. A *Web-based portal system* (WBPS) is developed aiming to streamline polishing process planning as well as addressing the hindrance problem of subjective human determination of knowledge discovery using data-mining techniques. In order to integrate process information with geometry data, Chapter 3 introduces a rule-based *IMPlanner* for process selection that considers part geometry and tolerance. Each selected process is then verified through a virtual machining module using Java 3D to identify any undesired part–tool collisions.

Recognising the importance of process planning in distributed manufacturing, Chapter 4 reports on a *CyberCut* system, being a coordinated pipeline of design, process planning and manufacturing. It is an integrated system that targets the automation of a design-to-manufacturing process, including a *macroplanner* and a *microplanner*, for Web-based rapid part machining. With mass customisation in mind, Chapter 5 continues the integrated efforts on producing one-of-a-kind products (OKP). Being a real industrial application, the reported work adopts the one-piece flow method to produce variations on a number of different kinds of doors and windows.

Linking to process planning, Chapter 6 introduces a tolerance analysis method based automated setup planning strategy that can be applied in the mass-customisation environment. It is based on the concept of production and process similarity and the best-practice knowledge in the automotive industry.

Extending the scope to scheduling, Chapter 7 addresses the issue with holonic manufacturing systems in mind. The reported work is based on task and resource holons that cooperate with each other based on a variant of the contract net protocol. This idea is implemented in a *Fabricare* holonic system. Similar to the holonic approach, Chapter 8 utilises an agent-based approach to achieve autonomy, while tackling dynamic scheduling problems in distributed manufacturing. An *iShopFloor* framework has been developed to support the agent-based dynamic scheduling. The discussions are carried on in Chapter 9 to include job-shop scheduling, combining a multi-agent system with an evolutionary approach. It is thus able to evolve the rules by which a schedule is created rather than the schedule itself.

Targeting distributed scheduling issues in multi-factory production with machine maintenance, Chapter 10 uses a *GADG* (genetic algorithm with dominant genes) approach to generate schedules that can minimise the interruption of maintenance during production. Moving to the enterprise level, the problem of distributed resources scheduling is addressed in Chapter 11. A *virtual CIM* concept is proposed as a solution to satisfy the emerging technological application of virtual enterprises, whereas the resource scheduling is dealt with using an agent-based approach.

While process planning and scheduling are crucial in distributed manufacturing, their integration can offer added value in achieving global optimisation. Chapter 12 introduces a unified representation model and a simulated annealing based approach to facilitate the integration and process optimisation. Moreover, in Chapter 13, a different integration approach is applied to holonic manufacturing systems, in which a basic architecture is proposed to determine both suitable sequences of machines needed to fabricate a part and suitable production schedules of the machines.

In addition to process planning and scheduling, how to manage dynamic demand events and how to replan current operations are equally important in winning global competitions in distributed manufacturing. In view of increasing dynamic demands in the semiconductor industry, Chapter 14 describes a planning method proactively responding to dynamic events that pose a threat to the services of manufacturing chains. A dynamic system model for manufacturing chains under exogenous demand shocks is also provided, together with an optimal control model. In Chapter 15, a parameter-perturbation approach to replanning operations is presented. Several planning models are developed and used to show how the operating conditions can be perturbed so that optimal short-term plans can be developed.

Finally, in Chapter 16, an emerging ISO standard, informally known as STEP-NC, is introduced, which points out one of the potential future directions leading to an interoperable manufacturing paradise. A three-tiered and Internet-based prototype system demonstrates how design and manufacturing data can be exchanged over the Internet during process planning in a distributed manufacturing environment.

All together, the sixteen chapters provide an overview of some recent research and development efforts on process planning and scheduling for distributed manufacturing, and are believed to make significant contributions to the literature. With the rapid advancement of information and communication technologies, particularly Internet- and Web-based technologies, we believe that this will continue to be a very active research field for many years.

The editors would like to express their deep appreciation to all the authors for their significant contributions to this book. Their commitment, enthusiasm, and technical expertise are what made this book possible. We are also grateful to the publisher for supporting this project, and would especially like to thank Mr Anthony Doyle, Senior Editor for Engineering, and Mr Simon Rees, Editorial Assistant, for their constructive assistance and earnest cooperation, both with the publishing venture in general and the editorial details. We hope the readers find this book informative and useful.

London, Ontario, Canada   Lihui Wang
October 2006   Weiming Shen

Contents

**3   Integration of Rule-based Process Selection with Virtual Machining
      for Distributed Manufacturing Planning**...................................... 61

Dusan N. Sormaz, Jaikumar Arumugam, Chandrasekhar Ganduri

**4   CyberCut: A Coordinated Pipeline of Design, Process Planning
      and Manufacture**... 91

V. Sundararajan, Paul Wright

**5   Process Planning, Scheduling and Control for
One-of-a-Kind Production** 109

Paul Dean, Yiliu Tu, Deyi Xue

6   Setup Planning and Tolerance Analysis 137

Yiming (Kevin) Rong

10   Distributed Scheduling in Multiple-factory Production with Machine Maintenance

Felix Tung Sun Chan, Sai Ho Chung

11   Resource Scheduling for a Virtual CIM System

Sev Nagalingam, Grier Lin, Dongsheng Wang

List of Contributors

Jaikumar Arumugam
Department of Industrial and
Manufacturing Systems Engineering
Ohio University
Athens, OH 45701
USA

Robert W. Brennan
Schulich School of Engineering
University of Calgary
Calgary, AB T2N 1N4
Canada

Ningxu Cai
Department of Mechanical and Materials
Engineering
The University of Western Ontario
London, ON N6A 5B9
Canada

Felix Tung Sun Chan
Department of Industrial & Manufacturing
Systems Engineering
The University of Hong Kong
Hong Kong
China

K.C. Cheng
PME International Company Limited
Hong Kong
China

Yon-Chun Chou
Institute of Industrial Engineering
National Taiwan University
Taipei 106
Taiwan

Sai Ho Chung
Department of Industrial & Manufacturing
Systems Engineering
The University of Hong Kong
Hong Kong
China

Paul Dean
Gienow Windows and Doors LLP
Calgary, AB T2C 2B6
Canada

Hsi-Yung Feng
Department of Mechanical and Materials
Engineering
The University of Western Ontario
London, ON N6A 5B9
Canada

Chandrasekhar Ganduri
Department of Industrial and
Manufacturing Systems Engineering
Ohio University
Athens, OH 45701
USA

Qi Hao
Integrated Manufacturing Technologies
Institute
National Research Council of Canada
London, Ontario N6G 4X8
Canada

George Q. Huang
Department of Industrial and
Manufacturing Systems Engineering
The University of Hong Kong
Hong Kong
China

Koji Iwamura
Department of Mechanical Engineering
Osaka Prefecture University
Sakai, Osaka 599-8531
Japan

Wei Jin
Department of Mechanical and Materials
Engineering
The University of Western Ontario
London, ON N6A 5B9
Canada

Weidong Li
School of Applied Science
Cranfield University
Cranfield, Bedfordshire, MK43 0AL
UK

Grier Lin
Centre for Advanced Manufacturing
Research
University of South Australia
Mawson Lakes, SA 5096
Australia

V.H.Y. Lo
Department of Industrial and
Manufacturing Systems Engineering
The University of Hong Kong
Hong Kong
China

Michael Masin
Department of Industrial Engineering and
Management
Technion-Israel Institute of Technology
Haifa 32000
Israel

Sev Nagalingam
Centre for Advanced Manufacturing
Research
University of South Australia
Mawson Lakes, SA 5096
Australia

Andrew Y.C. Nee
Department of Mechanical Engineering
National University of Singapore
Engineering Drive 1, 117576
Singapore

José Neves
Escola de Engenharia
Universidade do Minho
Braga
Portugal

B.K.K. Ngai
Department of Industrial and
Manufacturing Systems Engineering
The University of Hong Kong
Hong Kong
China

Douglas H. Norrie
Schulich School of Engineering
University of Calgary
Calgary, AB T2N 1N4
Canada

S.K. Ong
Department of Mechanical Engineering
National University of Singapore
Engineering Drive 1, 117576
Singapore

Carlos Ramos
GECAD R&D Group
School of Engineering
Polytechnic Institute of Porto
Porto 4200-072
Portugal

Yiming (Kevin) Rong
Department of Mechanical Engineering
Worcester Polytechnic Institute
Worcester, MA 01609
USA

Nazrul I. Shaikh
Marcus Department of Industrial and
Manufacturing Engineering
The Pennsylvania State University
University Park, PA 16802
USA

Weiming Shen
Integrated Manufacturing Technologies
Institute
National Research Council of Canada
London, Ontario N6G 4X8
Canada

Rajesh Shrestha
Osaka Prefecture University
Sakai, Osaka 599-8531
Japan

Dusan N. Sormaz
Department of Industrial and
Manufacturing Systems Engineering
Ohio University, Athens, OH 45701
USA

Paulo Sousa
GECAD R&D Group
School of Engineering
Polytechnic Institute of Porto
Portugal

Nobuhiro Sugimura
Osaka Prefecture University
Sakai, Osaka 599-8531
Japan

V. Sundararajan
Department of Mechanical Engineering
University of California at Riverside
Riverside, CA 92521
USA

Yoshitaka Tanimizu
Osaka Prefecture University
Sakai, Osaka 599-8531
Japan

V.Y.M. Tsang
Department of Industrial and
Manufacturing Systems Engineering
The University of Hong Kong
Hong Kong, China

Yiliu Tu
Departments of Mechanical and
Manufacturing Engineering
University of Calgary
Calgary, AB T2N 1N4
Canada

Scott S. Walker
Schulich School of Engineering
University of Calgary
Calgary, AB T2N 1N4
Canada

Dongsheng Wang
Centre for Advanced Manufacturing
Research
University of South Australia
Mawson Lakes, SA 5096
Australia

Lihui Wang
Integrated Manufacturing Technologies
Institute
National Research Council of Canada
London, Ontario N6G 4X8
Canada

Paul Wright
Department of Mechanical Engineering
University of California at Berkeley
Berkeley, CA 94720-1740
USA

Richard A. Wysk
Marcus Department of Industrial and
Manufacturing Engineering
The Pennsylvania State University
University Park, PA 16802
USA

Xun Xu
Department of Mechanical Engineering
University of Auckland
Auckland
New Zealand

Deyi Xue
Departments of Mechanical and
Manufacturing Engineering
University of Calgary
Calgary, AB T2N 1N4
Canada

1

An Effective Approach for Distributed Process Planning Enabled by Event-driven Function Blocks

Lihui Wang[1], Hsi-Yung Feng[2], Ningxu Cai[2] and Wei Jin[2]

[1] Integrated Manufacturing Technologies Institute
National Research Council of Canada, London, ON N6G 4X8, Canada
Email: lihui.wang@nrc.gc.ca

[2] Department of Mechanical and Materials Engineering
The University of Western Ontario, London, ON N6A 5B9, Canada
Email: sfeng@eng.uwo.ca

Abstract
This chapter presents a function block enabled approach towards distributed process planning. It covers the basic concept, generic machining process sequencing using enriched machining features, process plan encapsulation in function blocks, and process monitoring through event-driven function blocks. A two-layer structure of *supervisory planning* and *operation planning* is proposed to separate generic data from machine-specific ones. The supervisory planning is only performed once, in advance, at the shop level to generate machine-neutral process plans, whereas the operation planning is carried out at runtime at the machine level to determine machine-specific operations. This dynamic decision making is facilitated by resource-driven algorithms embedded in the function blocks. The internal structures of typical function blocks are also introduced in the chapter. Our approach and algorithms are verified through case studies before drawing conclusions. It is expected that the new approach can greatly enhance the dynamism of fluctuating job-shop operations.

1.1 Introduction

Manufacturing processes involved in job shop operations are complicated, especially on machining shop floors where a large variety of products, usually in small batch sizes, are handled dynamically. Such a dynamic environment of job-shop operations is usually characterised with operation uncertainty, including job delay, urgent-job insertion, fixture shortage, missing tools, and even machine breakdown. It thus requires an adaptive means that enables distributed process planning, dynamic job dispatching, and process monitoring. It should be responsive to unpredictable changes of distributed production capacity and functionality. An ideal shop floor should be one that effectively uses real-time manufacturing intelligence to achieve the best overall performance with the least unscheduled machine downtime. However, most available CAPP (computer-aided process planning) systems were developed mainly for flow-shop operations, and are inflexible if applied directly to the dynamic job-shop environment. In response to the above requirements and in order to adapt to the dynamic activities in job-shop operations, a function block enabled distributed process planning (DPP) approach is introduced in this chapter to achieve the dynamism during job-shop planning and plan execution.

Process planning is a knowledge-intensive and complex task that transforms design information into manufacturing processes and determines the optimal sequence of operations. It could also be defined as the act of preparing detailed work instructions to produce a part [1.1]. There are many factors that affect the process planning task. Part geometry, tolerance, surface finish, raw material, lot size, and the available resources (machines, fixtures, and tools, *etc.*) all contribute to the decision making during process planning. The variety of resource and operation selections together with their combinations makes the task of process planning complex and time consuming. This combinatorial optimisation problem has been proven to be NP-complete [1.2]. Maintaining the consistency of all process plans and keeping them optimised is difficult, especially for dynamic job-shop operations.

Targeting the uncertainty of job-shop operations, the objective of our research is to develop a methodology for distributed process planning. Within the DPP context, function blocks are used for machining-data encapsulation and event-driven process control, while machining features are used for process sequencing. The ultimate goal of DPP is to achieve the adaptability of a process plan for job-shop operations. The rest of the chapter is organised as follows: Section 1.2 briefly reviews the literature of process planning; Section 1.3 outlines the fundamentals of the DPP approach; Section 1.4 presents decision-making algorithms used in DPP; Section 1.5 gives details on function block enabled plan execution and process monitoring; finally, Section 1.6 concludes the chapter and identifies our contributions.

1.2 Brief Literature Review

For decades, since Nieble reported the idea of using the power of computers to assist process planning in 1965 [1.3], subsequent research efforts have been numerous. In order to transform product data in design to numerical control data in machining, a number of steps must be followed, including setup planning [1.4][1.5], process sequencing [1.6][1.7], tool selection [1.8]–[1.10], tool-path planning [1.11][1.12], machining-parameters selection [1.13][1.14], and fixture design [1.15][1.16]. These activities are knowledge-intensive, complex and dynamic in nature [1.17]. These research achievements can be classified into three categories: variant, generative, and AI (artificial intelligence)-based CAPP systems. The variant approach is a data-retrieval and editing method, whereas the generative approach is a knowledge-based method that can automatically generate the process plan of a part according to its features and manufacturing requirements. Most AI-based CAPP systems can be categorised into the generative approach.

Among many others, previous achievements on process planning include object-oriented approaches [1.18][1.19], genetic algorithm based approaches [1.20][1.21], neural network based approaches [1.22][1.23], Petri net based approach [1.6], feature recognition or feature-driven approaches [1.24][1.25], and knowledge-based approaches [1.26][1.27].

The early hybrid approaches (variant + generative) include interactive process planning [1.19], rapid process planning [1.28], micro/macroplanning [1.29][1.30], integrated process planning [1.31], dynamic process planning [1.32], incremental process planning [1.33], Web-based process planning in CyberCut [1.34], and agent-based process planning [1.35]–[1.40]. No attempt is made in this chapter to compare and evaluate these different approaches, although some details of certain approaches are given below.

As decentralisation of business grows, the research focus of process planning is moving towards solving problems in distributed manufacturing environments. Tu *et al.* [1.33] introduced a method called IPP (incremental process planning) for one-of-a-kind production (OKP). The IPP approach is used to extend or modify a primitive plan (a skeletal process plan) incrementally, according to the new features that are identified from a product design until no more new features can be found. A complete process plan generated by the IPP may include alternative processes. This means that a given part can also be processed by alternative machines in alternative sequences in a different plant.

A distributed manufacturing environment also changes the way of applying AI techniques to process planning. In addition to centralised AI approaches (*e.g.* genetic algorithms, neural networks, fuzzy logic, knowledge-based or expert systems, *etc.*), agent technology, being one type of distributed AI approaches, has attracted wide attention. Instead of being one large expert system, cooperative intelligent agents are being used in developing distributed CAPP systems. Among others, CoCAPP [1.37] (cooperative computer-aided process planning) attempts to distribute complex process planning activities to multiple specialised problem solvers and to coordinate them to solve complex problems. The CoCAPP attempted to satisfy five major requirements: autonomy, flexibility, interoperability, modularity, and scalability. It utilises cooperation and coordination mechanisms built into distributed agents with their own expert systems. Each agent deals with a relatively independent domain of process planning. Collectively, the multiple agents can solve complex problems.

Shih and Srihari [1.41] proposed a distributed AI-based framework for process planning. Their approach decomposes the entire production control task into several subtasks, each of which is implemented by an intelligent agent. By working collectively, the agents can arrive at a solution for the problem. Similarly, Sluga *et al.* [1.38] introduced a VWS (virtual work system) as the essential building block for decision making in a distributed manufacturing environment. The VWS represents a manufacturing work system in the information space, and is structured as an autonomous agent. It is a constituent entity of an agent network in which dynamic clusters of cooperating agents are solving manufacturing tasks. The decision making in process planning is based on a market mechanism consisting of bidding–negotiation–contracting phases. The VWS approach aims at enabling dynamic decision making based on the actual state of a given environment. The bidding-based approach is also useful to integrate product design, process planning, and scheduling [1.39].

CyberCut is a research project that aims to develop a networked manufacturing service for rapid part design and fabrication on the Internet [1.34]. A critical part of this service is an automated process planning module that is capable of generating process plans to satisfy the desired geometries and specified requirements. Three types of agents are designed to facilitate the CyberCut: primary process-planning agent, environmental planning agent, and burr minimisation tool path planning agent [1.40]. The multi-agent planning module incorporates conventional and specialised planning agents for environmental consideration and burr minimisation. However, the interactions between those agents are based on human decisions.

The agent-based approach is also being recognised as one of the effective ways to realise adaptability and dynamism of process planning. Zhang *et al.* proposed an AAPP (agent-based adaptive process planning) system on top of an OOMRM [1.19] (object-oriented manufacturing resources modelling) framework. The OOMRM is used to model manufacturing resources capability and capacity in an object-oriented manner, which intends to encapsulate manufacturing system knowledge and the methods of using the knowledge, while the AAPP is implemented as a man-machine integrated process planning platform. Instead of automating process-planning tasks completely, the AAPP system provides an interactive mode for experienced manufacturing engineers to map out more reasonable and flexible manufacturing processes for a realistic manufacturing environment. Five agents are used in the AAPP system to carry out part information classification, manufacturing resources mapping, process planning, human planning, and machining parameter retrieval. A contract net-based scheme is utilised as the coordination protocol between agents.

As partial trends in CAPP, as pointed out by ElMaraghy [1.42], integrations of CAPP with either product design or manufacturing scheduling or both remain attractive to researchers and practitioners. Previous studies in this area include design-planning integration [1.28], design-to-control [1.43], a reactive planning environment [1.32], machining-feature-based product design and process planning [1.44][1.45], and planning-scheduling integration [1.21][1.46][1.47].

While the reported process planning algorithms were claimed to be effective in achieving their optimal solutions, most techniques focus on enhancing the system functionality in their respective domains. They attempted to generate a narrowly targeted optimal solution for a specific set of tools and machine, or to suggest alternative solutions for various operation scenarios. In both cases, the generated process plans are static in nature and specific to dedicated resources. The dynamic shop-floor environment and disruptions caused by stochastic bottlenecks, unavailability of cutting tools, or breakdown of equipment are often overlooked during process processing.

Although there have been many efforts devoted to the generative approaches in CAPP, their adaptability to unpredictable changes in shop-floor operations remains insufficient. This may be due to the fact that most of the developments attempt to recognise very detailed design information and lack adaptive-learning and decision-making capability. Moreover, most CAPP systems available today are centralised in architecture, vertical in sequence, and offline in knowledge processing. It is difficult (if not impossible) for a centralised offline system to make adaptive decisions in advance, without the knowledge of the actual status of machines on shop floors. The changing shop-floor environment demands a dynamic approach for process planning that is adaptive and responsive to a sudden change in production. In response to the demands, we selected a fundamentally different approach to process planning. A novel DPP (distributed process planning) concept was therefore proposed by the authors [1.25][1.48] to generate adaptive process plans so as to address these dynamic issues, such as product changeover, job delay, urgent job insertion, unavailable fixtures, missing tools, and even a machine breakdown that happen unpredictably in today's manufacturing shop floors. Within the context of DPP, the benefits of a so-generated process plan are generic to different machines and adaptive to environmental changes.

Table 1.1 provides a comparison between conventional CAPP approaches and DPP in terms of the processing tasks involved. While conventional process plans are normally associated with specific resources at an early stage, our approach is designed to generate machine-neutral process plans during high-level supervisory planning and consider machine-specific operation planning at the CNC controller level. More details on DPP are given in the next section.

Table 1.1. Conventional CAPP approaches vs. DPP

Conventional CAPP approaches	DPP approach

– Interpretation of part design data
– Selection of machining processes (drilling, milling, grinding, *etc.*)
– *Selection of machines, tools, and fixtures*
– Process optimisation
– Decomposition of material volume to be removed
– Selection of machining operations
– Generation of precedence constraints
– Sequencing of machining operations
– Cutting parameter selection
– Tool-path planning
– Operation optimisation
	Supervisory planning:
	  – Product data analysis/EMF parsing
	  – Selection of machining processes
	  – EMF grouping/setup planning
	  – Multiple setup sequencing and EMF sequencing
	  – Function block generation
	  – Generic process plan dispatching
	Operation planning:
	  – *Tool/jig/fixture selection*
	  – Cutting parameter selection
	  – Tool-path planning
	  – Machine-specific local operation optimisation

EMF: enriched machining feature

1.3 Distributed Process Planning

1.3.1 Fundamentals of DPP

According to the literature survey, most process plans generated using existing CAPP systems are tied to specific resources (machines, fixtures and cutters, *etc.*) in the first place. They are inflexible, unportable, and not responsive to unexpected changes. When a resource becomes unavailable, its dedicated process plan is subject to a major change or even regeneration. Such repetitive planning tasks may happen frequently in job-shop operations. DPP, however, is used to generate adaptive process plans that can also be integrated with dynamic scheduling functions. The ultimate goal of DPP is to improve flexibility, responsiveness, agility, adaptability, and real-time manufacturing intelligence of shop floors. A process plan generally consists of two parts: *generic data* (machining method, machining sequence, and machining strategy) and *machine-specific data* (tool data, cutting conditions, and tool paths). A two-layer hierarchy is considered suitable to separate the generic data from those machine-specific ones in DPP. As shown in Figure 1.1, the tasks of DPP can be divided into two groups and accomplished at two different levels: shop-level *supervisory planning* and controller-level *operation planning*. The former focuses on product data analysis, machining feature decomposition, setup planning, machining process sequencing, jig/fixture selection, and machine selection. The latter considers the detailed working steps for each machining operations, including cutting-tool selection, cutting-parameter assignment, tool-path planning, and control code generation. Between supervisory planning and operation planning, scheduling functions can be integrated with DPP by means of function blocks. (The integration of DPP with scheduling is beyond the scope of this chapter. Readers are referred to [1.47][1.49] for more details.)

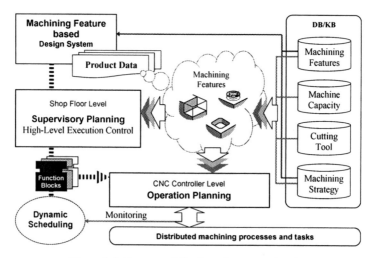

Figure 1.1. Concept of distributed process planning

1.3.2 Basic Requirements

Being an adaptive process planning tool, DPP is required to be able to process and transform product design data to NC machining data, and optimise the generated technological data (cutting parameters and tool paths, *etc.*) on the fly. The basic requirements for DPP can be summarised as follows.

- *Intelligent*: to be able to handle uncertainty and maximise the probability of success of controlled machining process;
- *Adaptive*: to be able to adjust its machining strategy/data when alternative resources are used through embedded algorithms;
- *Fault tolerant*: to be able to react to a failure and resume properly on the same machine, or find alternative plans on a different machine;
- *Portable*: to be able to run in different machine controllers without control code modification or regeneration; and
- *Reusable*: to be able to reuse the basic function blocks of a process plan for other machining operations.

1.3.3 System Architecture

The system architecture of DPP is illustrated in Figure 1.2, which consists of five modules across supervisory planning and operation planning: (1) machining feature parsing, (2) machining sequence generation, (3) function block design, (4) execution control, and (5) function block (FB) processing. They are supported by a group of shared databases and knowledge bases for decision making.

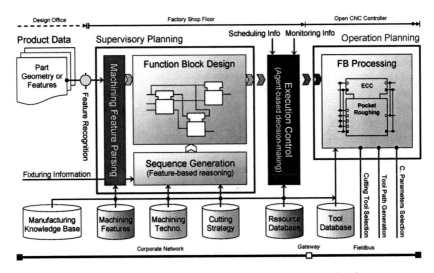

Figure 1.2. System architecture for distributed process planning

At the supervisory planning level, a list of enriched machining features (EMFs) is first identified after feature recognition and feature parsing. The machining processes are then sequenced using an EMF-based reasoning approach, before being embedded in a set of function blocks. The system outputs of supervisory planning are a set of function blocks with sequenced process, machining data, and algorithms embedded. It is the set of algorithms that make adaptive decisions at runtime during operation planning after a set of function blocks is dispatched to an appropriate machine for execution. During the operation planning, the function block is finalised with actual tool data, cutting conditions and tool path. The resultant function block is thus able to describe the detailed machining operations. Runtime optimisation is performed at the latter stage, when specific resources (machine, tool and fixture) are known, and within a relatively small search space, because the machine controller is knowledgeable of the machine in terms of its dynamics and runtime status.

On the other hand, because of the event-driven nature, the execution status of a function block is sent back to the execution control module together with the runtime machine status for dynamic scheduling, whenever necessary.

This system architecture also considers extensions to Web-based manufacturing. The major decision logics are deployed on an application server. A client conducts distributed process planning through an applet-based user interface running in a Web browser. This function is currently under development.

1.3.4 Enabling Technologies

As can be seen in Figure 1.2, machining features, function blocks and intelligent agents are the three enabling technologies, with the first two used extensively in the current implementation. In this section, only machining features and function blocks are introduced in detail.

1.3.4.1 Machining Features

As shown in Figure 1.3, machining features are those shapes such as *step*, *slot*, *pocket*, and *hole* that can be easily achieved by the available resources and defined machining technologies [1.44]. Being standard shapes that can be machined, the machining features are different from design features (geometric standard shapes such as vertex, curve, and face). Each machining feature holds a set of loosely coupled information about how to fabricate it, including cutter type, operation sequence, tool path generation logic, and suggested cutting parameters, which provide an indication as to what kind of operation and tools will be required to manufacture the feature. Since milling and drilling operations are dominant in machining operations, only milling and drilling features on prismatic workpiece are considered here. Today, these machining features are widely accepted in design and machining for the ease of information retrieval and processing.

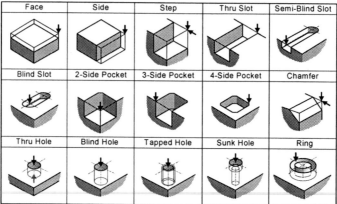

↓ Tool access direction

Figure 1.3. Typical machining features

Within the context, each machining feature can be represented by its geometric feature, surface feature, volume feature, and loosely coupled cutting information. A geometric feature is a topological unit that holds the main information of the machining feature itself, such as geometry, dimension, and tolerance; a surface feature captures the attributes and the relationship of faces defining the surface of the machining feature; and a volume feature is the solid volume enclosed in the machining feature. Figure 1.4 illustrates the combined feature model of a machining feature *step*.

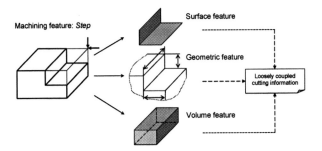

Figure 1.4. Combined feature model of machining feature *step*

This combined feature model can be expressed by the following equations:

$$MF = \bigcup_{i=1}^{I} MF_i \tag{1.1}$$

$$MF_i = \left[\bigcup_{l=1}^{L} f_{il}^{MF}(SF_i), \bigcup_{m=1}^{M} f_{im}^{MF}(GF_i), \bigcup_{n=1}^{N} f_{in}^{MF}(VF_i), \bigcup_{r=1}^{R} f_{ir}^{MF}(CI_i) \right] \tag{1.2}$$

where

MF	the set of all machining features of a given part; I is the total number of the machining features;
MF_i	the ith machining feature of the set;
$\bigcup_{l=1}^{L} f_{il}^{MF}(SF_i)$	the attributes of the ith surface feature; L is the total number of attributes used to describe this surface feature;
$\bigcup_{m=1}^{M} f_{im}^{MF}(GF_i)$	the attributes of the ith geometric feature; M is the total number of attributes used to describe this geometric feature;
$\bigcup_{n=1}^{N} f_{in}^{MF}(VF_i)$	the attributes of the ith volume feature; N is the total number of attributes used to describe this volume feature;
$\bigcup_{r=1}^{R} f_{ir}^{MF}(CI_i)$	the machining information loosely coupled with the ith machining feature, R is the total amount of machining information used to fabricate this machining feature.

Within DPP, it is assumed that the machining feature list of a part is given. Such a feature list can be obtained either by adopting third-party feature-recognition solutions [1.24][1.50]–[1.52] or by incorporating machining feature based design methodology, *i.e.* designing a part in the same way of "machining" by subtracting machining features from its blank [1.44].

1.3.4.2 Function Blocks

The concept of function blocks is described in the IEC-61499 specification [1.53], as an IEC standard for distributed industrial processes and control systems, particularly for PLC control. It is based on an explicit event-driven model and provides for data flow and finite-state automata-based control. It is relevant to the DPP in machining data encapsulation and process plan execution. The event-driven model of a function block gives an NC machine more intelligence and autonomy to make decisions on how to adapt a process plan to match the actual machine capacity and dynamics. It enables dynamic resource scheduling, execution control, and process monitoring.

Figure 1.5 illustrates the internal structure of a basic (left) and a composite (right) function blocks. A basic function block can have multiple outputs and can maintain its internal hidden-state information. This means that a function block can generate different outputs even if the same inputs are applied. This fact is of vital importance for automatic cutting-parameter modification, after a function block has been dispatched to a machine, by changing the internal hidden state of the function block. For example, a function block of *pocket_milling* can be used for roughing and finishing at the same machine, or at different machines, with different cutting parameters and tool paths by adjusting the internal state variables of the function block to fine tune the algorithms in use. The behaviour of a basic function block is controlled by an internal finite-state machine, whose operation is represented by an execution control chart (ECC) as shown in Figure 1.6.

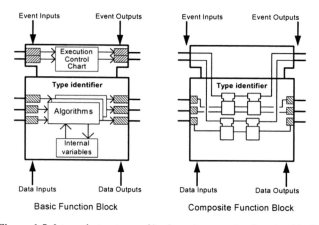

Figure 1.5. Internal structures of basic and composite function blocks

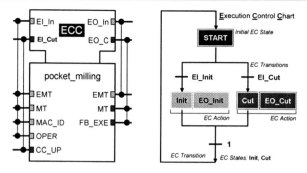

Figure 1.6. Execution control mechanism of a basic function block

Being a (representation of a) finite-state machine, an ECC is made up of EC states, EC transitions, and EC actions. The initial EC state, START in this case, cannot have any EC actions associated with it. The occurrence of an event input, such as EI_Init and EI_Cut, causes the ECC to be invoked and the input variables (EMT, MAC_ID, OPER, *etc.*) to be mapped. The EC transitions use a Boolean combination of conditions that may be comprised of event inputs, data inputs, and internal state variables. A triggered EC transition causes a change of EC state and this leads to the execution of an associated EC action, *Init* or *Cut* in this case. The EC action then sends out an event, EO_Init or EO_Cut, upon completion. In the DPP, each machining feature (hole, step, slot, or pocket, *etc.*) is mapped to a basic function block that defines fundamental relationships between events and data. The needed processing algorithms are encapsulated in the basic function block and they can only be accessed by the basic function block itself. A setup of machining features forms a composite function block (combination of basic function blocks) connected by events and data. Within the context, event inputs of a function block are used to trigger appropriate machining strategies (algorithms) encapsulated inside the function block, while the data inputs are used for algorithm processing. The event flow between function blocks determines both the machining sequence and the type of machining operation (roughing or finishing, *etc.*). How a function block is designed is presented in Section 1.4.4.

1.4 Decision Solutions for Supervisory Planning

This section introduces the decision-making algorithms for supervisory planning and execution control. Note that the supervisory planning needs only to be done once, the result of which is a resource-neutral generic process plan. This generic process plan is embedded in a set of predefined basic function blocks in the unit of setup, which are ready for merging and dispatching to a specific machine.

1.4.1 EMF for Machining Process Sequencing

One critical task in DPP is machining process sequencing. Since a part design can be represented by machining features either through feature-based design or applying a third-party feature-recognition solution, machining process sequencing in DPP is treated as the task of putting machining features into proper setups and in the proper sequence. However, for the purpose of effective machining process sequencing, only the information of machining features is insufficient. We combine the *intermediate machining volume* (IMV) with each machining feature to reflect the dynamic change of its shape during machining. A combined machining feature is called an *enriched machining feature* (EMF) in this research.

The IMV of a machining feature is the intersection of its *maximum machining volume* (MMV) and the current workpiece; whereas the MMV of the machining feature is the volume to be removed to create the machining feature directly from the surface of its raw material along the defined tool-access direction without destroying the part. Figure 1.7 shows the concept of IMV through a hole. The IMV of the machining feature *Hole* varies between its MMV and its *actual machining volume* (AMV) along the machining process of the part. The upper limit of the IMV is the volume (or MMV) to be removed from the raw material; the lower limit of the IMV is the volume feature (or AMV) of this machining feature. Collectively, the change of IMVs of machining features demonstrates the change of a workpiece while the workpiece is gradually taking its shape during the machining operation.

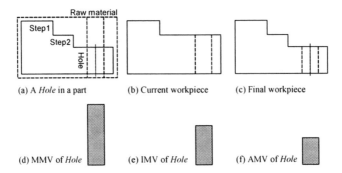

Figure 1.7. Intermediate machining volume in machining

Adding the current IMVs into Equation (1.2), the enriched machining features of a given part can be denoted as:

$$EMF = \bigcup_{i=1}^{I} EMF_i \qquad (1.3)$$

$$EMF_i = (MF_i, IMV_i) = \left[\bigcup_{l=1}^{L} f_{il}^{MF}(SF_i), \bigcup_{m=1}^{M} f_{im}^{MF}(GF_i), \bigcup_{n=1}^{N} f_{in}^{MF}(VF_i), \right.$$

$$\left. \bigcup_{r=1}^{R} f_{ir}^{MF}(CI_i), \bigcup_{t=1}^{T} f_{it}^{MF}(IMV_i) \right] \qquad (1.4)$$

where,

EMF the set of all enriched machining features of a given part;

EMF_i the ith enriched machining feature of the set;

$\bigcup_{t=1}^{T} f_{it}^{MF}(IMV_i)$ the attributes of the ith intermediate machining volume; T is the total number of attributes used to describe this IMV.

The detailed representation of an EMF, especially its surface feature, volume feature, and IMV, is formulated using the basic geometric entity – *surface*. Here, a surface refers to a basic individual face shape, such as planar surface, cylindrical surface, *etc.* Jointly, they define the geometry of the EMF. A surface is termed *real* when the inside of its boundary is solid, or *imaginary* when the boundary is enveloping an empty area [1.54]. The surfaces of a given part can be defined as:

$$S = RS \cup IS = \bigcup_{j=1}^{J} RS_j \cup \bigcup_{k=1}^{K} IS_k \qquad (1.5)$$

where,

S the set of all surfaces of a given part;

RS the set of all real surfaces of the part; J is the total number of real surfaces;

IS the set of all imaginary surfaces of the part; K is the total number of imaginary surfaces;

$RS_j = f_j^{RS}(R_{j1}, R_{j2}, ..., R_{jP})$ the jth real surface described by a function of a set of parameters; P is the total number of parameters;

$IS_k = f_k^{IS}(I_{k1}, I_{k2}, ..., I_{kQ})$ the kth imaginary surface of the part described by a function of a set of parameters; Q is the total number of these parameters.

The surface feature of the ith enriched machining feature can be described by a set of real surfaces, denoted as:

$$\bigcup_{l=1}^{L} f_{il}^{MF}(SF_i) = \bigcup_{l=1}^{L} f_{il}^{MF}\left(f_i^{SF}\left(\bigcup_{j=1}^{J} f_j^{RS}(R_{j1}, R_{j2}, ..., R_{jP}) \right) \right) \tag{1.6}$$

Similarly, the volume feature and the IMV of the ith EMF can be described by a set of surfaces, including the real surfaces and the imaginary surfaces:

$$\bigcup_{n=1}^{N} f_{in}^{MF}(VF_i) = \bigcup_{n=1}^{N} f_{in}^{MF}\left(f_i^{VF}\left(\bigcup_{j=1}^{J} f_j^{RS}(R_{j1}, R_{j2}, ..., R_{jP}) \cup \bigcup_{k=1}^{K} f_k^{IS}(I_{k1}, I_{k2}, ..., I_{kQ}) \right) \right) \tag{1.7}$$

$$\bigcup_{t=1}^{T} f_{it}^{MF}(IMV_i) = \bigcup_{t=1}^{T} f_{it}^{MF}\left(f_i^{IMV}\left(\bigcup_{j=1}^{J} f_j^{RS}(R_{j1}, R_{j2}, ..., R_{jP}) \cup \bigcup_{k=1}^{K} f_k^{IS}(I_{k1}, I_{k2}, ..., I_{kQ}) \right) \right) \tag{1.8}$$

Substituting Equations (1.6)–(1.8) to Equation (1.4), a comprehensive representation of an EMF can be obtained as the ith EMF of a given part:

$$EMF_i = \begin{bmatrix} \bigcup_{l=1}^{L} f_{il}^{MF}\left(f_i^{SF}\left(\bigcup_{j=1}^{J} f_j^{RS}(R_{j1}, R_{j2}, ..., R_{jP}) \right) \right) \\ \bigcup_{m=1}^{M} f_{im}^{MF}(GF_i) \\ \bigcup_{n=1}^{N} f_{in}^{MF}\left(f_i^{VF}\left(\bigcup_{j=1}^{J} f_j^{RS}(R_{j1}, R_{j2}, ..., R_{jP}) \cup \bigcup_{k=1}^{K} f_k^{IS}(I_{k1}, I_{k2}, ..., I_{kQ}) \right) \right) \\ \bigcup_{r=1}^{R} f_{ir}^{MF}(CI_i) \\ \bigcup_{t=1}^{T} f_{it}^{MF}\left(f_i^{IMV}\left(\bigcup_{j=1}^{J} f_j^{RS}(R_{j1}, R_{j2}, ..., R_{jP}) \cup \bigcup_{k=1}^{K} f_k^{IS}(I_{k1}, I_{k2}, ..., I_{kQ}) \right) \right) \end{bmatrix} \tag{1.9}$$

This EMF_i, $i \in [1, I]$, is first grouped into a setup and then sequenced in the setup using an EMF-based reasoning approach described in the subsequent sections.

1.4.2 EMF Grouping

In a mechanical design, the functional requirements of a part are normally expressed by geometrical dimensions and tolerances. To eliminate as much machining error stackup as possible, it is suggested that the machining features with certain functional relationships should be grouped together and machined in one single setup [1.15], based on an appropriate datum reference frame. A datum reference frame is a reference coordination system used to secure other machining features in the same part [1.55], and is determined by the functional relationships (*e.g. //, ⊥, ⊕, etc.*) among the machining features. The EMF grouping in DPP follows three steps: (1) choosing datum references; (2) finding a primary locating surface and direction; and (3) grouping EMFs into appropriate setups.

Step 1: Choosing Datum References
   One of the relationships among EMFs is the datum dependency precedence given in the representation of an EMF as a *reference feature* and/or *reference face*, which expresses the position, orientation or profile tolerance requirements of the EMF. By tracing the reference feature/face of each EMF, a primary datum reference frame and its dependency precedence of multiple datum references (if any) can be identified as follows,

> for $\forall i$ that $EMF_i \in EMF$ do
>   search an RF_i $(RF_i \subset EMF_i \wedge RF_i \neq none)$;
>   for $\forall j_{(j \neq i)}$ that $EMF_j \in EMF$ do
>   search an RF_j $(RF_j \subset EMF_j \wedge RF_j \neq none)$;
>   if RF_i depends on RF_j
>   switch i and j;
>   end if
>   end for
> end for

where RF is the set of all reference faces. The first item of the sorted results of datum dependency is the primary datum reference. The EMF grouping must be arranged according to the datum reference frame and their dependency.

Step 2: Finding a Primary Locating Direction
   A primary locating direction is the surface normal $\vec{V}$ of the primary locating surface (*PLS*), which usually serves as the primary datum reference for determining the spatial position and orientation of a workpiece and constrains at least three degrees of freedom. It should be aligned with or be orthogonal to the Z-axis of a machine tool, depending on the configuration of the machine. The primary locating surface and its locating direction can be determined by the following equations:

$$PLS = \left\{ f\left(A^*, T^*\right) \middle| W_A \times \frac{A^*}{A_{max}} + W_T \times \frac{T^*}{T_{max}} = max\left(W_A \times \frac{A}{A_{max}} + W_T \times \frac{T}{T_{max}} \right) \right\} \qquad (1.10)$$

$$\vec{V} = \left[\frac{\partial f}{\partial x}, \frac{\partial f}{\partial y}, \frac{\partial f}{\partial z} \right] \qquad (1.11)$$

where, A^* and T^* are the surface area and the generalised accuracy grade of the *PLS*; W_A is the weight factor of surface area; W_T is the weight factor of the surface accuracy grade; A_{max} and T_{max} are the maximum values of surface area and generalised accuracy grade of all candidate locating surfaces. A generalised accuracy grade T can be obtained by applying the algorithms described in [1.16] [1.56][1.57].

Step 3: Grouping EMFs into Appropriate Setups

Based on the primary locating direction $\vec{V}$ (setup orientation) determined in Step 2, the EMF grouping can be accomplished by searching for those EMFs whose tool-access directions $\vec{T}_{EMF}$ are opposite to $\vec{V}$, and grouping them into setup $ST_{\vec{V}}$. The procedure of EMF grouping is given below.

> for $\forall i$ that $EMF_i \in EMF$ do
>   search a $\vec{T}_{EMF}$ ($\vec{T}_{EMF} \subset EMF_i$);
>   if ($\vec{T}_{EMF}$ is a primary access direction $\wedge$ $\vec{T}_{EMF}$ is opposite to $\vec{V}$)
>   group EMF_i into the setup $ST_{\vec{V}}$;
>   else if ($\vec{T}_{EMF}$ is a secondary access direction $\wedge$ EMF_i is ungrouped $\wedge$ $\vec{T}_{EMF}$
>   is opposite to $\vec{V}$)
>   group EMF_i into the setup $ST_{\vec{V}}$;
>   end if
> end for

Following the procedure, a setup is formed, which can also be denoted as:

$$ST_{\vec{V}} = \left\{ EMF \mid \vec{T}_{EMF} = -\vec{V} \right\} \qquad (1.12)$$

Note that the remaining EMFs are grouped by repeating Steps 2 and 3, but based on the secondary locating direction and so on until all the EMFs are properly grouped.

To be generic, the EMF grouping (into setups) at this stage is done for 3-axis machines, as their configurations form the basis of other machines with more axes. In other words, the 3-axis-based EMF grouping makes a process plan generic and applicable to other machines with varying configurations. However, a setup merging is required for 4- or 5-axis machines, after a specific CNC machine is selected. This setup merging is straightforward if a machine's configuration is known [1.58].

1.4.3 EMF Sequencing

EMF sequencing normally consists of two parts: (1) multiple setup sequencing, and (2) EMF sequencing within each setup. The issue of multi-setup sequencing is addressed implicitly when selecting locating directions (primary, secondary, *etc.*) for the EMF grouping, in terms of the generalised accuracy grade and critical datum reference. The true challenge of EMF sequencing is now shifted to how to sequence EMFs within each setup, when their machining sequence cannot be determined simply by the datum relationships and manufacturing constraints among the EMFs. In DPP, an EMF-based geometric reasoning approach is proposed by tracking and comparing the IMV against the AMV (or volume feature) of each EMF. By applying the following five reasoning rules sequentially, a machine-neutral sequence plan with multiple setups can be created. For example, in the case shown in Figure 1.7, the IMV of the *Hole* varies between its MMV (Figure 1.7(d)) and its AMV (Figure 1.7(f)) along the machining process. As a rule of thumb, if the IMV of an EMF equals the AMV of the EMF, it is the time to machine the EMF.

Rule 1: *During sequencing, when the IMV of an EMF equals the AMV of the EMF, or IMV=AMV, this machining feature is ready for machining.*

Applying Rule 1 to the case shown in Figure 1.7, it is easy to conclude a sequence of *Step1* → *Step2* → *Hole* for machining. Figure 1.8 shows 30 typical cases after applying Rule 1.

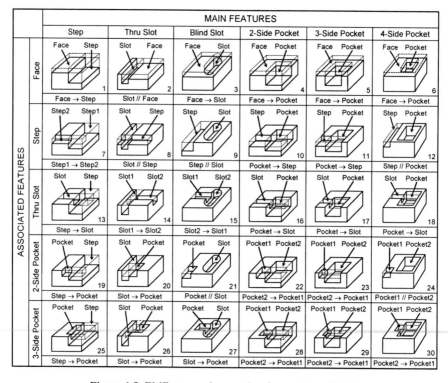

Figure 1.8. EMF sequencing results after applying Rule 1

This reasoning rule works effectively for EMF sequencing in the same setup and with feature interactions. However, after applying Rule 1, there still exist some cases that cannot be handled by this rule, in which the sequence of two machining features remains in parallel (shown as // in Figure 1.8), such as Case 8: Thru Slot + Step. In this case, if the *Thru Slot* is cut first, the *Step* will be divided into two smaller ones, which is contrary to the definition of a machining feature being a *basic single machinable shape*.

Rule 2: *If the IMF of machining feature A is to be divided into more than one piece as a result of the machining operation of machining feature B, the machining feature A should be cut first.*

In addition to the feature-splitting case encountered in Rule 2, there are cases where incorrect sequences may result in different types of machining features, *e.g.* Case 2: Thru Slot + Face. In this case, if the *Face* is milled first, the *Face* feature in machining is actually changed to a *Step*. This is not allowed, as different EMF types require different machining data (tool type, tool-access direction, and tool-path pattern, *etc.*). Rule 3 is therefore established to prevent such ill cases.

Rule 3: *If a machining feature is to be changed to another feature type as a result of its own machining operation, this machining feature is not ready and should be cut later.*

The remaining parallel cases after applying Rule 1 to Rule 3 do not have feature interactions and their machining sequences are not critical. They are further handled by adopting the knowledge of best practice or know-how of operators. One rule commonly used by machinists is that the bigger volume is to be removed first, because removing a bigger volume generally produces more cutting force and cutting heat that may result in more deformation and poor surface quality, especially for large workpieces.

Rule 4: *A bigger machining volume is to be cut first.*

Figure 1.9 shows the EMF sequencing results after applying the reasoning rules 2–4 to those parallel cases remaining in Figure 1.8.

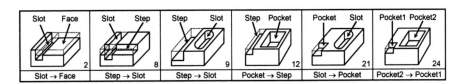

Figure 1.9. Sequenced results of the 6 parallel cases in Figure 1.8

Although Rules 1–4 are applied sequentially during EMF sequencing, *Face* and *Side* features are handled differently, except for Case 2. These two types of EMFs usually cover large surface areas and are frequently used as datum references. They are normally removed first in each setup. In addition, the tool-type information embedded in each EMF is used to group the sequenced EMFs into clusters (within each setup) to minimise the tool-change time.

Rule 5: *In a setup, the machining features sharing the same tool types are grouped into clusters.*

By applying the five rules, a machine-neutral sequence plan can be created. These rules cover all critical EMF sequences of a prismatic part. The remaining parallel sequences, if any, are not critical and will be up to the controller-level operation planning (see Figure 1.2) to determine.

1.4.4 Function Block Design

Three basic function block types are defined in the DPP: (1) *machining feature function block* (MF-FB), (2) *event switch function block* (ES-FB), and (3) *service interface function block* (SI-FB).

1.4.4.1 Machining Feature Function Block

In DPP, each machining feature can be mapped to one machining feature function block (MF-FB). Here, we use a 4-side pocket feature to demonstrate the function block design process. Figure 1.10(a) gives the graphical definition of a 4-side pocket function block, and Figure 1.10(b) depicts its execution control chart.

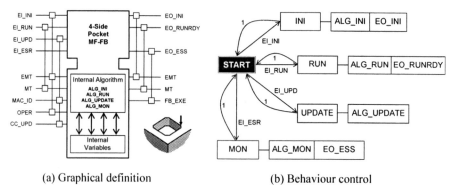

(a) Graphical definition   (b) Behaviour control

Figure 1.10. A 4-side pocket machining feature function block

It can also be expressed in a textual way that is more readable by human designers:

```
FUNCTION_BLOCK
   4-SIDE POCKET
   EVENT_INPUT
      EI_INI WITH EMT, OPER;
      EI_RUN WITH MT;
      EI_UPD WITH MAC_ID, CC_UPD;
      EI_ESR;
   END_EVENT
   VAR_INPUT
      EMT:     FLOAT;
      MT:      FLOAT;
      MAC_ID:  INT;
      OPER:    STRING;
      CC_UPD:  VECTOR;
   END_VAR
   EVENT_OUTPUT
      EO_INI WITH EMT;
      EO_RUNRDY WITH MT;
      EO_ESS WITH FB_EXE;
   END_EVENT
```

```
VAR_OUTPUT
   EMT:     FLOAT;
   MT:      FLOAT;
   FB_EXE:  VECTOR;
END_VAR
ALGORITHM
   ALG_INI;
   ALG_RUN;
   ALG_UPDATE;
   ALG_MON;
END_ALGORITHM
END_FUNCTION_BLOCK
```

where, EMT is the estimated machining time based on the suggested machining data, which is accumulated and relayed along an MF-FB chain; MT is used to store the actual machining time accumulated during function block execution; MAC_ID passes the selected machine ID to the MF-FB for machine-specific local optimisation; OPER tells the MF-FB the type of machining operation such as roughing, semi-finishing or finishing; FB_EXE is a vector storing the execution status and cutting parameters of the function block for monitoring; and CC_UPD is another vector that can be used by an operator to override the auto-generated cutting parameters. Based on the external variables and embedded internal variables (not shown in Figure 1.10(a), such as machining feature ID, workpiece material), the four defined algorithms can provide the needed functions upon request.

In Figure 1.10(b), the START state is an initial idle state ready for receiving event inputs. EI_INI triggers the state transition from START to INI, and when state INI is active, algorithm ALG_INI is being executed for initialisation. Upon its completion, ALG_INI will fire an event EO_INI indicating the success of the initialisation. Similarly, for other state transitions to RUN, UPDATE and MON, different embedded algorithms ALG_RUN (MF-FB execution), ALG_UPDATE (cutting-parameter update), and ALG_MON (MF-FB monitoring) are triggered, correspondingly. An event "1" means a state transition is always true. That is to say, the state will transit back to the START state and be ready for receiving the next event input. If a START state is not ready, any arrival events will be ignored.

Table 1.2 lists the required machining information embedded in an MF-FB.

Table 1.2. MF-FB embedded machining information

Feature type		Operation	Cutter type
	f: Feed per tooth n: Flute number L: Tool-path length	Roughing Finishing	Square end mill Square end mill (Diameter smaller than twice the corner radius)
Suggested tool-path patterns			
(1)		(2)	
Machining time estimation: $T = L / (f \cdot n \cdot rpm)$			

With the required machining data and the embedded algorithms, an MF-FB like this is empowered to make adaptive decisions at runtime with respect to the assigned resources (*e.g.* machine and cutters). For an MF-FB, the initialisation algorithm ALG_INI can fulfil the following tasks before actual fabrication:

- calculating optimal cutting conditions,
- generating tool path according to suggested tool-path patterns, and
- estimating or accumulating machining time.

1.4.4.2 Event Switch Function Block

As mentioned in Section 1.3.4.2, while basic MF-FBs can define the functional relationships between events, data and algorithms for individual machining features fabrication, their combination can form a composite function block representing a setup. A composite function block may consist of several basic and/or composite function blocks with partially sequenced connections via events and data. The event flow among MF-FBs also determines their machining sequence. Figure 1.11 shows a composite function block, where the event flow (or sequence) among three MF-FBs is facilitated at runtime by an event switch function block (ES-FB).

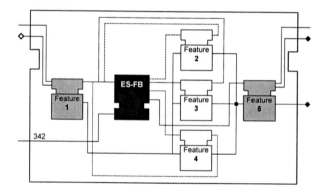

Figure 1.11. An ES-FB in a composite function block

For example, if a machining sequence of "342" is given to the three machining features (*Features 2–4*), the ES-FB will fire events accordingly to appropriate MF-FBs for feature fabrications in the order of 3→4→2. It thus adds flexibility for the composite function block to dynamically adjust the machining sequence of non-critical machining features. Figure 1.12 gives the graphical definition of the ES-FB, where ROUTE is the only data input to the function block. It is used as a reserved port for controller-level operation planning to do the local optimisation of machining sequence. Once the final sequence becomes explicit for those parallel features, a string of integer numbers indicating the sequence is applied to the port. The function of event switching is realised by an internal algorithm ALG_SWITCH, which parses the input data string and triggers one execution event at a time until the entire string is exhausted. An acceptable string must (1) only consist of non-repetitive numbers, *e.g.* "342", and (2) represent all parallel features needed for switching.

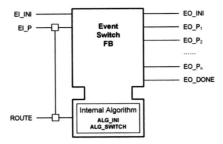

Figure 1.12. Event switch function block

Denoting an acceptable string as S_{tr} for an event-switch state machine M yields

$$S_{tr}(M) = \left\{ x \in I \,\middle|\, \begin{matrix} \delta * (START, x) = DONE, \\ I = \{(I_e, I_d)\} \end{matrix} \right\} \qquad (1.13)$$

where, START and DONE are the initial and final states; I is the set of both event input I_e and data input I_d.

$$I_d = (x_1, x_2, \ldots, x_i, \ldots, x_p), \ \forall x_i \,\big|\, (x_i \in I^+, i \le p) \qquad (1.14)$$

where I^+ is a set of positive integers; p is the number of all parallel features; and $x_i \ne x_j$ when $i \ne j \, (i, j = 1, 2, 3, \ldots, p)$. As we are only interested in data inputs when forming an acceptable string for an ES-FB, Equation (1.13) can be rewritten as

$$S_{tr}(M) = \{ x \in I_d \,|\, \delta * (START, x) = DONE \} \qquad (1.15)$$

1.4.4.3 Service Interface Function Block

In addition to MF-FBs and ES-FB, a service interface function block (SI-FB) is designed to facilitate the execution control of MF-FBs in DPP. It also enables machining-process monitoring during function block execution. As mentioned earlier, all MF-FBs are grouped in setups before being dispatched to appropriate machines. Each setup is a composite function block. An SI-FB is plugged to each composite function block with the following assigned duties: (1) to collect the runtime *execution status* of an MF-FB including its ID, cutting parameters, and job-completion rate; (2) to collect *machining status* (cutting force, cutting heat, and vibration, *etc.*) if made available; and (3) to report any *unexpected situations* to the execution control module of DPP, *e.g.* security alarms and tool breakage, *etc.*

Similar to other function block types, an SI-FB is designed (as illustrated in Figure 1.13) with five embedded algorithms for requesting and reporting execution status (ES), machining status (MS), and unexpected situation (US) from MF-FBs and to the Execution Control module (see Figure 1.2), respectively. In order to monitor the machining process during execution, an SI-FB can be connected to a composite function block as shown in Figure 1.14. At each request of the Execution Control module, the SI-FB will pass the request (EI_ESR, execution status request) to the composite function block, which will then return an array of FB_EXE containing runtime execution status back to the SI-FB and finally to the Execution Control module. The SI-FB is of vital importance for machining-process monitoring and dynamic rescheduling in case of machine failure.

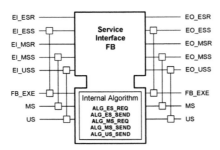

Figure 1.13. Service interface function block

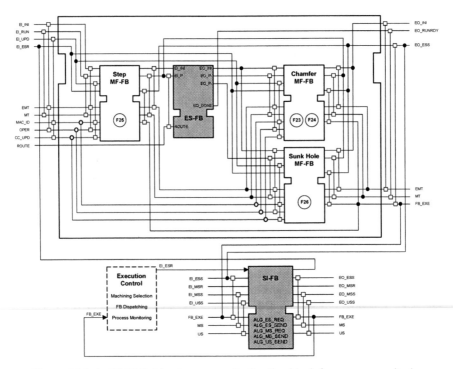

Figure 1.14. An SI-FB linking to a composite function block for process monitoring

1.5 Setup Merging and Monitoring

The Execution Control module shown in Figure 1.2 integrates supervisory planning with operation planning. It also links to an agent-based dynamic scheduling system. Due to page limitation, this section only presents the algorithms for setup merging and its execution monitoring through examples.

1.5.1 Setup Merging

As mentioned in Section 1.4.2, a DPP-generated sequence plan is 3-axis based. In the case that a 4-/5-axis machine is selected, proper setup merging is required for the best utilisation of the machine. According to the five EMF-based reasoning rules, a 3-axis-based generic setup plan of a test part (shown in Figure 1.15(a)) with 26 machining features can be generated. It consists of 5 setups, each of which contains a set of partially sequenced machining features, as shown in Figure 1.15(b). The light grey areas are setups and the dark grey areas indicate the feature groups sharing the same tools. Each 3-axis-based setup can be represented by a unique unit vector u indicating its tool-access direction (TAD).

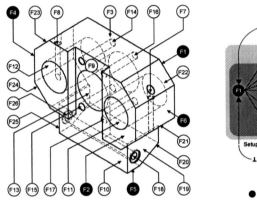 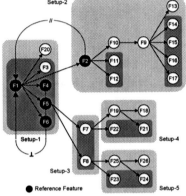

(a) A test part with 26 machining features   (b) Partially sequenced machining features

Figure 1.15. A test part with 5 setups after applying EMF-based reasoning rules

In the case that a 5-axis machine tool $\{X, Y, Z, A$ (around X), B (around Y)$\}$ is selected, more than one setup of the test part may have the chance to be machined in one *base* setup through setup merging. The setup merging examines whether other setups can be included in the base setup by checking the unit vector u of each setup against the tool-orientation space (TOS) of the selected machine. The procedure is straightforward by following two steps and their iterations, *i.e.* (1) aligning the locating direction of a base setup to the spindle axis Z, and (2) searching for a position that includes a maximum number of 3-axis-based setups by rotating the part around the locating direction (or spindle axis Z). This merging process is repeated for all setups until a minimum number of 5-axis-based setups can be reached. Since the first step can be done easily using matrix transformation, we only provide more details on the second step.

Figure 1.16(a) shows a typical scenario, where a base setup has been aligned with $-Z$ axis and another 3-axis-based setup with a tool-access direction u_i (x_i, y_i, z_i) is under consideration. The goal is to rotate the vector u_i (or the test part) around Z and at the same time determine a mergable range (or ranges) within 2π, that u_i can fit in the TOS of the machine. The TOS is represented as a spherical surface patch denoted by **EFGH** in Figure 1.16(a).

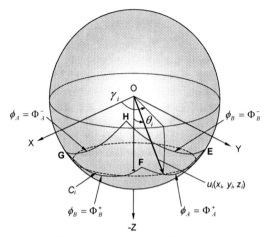

(a) Searching for setup mergability in TOS

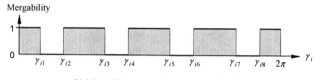

(b) Mergable range of a setup with TAD u_i

Figure 1.16. Setup merging for a 5-axis machine

As shown in Figure 1.16(a), the spherical coordinates of u_i are $(1, \gamma_i, \theta_i)$. By rotating u_i around Z, a circle C_i is obtained.

$$\begin{cases} x_i = \sin\theta_i \cos\gamma_i \\ y_i = \sin\theta_i \sin\gamma_i \\ z_i = -\cos\theta_i \end{cases} \tag{1.16}$$

where, θ_i is a constant and $\gamma_i \in [0, 2\pi]$. The C_i may intersect with the spherical surface patch **EFGH** defined by

$$\textbf{EF:} \quad \phi_A = \Phi_A^+, \quad \phi_B \in [\Phi_B^-, \Phi_B^+] \tag{1.17}$$

$$\textbf{FG:} \quad \phi_B = \Phi_B^+, \quad \phi_A \in [\Phi_A^-, \Phi_A^+] \tag{1.18}$$

$$\textbf{GH:} \quad \phi_A = \Phi_A^-, \quad \phi_B \in [\Phi_B^-, \Phi_B^+] \tag{1.19}$$

$$\textbf{HE:} \quad \phi_B = \Phi_B^-, \quad \phi_A \in [\Phi_A^-, \Phi_A^+] \tag{1.20}$$

where, $[\Phi_A^-, \Phi_A^+]$ and $[\Phi_B^-, \Phi_B^+]$ are the motion ranges of axes A and B, respectively. For $\phi_A = \Phi_A^+$ and $\phi_B \in [\Phi_B^-, \Phi_B^+]$,

$$|z| = \sqrt{\frac{(\cos(\Phi_A^+))^2}{1 + (\cos(\Phi_A^+) * \tan(\phi_B))^2}}, \quad \phi_B \in [\Phi_B^-, \Phi_B^+] \tag{1.21}$$

If $|z_i| < |z|_{min}$, the segment **EF**:$\{ \phi_A = \Phi_A^+ ,\ \phi_B \in [\Phi_B^- , \Phi_B^+]\}$ and the circle C_i has no intersection. If $z_i < 0$ and $|z_i| > |z|_{max}$, the segment **EF** and circle C_i intersect over the entire range of $[0, 2\pi]$. Otherwise, if $z_i < 0$ and $|z|_{min} < |z_i| < |z|_{max}$, **EF** and C_i intersect with each other along the edge of the TOS. Figure 1.16(b) gives the mergable range of the case shown in Figure 1.16(a). This mergable range can be calculated for every 3-axis-based setup. A pose (position and orientation) of the test part that provides the most overlapping mergable range determines a 5-axis-based setup. Figure 1.17 depicts one case of setup merging of the test part after the generic sequence plan in Figure 1.15(b) has been combined for a 5-axis machine.

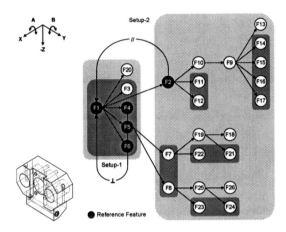

Figure 1.17. Results of setup merging for a 5-axis machine

1.5.2 Detailed Operation Planning

After a merged composite function block (*e.g.* Setup-2 in Figure 1.17) has been dispatched to its dedicated machine, detailed operation planning is performed. The algorithm ALG_INI in each MF-FB can choose a cutter, determine a set of cutting parameters, plan tool path, and generate optional G-code for conventional machines. A knowledge base that contains suggested tools and tool-path patterns for each MF-FB is used to facilitate operation planning. Although the runtime initialisation runs transparently in a controller, a user interface is implemented to visualise the process and to verify the concept, as shown in Figure 1.18. The machining sequence ❶ and setup ❷ are derived based on critical datum references, manufacturing constraints, and the EMF-based reasoning rules, while other detailed machining data ❸–❺ are derived by individual MF-FBs. The data in ❸ and ❹ is used to cut corresponding machining features. The G-code ❺ of a setup (setup-5 merged into base Setup-2 in this case) is generated for the selected machine, by assembling blocks of G-code of each machining feature in the order of the defined sequence. Note that the function of G-code generation is designed into function blocks to best utilise legacy machines. It is triggered when the process plan is dispatched to a non-OAC (open architecture controller)-based machine.

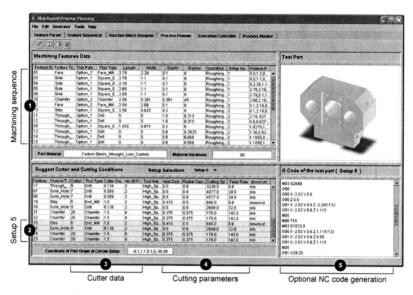

Figure 1.18. Detailed machining data derived by function block embedded algorithms

1.5.3 Function Block Execution Control and Monitoring

Unlike conventional CAPP systems, our function block enabled DPP approach can provide two-way information flow. The monitoring information from bottom up adds value to adaptive process planning and is of vital importance for shop-floor execution control and dynamic scheduling. The current DPP implementation enables both remote monitoring through the Execution Control module and local monitoring beside a machine using the Operation Planning module. TCP/IP through sockets is used for data communication. Figure 1.19 demonstrates one scenario for remote monitoring of another test part. Once a request is sent to the composite function block that runs on a specific machine, its runtime status including current cutting conditions and job completion rate (%) will be sent back to the requester.

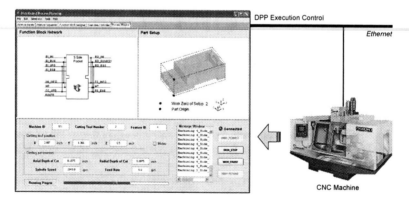

Figure 1.19. Real-time function block execution monitoring

Finally, the two sample parts machined using DPP-generated process plans are shown in Figure 1.20. It is worthy of mention that rather than "*how to do*" defined by ISO-6983 in terms of G-code, the function blocks in DPP only define "*what to do*", which is independent of machines. The detail instructions on "*how to do*" a job, however, is up to the embedded algorithms to determine at runtime.

Figure 1.20. Machined test parts for DPP concept validation

1.6 Conclusions

This chapter presents an innovative *distributed process planning* (DPP) approach to meet the challenges of process planning for dynamic job-shop operations, including the DPP concept, architecture, and algorithms for machining process sequencing, setup merging and function block design. A two-layer hierarchy of shop-level *supervisory planning* and CNC controller-level *operation planning* are considered suitable for separating generic machining data from machine-specific ones. Enabled by function blocks, the machine-specific data are generated at runtime, before plan executions, to better adapt to shop-floor uncertainty.

Within DPP, a process plan of a given part is divided into setups, each of which is a composite function block. In order to achieve generality of a process plan, 3-axis machines and partial sequencing are considered for feature grouping and sequencing using five EMF-based reasoning rules. Setup merging is then performed when a specific machine is selected based on a mergability analysis. Three basic function block types are defined in DPP to facilitate job execution control and machining process monitoring. The DPP concept and algorithms are demonstrated through two test parts. It is expected that the DPP approach will largely increase the flexibility and adaptability of process plans, especially in fluctuating job-shop environments.

Acknowledgments

This research is jointly supported by the Natural Sciences and Engineering Research Council of Canada (NSERC), Materials and Manufacturing Ontario (MMO), and the National Research Council of Canada (NRC). The authors are also grateful to the research assistance provided by Mr. Changjin Song during his graduate study at the University of Western Ontario.

References

[1.1]   Chang, T.-C., Wysk, R.A. and Wang, H.-P., 1991, *Computer-Aided Manufacturing*, Prentice Hall, New Jersey.

[1.2]   Garey, M.R. and Johnson, D.S., 1979, *Computers and Intractability: a Guide to the Theory of NP-Completeness*, W.H. Freeman, San Francisco.

[1.3]   Nieble, B., 1965, *Mechanised Process Selection for Planning New Designs*, ASTME paper 737.

[1.4]   Ong, S.K. and Nee, A.Y.C., 1996, "Fuzzy-set-based approach for concurrent constraint set-up planning," *Journal of Intelligent Manufacturing*, 7(2), pp. 107–120.

[1.5]   Wu, R.-R. and Zhang, H.-M., 1998, "Object-oriented and fuzzy-set-based approach for set-up planning," *International Journal of Advanced Manufacturing Technology*, 14, pp. 406–411.

[1.6]   Lee, K.Y. and Jung M.Y., 1995, "Flexible process sequencing using Petri net theory," *Computers & Industrial Engineering*, 28(2), pp. 279–290.

[1.7]   Yeo, S.H., Ngoi, B.K.A. and Chen, H., 1998, "Process sequence optimisation based on a new cost-tolerance model," *Journal of Intelligent Manufacturing*, 9, pp. 29–37.

[1.8]   Lin, A.C. and Wei, C.-L., 1997, "Automated selection of cutting tools based on solid models," *Journal of Materials Processing Technology*, 72, pp. 317–329.

[1.9]   Edalew, K.O., Abdalla, H.S. and Nash, R.J., 2001, "A computer-based intelligent system for automatic tool selection," *Materials and Design*, 22, pp. 337–351.

[1.10] Lim, T., Corney, J., Ritchie, J.M. and Clark, D.E.R., 2001, "Optimising tool selection," *International Journal of Production Research*, 39(6), pp. 1239–1256.

[1.11] Jung, J.Y. and Ahluwalia, R.S., 1993, "Prismatic part feature extraction and feature-based tool path selection," *Journal of Design and Manufacturing*, 3, pp. 1–19.

[1.12] Boogert, R.M., Kals, H.J. and van Houten, F.J., 1996, "Tool paths and cutting technology in computer-aided process planning," *International Journal of Advanced Manufacturing Technology*, 11, pp. 186–197.

[1.13] Hashmi, K., El Baradie, M.A. and Ryan, M., 1998, "Fuzzy logic based intelligent selection of machining parameters," *Computers & Industrial Engineering*, 35(3–4), pp. 571–574.

[1.14] Arezoo, B., Ridgway, K. and Al-Ahmari, A.M.A., 2000, "Selection of cutting tools and conditions of machining operations using an expert system," *Computers in Industry*, 42, pp. 43–58.

[1.15] Rong, Y. and Zhu, Y., 1999, *Computer-Aided Fixture Design*, Marcel Dekker, New York.

[1.16] Ma, W., Li, J. and Rong, Y., 1999, "Development of automated fixture planning systems," *International Journal of Advanced Manufacturing Technology*, 15(3), pp. 171–181.

[1.17] Ming, X.G., Mak, K.L. and Yan, J.Q., 1999, "A hybrid intelligent inference model for computer aided process planning," *Integrated Manufacturing Systems*, 10(6), pp. 343–353.

[1.18] Sormaz, D.N. and Khoshnevis, B., 1997, "Process planning knowledge representation using an object-oriented data model," *International Journal of Computer Integrated Manufacturing*, 10(1–4), pp. 92–104.

[1.19] Zhang, Y., Feng, S.C., Wang, X., Tian, W. and Wu, R., 1999, "Object oriented manufacturing resource modelling for adaptive process planning," *International Journal of Production Research*, 37(18), pp. 4179–4195.

[1.20] Zhang, F., Zhang, Y.F. and Nee, A.Y.C., 1997, "Using genetic algorithms in process planning for job shop machining," *IEEE Transactions on Evolutionary Computation*, 1(4), pp. 278–289.

[1.21] Morad, N. and Zalzala, A., 1999, "Genetic algorithms in integrated process planning and scheduling," *Journal of Intelligent Manufacturing*, **6**, pp. 169–179.

[1.22] Devireddy, C.R. and Ghosh, K., 1999, "Feature-based modelling and neural network-based CAPP for integrated manufacturing," *International Journal of Computer Integrated Manufacturing*, **12**(1), pp. 61–74.

[1.23] Monostori, L., Viharos, Z.J. and Markos, S., 2000, "Satisfying various requirements in different levels and stages of machining using one general ANN-based process model," *Journal of Materials Processing Technology*, **107**, pp. 228–235.

[1.24] Tseng, Y.-J. and Joshi, S.B., 1994, "Recognising multiple interpretations of interacting machining features," *Computer-Aided Design*, **26**(9), pp. 667–688.

[1.25] Wang, L. and Norrie, D.H., 2001, "Process planning and control in a holonic manufacturing environment," *Journal of Applied Systems Studies*, **2**(1), pp.106–126.

[1.26] Vosniakos, G.C. and Davies, B.J., 1993, "Knowledge-based selection and sequencing of hole-making operations for prismatic parts," *International Journal of Advanced Manufacturing Technology*, **8**, pp.9–16.

[1.27] Stori, J.A. and Wright, P.K., 1996, "A knowledge-based system for machining operation planning in feature based, open-architecture manufacturing," In *Proceedings of ASME Design Engineering Technical Conference*, Irvine, CA.

[1.28] Kimura, F., 1989, "High-quality product realisation through the tight integration of product design and process planning," In *Proceedings of CIRP International Workshop on CAPP*, pp. 199–208.

[1.29] Srinivasan, M. and Sheng, P., 1999, "Feature based process planning in environmentally conscious machining – part 1: microplanning," *Robotics and Computer Integrated Manufacturing*, **15**(1), pp. 257–270.

[1.30] Srinivasan, M. and Sheng, P., 1999, "Feature based process planning in environmentally conscious machining – part 2: macroplanning," *Robotics and Computer Integrated Manufacturing*, **15**(1), pp. 271–281.

[1.31] Khoshnevis, B., Sormaz, D.N. and Park, J.Y., 1999, "An integrated process planning system using feature reasoning and space search-based optimisation," *IIE Transactions*, **31**, pp. 597–616.

[1.32] ElMaraghy, W.H., 1992, "Integrating assembly planning and scheduling – CAPP related issues," *Annals of CIRP*, **41**(1), pp. 11–14.

[1.33] Tu, Y., Chu, X. and Yang, W., 2000, "Computer-aided process planning in virtual one-of-a-kind production," *Computers in Industry*, **41**, pp. 99–110.

[1.34] Smith, C.S. and Wright, P.K., 1996, "CyberCut: a World Wide Web based design-to-fabrication tool," *Journal of Manufacturing Systems*, **15**(6), pp. 432–442.

[1.35] Park, H.G. and Baik, J.M., 1999, "Enhancing manufacturing product development through learning agent system over Internet," *Computers & Industrial Engineering*, **37**, pp. 117–120.

[1.36] Sun, J., Zhang, Y.F., and Nee, A.Y.C., 2001, "A distributed multi-agent environment for product design and manufacturing planning," *International Journal of Production Research*, **39**(4), pp. 625–645.

[1.37] Zhao, F.L., Tso, S.K. and Wu, P.S.Y., 2000, "A cooperative agent modelling approach for process planning," *Computers in Industry*, **41**, pp. 83–97.

[1.38] Sluga, A., Butala, P. and Bervar, G., 1998, "A multi-agent approach to process planning and fabrication in distributed manufacturing," *Computers & Industrial Engineering*, **35**(3–4), pp. 455–458.

[1.39] Gu, P., Balasubramanian, S. and Norrie, D.H., 1997, "Bidding-based process planning and scheduling in a multi-agent system," *Computers & Industrial Engineering*, **32**(2), pp. 477–496.

[1.40] Dornfeld, D., Wright, P.K., Wang, F.-C., Sheng, P., Stori, J., Sundararajan, V., Krishnan, N. and Chu, C.-H., 1999, "Multi-agent process planning for a networked machining service," *Technical Paper of North American Manufacturing Research Conference*, MS99–175, pp. 1–6.

[1.41]   Shih, W. and Srihari, K., 1995, "Distributed artificial intelligence in manufacturing systems control," *Computers & Industrial Engineering*, **29**(1–4), pp. 199–203.

[1.42]   ElMaraghy, H.A., 1993, "Evolution and future perspectives of CAPP," *Annals of CIRP*, **42**(2), pp. 739–751.

[1.43]   Wang, L., 2001. "Integrated design-to-control approach for holonic manufacturing systems," *Robotics and Computer Integrated Manufacturing*, **17**(1–2), pp. 159–167.

[1.44]   Wang, L., Zhao, W., Ma'ruf, A. and Hoshi, T., 1996, "Setup-less fabrication technology incorporated with machining feature-based CAD/CAM system for machining workshop," In *Proceedings of International Manufacturing Engineering Conference*, pp. 95–97.

[1.45]   Liu, X., 2000, "CFACA: Component framework for feature-based design and process planning," *Computer-Aided Design*, **32**(7), pp. 397–408.

[1.46]   Little, D., Rollins, R., Peck, M. and Porter, J.K., 2000, "Integrated planning and scheduling in the engineer-to-order sector," *International Journal of Computer Integrated Manufacturing*, **13**(6), pp. 545–554.

[1.47]   Wang, L., Hao, Q. and Shen, W., 2003, "Function block based integration of process planning, scheduling and execution for RMS," In *Proceedings of the 2nd CIRP International Conference on Reconfigurable Manufacturing*, (CD-ROM).

[1.48]   Wang, L., Feng, H.-Y. and Cai, N., 2003, "Architecture design for distributed process planning," *Journal of Manufacturing Systems*, **22**(2), pp. 99–115.

[1.49]   Wang, L., Hao, Q. and Shen, W., 2006, "A novel function block based integration approach to process planning and scheduling with execution control," *International Journal of Manufacturing Technology and Management*, in press.

[1.50]   Kim, Y.S. and Wang, E., 2002, "Recognition of machining features for cast then machined parts," *Computer-Aided Design*, **34**(1), pp. 71–87.

[1.51]   Li, W.D., Ong, S.K. and Nee, A.Y.C., 2002, "Recognising manufacturing features from a design-by-feature model," *Computer-Aided Design*, **34**(11), pp. 849–868.

[1.52]   Sundararajan, V. and Wright, P.K., 2004, "Volumetric feature recognition for machining components with freeform surfaces," *Computer-Aided Design*, **36**(1), pp. 11–25.

[1.53]   IEC 61499-1, 2005, *International Standard of Function Blocks – Part 1: Architecture*, International Electrotechnical Commission, pp. 1–111.

[1.54]   Gindy, N.N.Z., 1989, "A hierarchical structure for form features," *International Journal of Production Research*, **27**(12), pp. 2089–2103.

[1.55]   Kimura, F. (ed.), 1995, *Computer-Aided Tolerancing – Proceedings of the Fourth CIRP Design Seminar*, Chapman & Hall.

[1.56]   Boerma, J.R. and Kals, H.J.J., 1989, "Fixture design with FIXES: the automated selection of positioning, clamping and support features for prismatic parts," *Annals of CIRP*, **38**, pp. 399–402.

[1.57]   Rong, Y., Liu, X., Zhou, J. and Wen, A., 1997, "Computer-aided setup planning and fixture design," *International Journal of Intelligent Automation and Soft Computing*, **3**(3), pp. 191–206.

[1.58]   Cai, N., Wang, L. and Feng, H.-Y., 2005, "Adaptive setup planning of prismatic parts by tool accessibility examination," In *Proceedings of 2005 ASME International Mechanical Engineering Congress & Exposition*, Paper # IMECE2005-81055.

Web-based Polishing Process Planning Using Data-mining Techniques

V.Y.M. Tsang[1], B.K.K. Ngai[1], G.Q. Huang[1,*], V.H.Y. Lo[1] and K.C. Cheng[2]

[1] Department of Industrial and Manufacturing Systems Engineering
The University of Hong Kong, Pokfulam Road, Hong Kong, China
*Email: gqhuang@hkucc.hku.hk

[2] PME International Company Limited, Hong Kong, China

Abstract
A Web-based portal system (WBPS) is developed to implement process planning that aims to streamline polishing products and processes. This Web application system mixes its functions of providing intelligent decision support to polishing enterprises by facilitating the sharing of vast collective polishing knowledge, as well as addressing the hindrance problem of subjective human determination of knowledge discovery using data-mining techniques. WBPS will create an important knowledge base for parameter optimisation using fuzzy logic and genetic algorithms through laboratory experiments and field studies within collaborating companies. Another aim of developing WBPS is to cope with the vast collective polishing knowledge, information sharing across the companies and applicable case initialisation. Functionality of WBKP will be explained with provision of online interfaces, access integration to polishing expertise and values, and application embedment to serve as self-documenting activities.

2.1 Introduction

Polishing process is a mechanical finishing process regarded as a common practice to enhance product value [2.1][2.2] by removing a considerable amount of metal or non-metallic and to smooth a particular surface. Figure 2.1 illustrates a model to simply demonstrate how polishing generates additional value for products. It has been widely used as a finishing process for a wide variety of manufactured products, including kitchenware, watch belts, jewellery, automobile interiors and other souvenirs items.

In polishing industries, product designers and production engineers have developed a huge range of products for their customers or buyers, meeting diverse product requirements. They will also practice product customisation to design products for customer approval according to customers' specific requests and updated market trends. This kind of continuously renewed product knowledge can become a company's competitive advantage and further implement Customer

Relationship Management (CRM) well in developing their own marketing strategies. Polishing enterprises can make efficient use of their practicable past experiences in product development to swiftly generate polishing process plans after receiving customised feature specification from their customers.

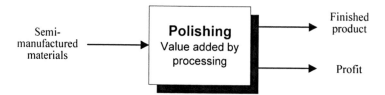

Figure 2.1. Basic model of polishing process adding product value

An ideal practice is for polishing enterprises to store successful and useful projects on their database system after the projects are completed. Key polishing information such as product structure and operation process parameters can be referred to in the future, provided that production reordering is launched or products of similar kinds are exploited. However, it is not unusual to see polishing technicians and operators making their own judgments to determine which projects to retrieve for use upon receiving a new product specification request. There is a lack of systematic approaches to case matching by impersonal assessment of polishing projects. Also, polishing enterprises seem to be unable to store this valuable information and share among their departments. These pieces of polishing experience are only focused in a small group of experienced polishing masters, which may encounter a possibility to be lost or to gradually vanish after these masters retire. Polishing enterprises seem to have no system for keeping track of this vital polishing information [2.2], storing key process parameters and interacting between parties and providing decision support [2.3][2.4]. Furthermore, another significance of process planning is to support the sharing of polishing operation knowledge and experience. However, the current situation is that it is not easy to extract this polishing knowledge because the processes are still operated on the basis of traditional subjective human determination, even if most polishing industries make use of automatic machinery nowadays. The quality of polished products can not be guaranteed.

In an attempt to relieve the information-sharing, case-retrieval and knowledge-extraction problem, our project is to develop a Web-based portal system aimed to propose a total solution to streamline the polishing product and process and to facilitate communication across the polishing enterprise. The system is supported by development of a knowledge base for optimisation of polishing process parameters. Up to now, the mechanism of polishing processes is still under investigation. It could be regarded as a black box by outsiders. Fuzzy modelling is one of the techniques currently used for extraction of knowledge of nonlinear, uncertain, and complex systems [2.5][2.6] and it has been applied successfully in many cases [2.7], especially in control engineering. Park et al. [2.8] showed that the introduction of genetic algorithms could improve the performance of fuzzy control systems. The introduction of genetic algorithms into fuzzy systems, which is known as genetic-fuzzy systems, was successfully applied to different areas [2.9][2.10]. However, it has not been applied in the polishing of stainless steel.

This chapter is organised as follows. A literature review is presented in Section 2.2. Then the concept of process planning for polishing is described in Section 2.3. Section 2.4 illustrates the framework of a Web-based portal system for polishing. Section 2.5 describes the methodologies of knowledge-base development. The results and discussion of the knowledge-base development are delivered in Section 2.6. Finally, conclusions are presented in Section 2.7.

2.2 Literature Review

2.2.1 Research Works in Polishing

In the study of polishing, research efforts have been focused on the polishing mechanism and modelling of the polishing process. In abrasive polishing, Samuel [2.11] stated that several researchers tried to explain the interactions between the abrasive and the substrate in the 19th century. However, it has recently been established that the concept is false. The polishing mechanism seems to be not fully understood and there are still many uncertainties. Currently, major polishing guidelines are provided. Davis [2.12] stated qualitative procedures to achieve different grades of finishes. Dickman [2.13] suggested wheel speeds for hand buffing for different materials and production techniques for achieving satin finishing, cutdown buffing and colour buffing for some materials. Reyers [2.14] recommended the work pressure and wheel speed for different applications. It seems to be too general to apply these features in specific materials.

2.2.2 Web Application for Knowledge-based Planning

Knowledge application and knowledge creation are at the core of any organisation's existence. Without knowledge, companies cannot survive. The relevance of various knowledge differs between organisations and changes over time. A professional organisation, that is, an organisation of so-called "knowledge workers" [2.15][2.16], leans more heavily on its intellectual assets than the average production firm.

Knowledge management is the management of corporate knowledge that can improve a range of organisational performance characteristics by enabling an enterprise to be more "intelligent acting". It is not a new movement *per se*, as the organisations have been trying to harness their internal processes and resources that have resulted in various movements over the years, such as total quality management, expert systems, business processes re-engineering, the learning organisation, core competencies, and strategy focus.

Most organisations already have a vast reservoir of knowledge in a wide variety of organisational processes. However, these organisations could not fully utilise their valuable recorded knowledge to further their business. Up to 2001, it did not seem evident that managerial users could focus on developing business applications that provide them with predefined planning reports giving them the information they need for decision-making purposes [2.17]. Even if the concept of a management information system (MIS) started to be developed by 1970, such systems are still not adequate to meet many of the decision-making needs of management [2.17]. This knowledge is quite diffused and mostly unrecognised. Often, organisational culture itself prevents people from sharing and disseminating their know-how in an effort to hold onto their individual powerbase and viability. Determining who knows what in an organisation itself could be a time-consuming and daunting task. This, in itself, justifies the need for a system to manage important knowledge assets for the organisations to allow them to identify and access workers' skills and expertise.

A Web-based portal system is a knowledge-based system developed with artificial intelligence that has many applications in the cognitive science area [2.17]. It enables e-business by provision of a unified application access, information and knowledge management within enterprises, and between enterprises and their trading partners, channel partner and customers. Users can add a knowledge base and some reasonable capability to the information system. A portal system is regarded as a key technical application to create an interface that presents polishing information to users and a gate to users to access the required data sources and knowledge acquisition [2.3][2.18]. Significant and invaluable polishing knowledge is collected and kept track of for future use [2.2], providing a way to shorten the product development time in later process planning and to smooth the knowledge management. It can be viewed as a way to access disseminated information within a company since information chunks can be stored in various systems using different formats. One of the major differences between a traditional Web site and a portal resides in the fact that the portal is usually tailored according to users' need. A portal is, consequently, a single point of access to Internet resources, an integration platform focusing on unification oriented towards the business processes of the companies. Therefore, portals synchronise knowledge and applications, creating a single view into the organisation's intellectual capital. The best example in this category is Yahoo. Today, the term "portal" is widely used to describe many different modules with corresponding interfaces for different purposes.

An enterprise portal can be defined as a single point of access (SPOA) for the pooling, organising, interacting, and distributing of organisational knowledge [2.19][2.20]. While it is true that enterprise portals are useful because this technology can bring about cost reduction, organised and structured information, and reduced access time, their competitive advantages are inherent in their abilities to filter, target, and categorise information so that users will get only what they need [2.21]. My Yahoo!, which was developed in 1998, was the first personalised Internet enterprise portal. Campus portals were pioneered by UCLA in 1999 [2.22]. Enterprise portals have fairly complex structures and features. However, their basic functions and elements are relatively easy to define. First, from an operational perspective, the strength of corporate portals lies in their ability to provide Web-based access to enterprise information, applications and processes. Second, in a functional view, they leverage existing information systems, data stores, networks, workstations, servers and applications as well as other knowledge bases to give each employee in every corporate site immediate access to invaluable sets of corporate data anytime and anywhere [2.20][2.23]. These capabilities are made possible because the generic framework is essentially focused on delivering information to the users from disparate databases.

2.2.3 Case-based Reasoning

Case-based reasoning stands for adapting the past solutions to resolve new problems or requests. It involves utilisation of precedent-applicable cases as paradigms to explain the new case. Sometimes, these selected existing cases even undergo revision in an attempt to match a new case's problem. A reasoning mechanism of case-based reasoning should be a learning system so that it has the capability to conduct evolution by itself, with the use of its own knowledge, to generate more feasible and consistent solutions to problem cases.

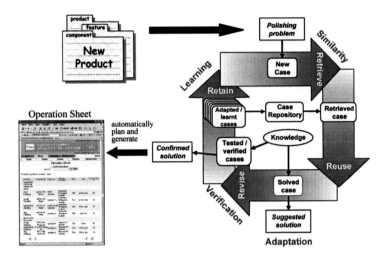

Figure 2.2. Artificial process planning with CBR to polishing problem

With adoption of the concept of case-based reasoning developed by Aamodt and Plaza [2.24], a polishing problem is studied to be resolved by optimal solutions derived from artificial process planning of WBPS, as demonstrated in Figure 2.2.

1. Case retrieval: The objective of this step is to retrieve the old cases stored in case libraries.
2. Case reuse: In case-based process planning, the selected old solution case is used as an inspirited candidate solution to solve the new problem case.
3. Case revise: Since a new problem case may not exactly match the old ones, the old knowledge may often need to be revised to generate a proposed solution case to fit the problem. The proposed solution will be verified as to whether it can solve the problem.
4. Case retain: Provided the new problem is solved, it is stored in case libraries for future use.

2.2.4 Fuzzy Modelling

Conventional approaches to system modelling mainly relied on mathematical functions for a precise description of a physical system [2.25]. One of the disadvantages of using conventional approaches is that the parameters obtained may not be easy to be interpreted by human beings. In addition, the approximation may not be good enough if a system that is to be modelled is highly nonlinear [2.26]. It was hard for users to extract knowledge if the mechanism of a system was unknown or highly nonlinear.

Fuzzy modelling is one of the techniques for nonlinear, complex systems. The underlying principle of this technique is based on fuzzy set theory, which was developed by Zadeh [2.27]. Details of the important concepts of fuzzy set theory were defined by [2.28]. Fuzzy set theory was the foundations of fuzzy logic (FL) systems, and the most important area of applications is fuzzy rule-based system (FRBs) [2.26].

In FRBs, fuzzy rules are presented in "IF-THEN" form. An "IF" part and a "THEN" part correspond to an antecedent part and a consequent part, respectively. Linguistic variables and their linguistic values are used in fuzzy rule representation and their meanings are defined by membership functions [2.29]. Mamdani and Takagi–Sugeno–Kang are two types of FRBs currently used for dealing with engineering problems. Mamdani-type FRB was the first FRB, developed by Mamdani in 1974, and is known as a fuzzy logic controller (FLC).

A FLC contains four major blocks. The FLC knowledge base consisted of two components: a database (DB) and a rule base (RB). The DB contains the membership functions and the RB contains a collection of fuzzy control rules. The operations of FLCs are well described by [2.26].

2.2.5 Genetic Algorithms

The formulation of genetic algorithms was started in the early 1950s for the simulation of biological system, and the work was done in the late 1960s and early 1970s by John Holland [2.30]. The operating principle is based on natural selection and natural genetics. It is summarised as follows: each chromosome represents a potential solution to a problem. Genetic algorithms start with chromosome initialisation. The fitness of each chromosome is evaluated by a user-defined fitness function. Chromosomes with higher fitness will have a higher chance to survive and *vice versa*. Crossover and mutation will be conducted so that a new search space can be exploited. The above processes are repeated until user-defined termination criteria such as the number of generations and fitness values are reached. After reaching the termination criteria, the best chromosome will be decoded into a real solution to the problem. Details of the mechanism can be found in [2.30][2.31].

The advantage of using genetic algorithms is that they are very effective in performing global searches. In addition, they can handle different kinds of objective functions and constraints [2.31].

2.2.6 GA-Fuzzy Systems

Park *et al.* [2.8] pointed out that the completeness of the fuzzy rule base, the subjective definition of a fuzzy subset and the choice of fuzzy implication operators are the challenging problems in the application of fuzzy modelling, and stated that the selection of membership functions have a significant effect on the output. The definition of fuzzy rules and the membership functions is one of the key questions in designing the FLCs. Genetic algorithms had been employed by many researchers to generate fuzzy rules and membership functions [2.32]. GA has demonstrated its power by producing promising results through different applications [2.10]. The approach is called a *genetic fuzzy system* [2.33]. The advantages of applying genetic algorithms into fuzzy models over other methods are stated in [2.8].

The automatic generation of fuzzy rules and membership functions using genetic algorithms has been investigated recently and had been categorised into four types [2.34]:

Type 1: learning fuzzy membership functions with fixed fuzzy rules
Type 2: learning fuzzy rules with fixed membership functions
Type 3: learning fuzzy rules and membership functions in stages
Type 4: learning fuzzy rules and membership functions simultaneously

Experiments have been conducted to compare the performance of DC motors using type 1, type 2 and type 4 approaches, and showed that the performance of fuzzy models was the best using type 4 [2.8].

In this chapter, type 4 is used to generate an optimal knowledge base for the polishing process of SS304 stainless steel.

2.3 Polishing Process Planning

Process planning can be defined as a systematic selection, determination and generation of detailed methods by which requested products or parts are manufactured from an initial form to a final form. For a polishing enterprise to proceed the polishing operations for products that meet the customers' appearance specification, the polishing processes of each feature part of the product must be thoroughly planned. Customers' requirements on polished products basically come from their satisfaction of products with respect to their appearance quality, aesthetics, reliability and durability. On the other hand, only guaranteeing that the products can meet the polishing quality specification is not enough. Like many other manufacturing processes, polishing processes of the products must be cost effective. In other words, the products must be polished to maximise the added value within the agreed processing time and production cost.

2.3.1 Purpose of Polishing Process Planning

Precision surface finishing makes remarkable finishing progress to complete the final product manufacturing. Precision finishing is mainly accomplished by lapping, grinding and polishing, according to Heiche and Spengler [2.36]. On the other hand, they mentioned that polishing can be considered the best finishing methods among the three, in the context of yielding the best precision and finish. Therefore, a scrupulous planning of the polishing process is a crucial stage of product manufacturing and requires defining the selection and proceeding of processes and operations to achieve the requested products. The main objective of polishing process planning is to identify the main tasks and corresponding detailed work instructions implemented to generate a well-designed polishing process plan.

Examples of the tasks include selection of polishing operations, equipments, tooling, wheels and compounds used. In some special requirements, selection of quality levels is also involved by determining the process parameters. It can also be used to avoid the costly changes due to manufacturability problems, one of which is the quality levels of products determined by factors such as product components, features and process parameters. Therefore, process planning becomes important in stipulating the polishing activities to provide the best products to customers. Several contexts of process planning are described as:

- identify the product and components;
- identify the polishing features corresponding to the products;
- identify the main operations undertaken during process planning;
- specify and describe process parameters of each operation;
- specify assets and resources required for each operation;
- identify the main process planning documentation;
- identify the main data libraries used to store key polishing information.

These tasks in polishing process planning are implemented for domain specification in processing a successful case with appropriate adoption of equipments, wheels, compounds, and other expected resources for a total polishing solution.

2.3.2 Design of Polishing Process Planning

As polishing process planning aims to meet the production need, which in turn determines the final product quality in the whole operation management, the input parameters of the process planning play an important role in minimal reachable quality control, such as the product's feature specification, sequencing and optimisation and so on. This can be determined through the identification of the product item and consultation of those customers. Figure 2.3 shows the stages of polishing process planning adopted in polishing enterprises.

Within the flow chart, identification of the product item is constantly referred to during process planning. It formally states the customer's quality requirement, as this requirement plays a pivotal role in settling the quantitative stages of the process plan afterwards. Process planning is constructed by selection and sequencing of polishing processes and operations to produce finished products and components from initial materials. It is a way to prepare detailed work instructions including selection of polishing equipments, resources such as wheels and compounds, polishing time and so on. Another essential procedure is to specify polishing process parameters whose values regulate the product quality. In the polishing industry, polished products, for example, stainless steel cooker, glasses, are complex in shape and structure. They involve many components that are composed of various kinds of polishing features. These product hierarchical structures raise the complexity of polishing processes. This is the most apparent issue for the enterprise when considering the appearance-polishing process to products.

As illustrated in Figure 2.4, polished products form a hierarchical structure to components, features, polishing operations with reference to each feature and process parameters to generate the polishing operation sheet. In general, this operation sheet is the output of polishing process planning. According to this figure, we can see that when the complexity level of a product's design structure increases, required efforts to make an appropriate process plan on a larger number of components, features of each component and the product assembly will improve correspondingly.

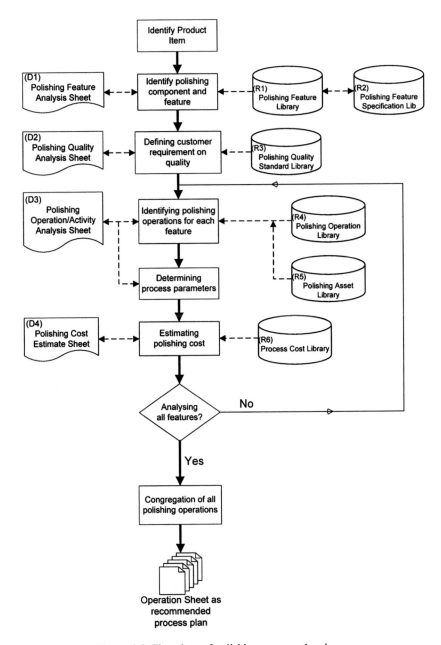

Figure 2.3. Flowchart of polishing process planning

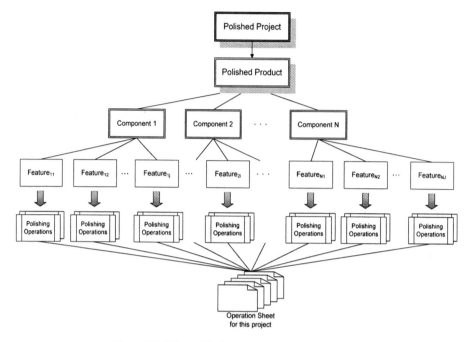

Figure 2.4. Hierarchical structure of polished product

2.4 Web-based Portal System for Polishing

This section presents an online system to streamline polishing products and processes through development of a Web-based portal system (WBPS). It is also used to communicate with users across the enterprise and even the whole industry. WBPS is a key technical application to create an interface that presents polishing information to users and a gate to users to access the required data sources and knowledge acquisition [2.3][2.18]. It can keep track of vital polishing information [2.2] by storing key process parameters, interacting between collaborating data and making decisions [2.3][2.4], and provides a benchmarking tool to smooth the knowledge management work by users, including disciplines of business intelligence management, information system resources [2.37]. Without adopting a knowledge-management approach, polishing enterprises might find it quite difficult to capture the key manufacturing knowledge and to retain recordable polishing information. Therefore, developing WBPS can benefit the polishing enterprises via important polishing knowledge integration and application, as well as its simpler circulation among departments and workers [2.4][2.37], which is expected to be shared in a multi-functional environment quickly and easily [2.38][2.39].

Figure 2.5 illustrates an overview of polishing knowledge sharing through the Web portal. An understanding of WBPS is dedicated to express its capability to locate and organise together the relevant polishing information involving a series of continuous processes through accessing structured datasets [2.4][2.37][2.40].

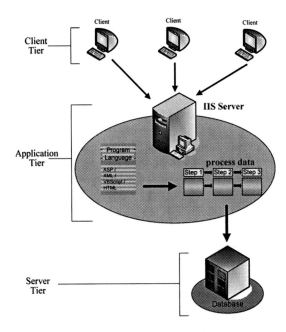

Figure 2.5. Knowledge integration in multi-functional environment by WBPS

2.4.1 Problem Definition

Current polishing practices are merely paper-based manual approaches, which are most likely unreliable and subject to human personal variations. WBPS has a purpose to develop an online portal system to streamline the polishing processes and products analysis (PPPA) such that a more scientific decision making can be generated with unambiguous expression of an innovative framework, and to build up a Web system to fulfil polishing product development analysis. Most often, users can even generate a higher value of polished products [2.1][2.2] through WBPS as intellectual and knowledge-based assets.

2.4.2 Objectives

The core points of introducing this system are the fixation of knowledge acquisition of parameters information in polishing processes and enabling that the key data of superior polishing features can be enrolled in recordable form [2.1]. While the portal system is attempting to build up a systematic knowledge-based support for polishing processes, it aims to achieve four main technological objectives:

- To identify and record polishing-incurring features of polished products and define quality requirements.
- To recognise relationships between polishing-process parameters and their corresponding quality levels through substantiated knowledge discovery.
- To collect information of polishing operations and specific quality levels they

could give rise to, and to prepare for building up of a discovery algorithm for optimisation of polishing-process parameters.

- To furnish a decision-support mechanism for polishing industries by WBPS's optimisation mechanism to produce appropriate and cost-effective polishing quality levels.

2.4.3 Design of Web-based Portal System

The framework of the portal system, its design and structure, as well as the organisational system are introduced as a tool to provide a platform to manage polishing information. Table 2.1 shows how knowledge application is conducted through WBPS.

Table 2.1. Knowledge application through WBPS

Activity	Selection criteria	System function
• Knowledge acquisition	Recognition procedure	Knowledge store and retrieval
• Recommendation	Algorithm procedure	Optimization analysis
• Product development	Product build-up	Case application

The Web-based portal system makes an important contribution to enable users to obtain efficient access to various types of structure and unstructured datasets (like features or operations) to streamline polishing process. The system also facilitates perspective-based combination and integration of information retrieval. Basically, WBPS is composed of 4 modules: Manual Planner, Repository, Business Cases and AutoPlanner. Figure 2.6 shows a graph to demonstrate the 4 architectural modules of the WBPS for polishing information sharing and how the process planning of polishing products and processes is accomplished.

2.4.3.1 Manual Planner

Planner is a starting user interface to allow them to process polishing projects. They either receive the enquiries or make stratified or "layer-to-layer" selection on project particularities. This kind of item-selection style provides a short time to give rise to a recommended polishing solution. It is designed to focus only on the work expected to be done by users planning to work out polished products structures and required operation sheets accordingly.

- Planner mainly builds a polishing project list for identification, management and accomplishment of product-oriented worksheets in Web page format.
- It helps look up the requested project, product, corresponding components and features, and operation processes quickly.
- It enables users to build up, delete or edit project or product details upon specific requirement of polishing process parameters.
- Mapping of product hierarchy helps manage its corresponding product structures and related process classes in a collaborative secured way.

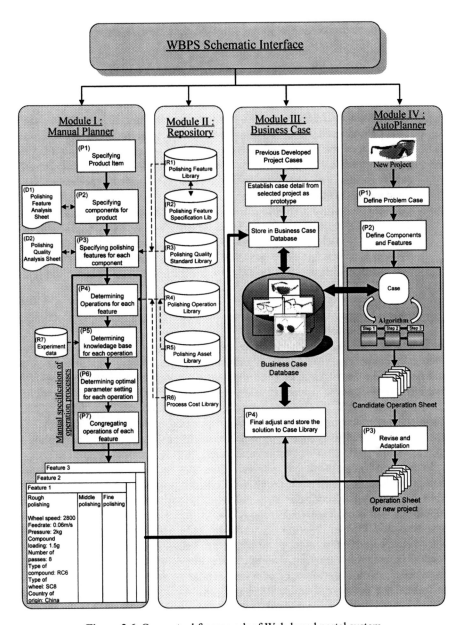

Figure 2.6. Conceptual framework of Web-based portal system

2.4.3.2 Repository

Repository consists of libraries for storing polishing information, related to the polishing processes. These data are self-contained information that can be used independently or shared with other components to satisfy the Planner requirement. Elementary data in the Repository module need to be customised in order to allow

Planner to handle the project planning specific to a new request of a conceptual product.

- Repository is managed to acquire the key polishing knowledge in the form of documentary, libraries, information record.
- It is referring to a quantified presentation of abstract polishing parameters in recordable form to users, so they can specify suitable and workable values to these parameters for streamlining the polishing processes and products.
- Definition of the content of polishing information can be supported by this package to store the key values. Products, components, polishing features, quality levels and operation process parameters can be generated for a standard meta database.
- Libraries are utilised to share polishing information categories for internal departments such as Project Team, Product Development Department, Operation Department and Quality Department, and for outside parties like customers or vendors. With a system to implement this strategy, every party can share significant information in order to achieve a better polishing product and process.

2.4.3.3 Business Cases

The Business Case module is another knowledge storage in WBPS. The Repository module contains libraries to record polishing process data. The Business Case module, on the other hand, is evolved to register the whole polishing projects in the format of cases. Each case consists of all elements of the polishing projects, such as products, hierarchical structures of products' components and features, polishing operations to make the qualities and corresponding process parameters. All the aspects are captured in Business Case databases to enable search, categorisation or other tasks in support of AutoPlanner, the fourth module of WBPS.

In terms of implementation, project cases can be built up from the Manual Planner module, in which the Operation Sheets for polished products are manually created and stored in databases as a practical case to achieve working memories. Another derivation of cases in Business Case is the adaptation of applicable cases from the AutoPlanner module. It involves many abstract case concepts and underlying product structures to mandate the knowledge-based mechanism. The bilateral case sharing between Business Case and AutoPlanner is shown in Figure 2.6, where it is clear that AutoPlanner searches and retrieves case records from the Business Case database for algorithm analysis. While the expected-to-be suitable case is extracted, it can be revised to solve the problems and for adaptation, then assembled to the database in the Business Case module.

2.4.3.4 AutoPlanner

AutoPlanner undertakes an essential contribution to develop an automatic processing algorithm to search for existing solutions, to revise the retrieved cases and to delve into applicable new solutions to the problem cases. It provides users with Operation Sheets of project solutions best resolving the new product prototype. This module will also help the users save valuable time in launching projects of similar products, so that a better time-to-market and quick return on its knowledge investment can be made.

- AutoPlanner provides a platform for users to define new product design and desired quality levels. Projects of similar product specifications or prototype styles can be looked up from the database of the Business Case module to search for existing solutions, where advantageous polishing features and process parameters can be traced and revised to new solutions.
- Polishing product and process are defined dynamically. New product structure and operation specification can be built on the basis of elite selected existing projects, followed by customisation of product details. This approach allows for a mapping of learning style to any successful product development.
- It is capable of building new projects from existing projects and implementing modifications. This facilitates product development improvement by driving knowledge exchange and brainstorming procedures across parties. It integrates the development of new products from past successful cases and advanced technical skills for products.
- AutoPlanner analyses the function gaps of existing products in terms of what functions the products can perform, what key features need to be developed and the cost. This can be managed and responded to through this module.

2.4.4 Implementation of Web-based Portal System

Implementation of the WBPS will be discussed in this section. Polishing process planning is detailed to show how a knowledge-based system is deployed through the Web application to conduct process planning for the Operation Sheet of problem cases. It is utilised as an identification to specialise every polishing project based on stating its key polishing properties or relevant product attributes. Figure 2.7 and Figure 2.8 show a simplified diagram to demonstrate how the Web application supports the streamlining procedures of polishing process planning and libraries management under different modules.

2.5 Knowledge-base Development Methodology

2.5.1 General Framework

A general framework of knowledge-base development is shown in Figure 2.9. The encoded chromosomes consist of fuzzy rule sets and membership functions. The internal structure of chromosomes is shown in Figure 2.10.

The generic format of fuzzy rule, R_i, is expressed in the following format.

R_i : IF x_1 is A_1 and x_2 is A_2 and ... and x_n is A_n
THEN y_1 is B_1 and ... y_m is B_m

where
$x_1, x_2, ... x_n$ = input variables
$y_1, y_2, ... y_m$ = output variables

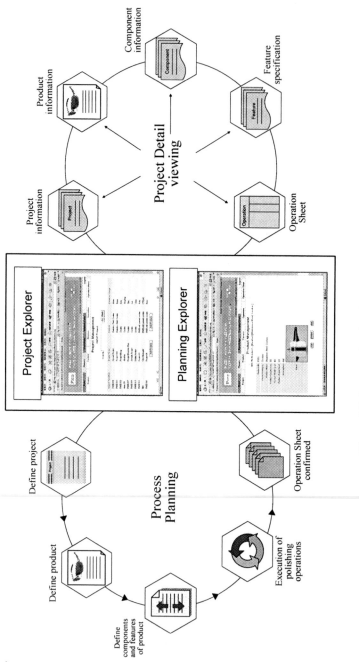

Figure 2.7. Process planning flow path of WBPS

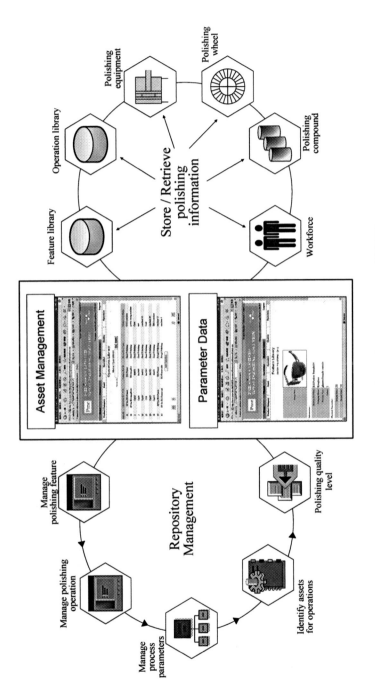

Figure 2.8. Repository information management of WBPS

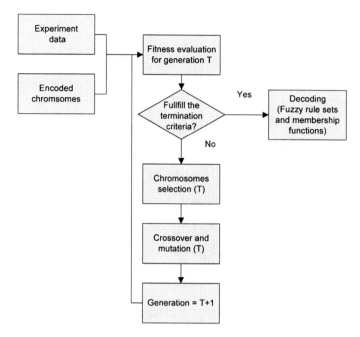

Figure 2.9. Framework of knowledge-base development

$A_1, A_2, \ldots A_n$ = fuzzy sets defined in the universe of discourse $U_1, U_2 \ldots U_n$
$B_1, B_2, \ldots B_m$ = fuzzy sets defined in the universe of discourse $U_1, U_2 \ldots U_m$

Parametric representation of a membership function of a fuzzy set A of a process parameter P is defined in the following.

$$MF_{PA} = (A_{(start)}, A_{(righttop)}, A_{(lefttop)}, A_{(finish)})$$

where

$A_{(start)}$ = the value on the left-hand side of the membership functions defined in the universe of discourse of fuzzy set A when $u(A_{(start)})$ is zero

$A_{(lefttop)}$ = the value on the left-hand side of the membership functions defined in the universe of discourse of fuzzy set A when $u(A_{(lefttop)})$ is one

$A_{(righttop)}$ = the value on the right-hand side of the membership functions defined in the universe of discourse of fuzzy set A when $u(A_{(righttop)})$ is one

$A_{(finish)}$ = the value on the right-hand side of the membership functions defined in the universe of discourse of fuzzy set A when $u(A_{(finish)})$ is zero

In fitness evaluation, the condition part of encoded chromosomes and past data will be compared to see whether they match or not. If their conditions match, the resultant parts will be compared. The fitness value will be assigned based on the proximity of the resultant part of chromosomes and past data. The higher the proximity of the resultant parts, the higher the fitness value of the chromosome. If the condition parts do not match, zero fitness values will be assigned. After fitness evaluation, chromosomes will be selected based on the fitness values for crossover and mutation to form a new population. The best chromosome, which contains the optimal fuzzy rule set and membership functions, will be selected and decoded after fulfilling the termination criteria. The decoded chromosome will be the optimal knowledge base for parameter optimisation.

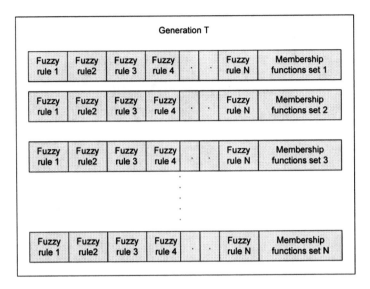

Figure 2.10. Internal structure of chromosomes in generation *T*

2.5.2 Case Study

2.5.2.1 Problem Description

The polishing process has been widely applied to different product categories. Products such as ornaments and jewels with polished surfaces could increase their attractiveness. A polished surface not only increases attractiveness, but also affects issues related to safety and health. However, the current practice is that the subjective judgement of polish masters determines polishing parameter settings. Parameter settings may be different from one to another and they are hard to express. Therefore, the quality of the surface finish could not be guaranteed. It is difficult to establish relationships between input parameters and the quality of the surface finish using the conventional approach as the polishing mechanism is not clear [2.35]. Fuzzy logic has been applied to model complex systems to determine outputs, which are known as fuzzy systems, since a knowledge base that consists of a fuzzy rules set and a membership function set affects the performance of the fuzzy system. The problem is to develop the best knowledge-base for the polishing process so that a proper surface finish quality would be determined with a given input. Intensive research efforts have been focused on the generation of knowledge bases using genetic algorithms. A number of papers have shown that a genetic algorithm is a powerful technique for selecting high performance FLC [2.10][2.41]. This has been applied in industrial processes [2.42], but not yet in the polishing industry.

The general framework proposed in Section 2.5.1 is used to generate the best knowledge base using fuzzy logic and genetic algorithms. The implementation of the proposed framework is described and the flow chart of the case study is shown in Figure 2.11.

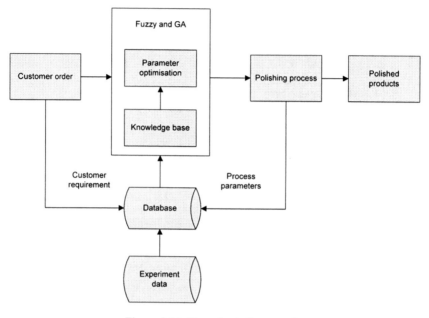

Figure 2.11. Flow chart of case study

2.5.2.2 Framework Implementation

Past data consists of three parts: customer requirement, process parameters and the output, which is shown in Table 2.2.

Table 2.2. Data structure

Process parameters	Values	
material	SS304	customer requirement
pressure	1 kg	
wheel speed	2100 rpm	
feed rate	0.06 m/s	
compound loading	1 g	process parameters
number of passes	10	
type of compound	RC6	
type of wheel	7X8	
country of origin	China	
roughness grade	6	surface roughness

Each chromosome consists of 10 fuzzy rules and membership functions. The genetic pattern of a chromosome is shown as follows:

Genetic pattern$_{chromosome}$ = FR1, FR2, ... FR10, MF$_{wheelspeed}$, MF$_{pressure}$, MF$_{feedrate}$, MF$_{compoundloading}$, MF$_{numberofpasses}$, MF$_{RGrade6}$, MF$_{RGrade7}$

where

FR1	= fuzzy rule 1
FR2	= fuzzy rule 2
FR10	= fuzzy rule 10
MF$_{wheelspeed}$	= membership functions for wheel speed
MF$_{pressure}$	= membership functions for pressure
MF$_{feedrate}$	= membership functions for feed rate
MF$_{compoundloading}$	= membership functions for compound loading
MF$_{numberofpasses}$	= membership functions for number of passes
MF$_{RGrade6}$	= membership functions for roughness grade 6
MF$_{RGrade7}$	= membership functions for roughness grade 7

For the format of fuzzy rules, input variables are customer requirement and process parameters and the quality of surface finish for rough polishing, and the output variables are the fuzzy terms defined in Table 2.3.

Table 2.3. Fuzzy terms for rough polishing

Process parameters	Range			
material	SS304(1)	SS316(2)	SS420(3)	customer requirement
pressure	low(1)	medium(2)	high(3)	
wheel speed	low(1)	high(2)		
feed rate	low(1)	medium(2)	high(3)	
compound loading	low(1)	medium(2)	high(3)	process parameters
number of passes	low(1)	medium(2)	high(3)	
type of compound	RC6(1)	SC8(2)	SC33(3)	
type of wheel	6X8(1)	7X8(2)	10X8(3)	
country of origin	China(1)	Korea(2)	Japan(3)	
roughness grade	6(6)	7(7)		surface roughness

For the format of membership functions, since all membership functions are expressed in triangular form, the value of A$_{(lefttop)}$ and A$_{(righttop)}$ will be the same. In this case study, fuzzy sets for each membership function and the universe of discourse of each fuzzy set is defined in Table 2.4.

Since each chromosome consists of a fuzzy rule region and a membership function region, genes in the fuzzy rule region stores the encoded values of input variables and output variables defined in Table 2.5. The encoded values of genes stored in the membership functions region are the values of the universe of discourse defined in Table 2.4.

Table 2.4. Fuzzy sets for membership functions

Membership functions	Fuzzy sets	Universe of discourse
$MF_{wheelspeed}$	high	2100–2800 rpm
	low	2100–2800 rpm
$MF_{pressure}$	high	1.5–2 kg
	medium	1–2 kg
	low	1–1.5 kg
$MF_{feedrate}$	high	0.06–0.08 m/s
	medium	0.04–0.08 m/s
	low	0.04–0.06 m/s
$MF_{compoundloading}$	high	1–1.3 g
	medium	1–2 g
	low	1.5–2 g
$MF_{numberofpasses}$	high	10–12
	medium	8–12
	low	8–10
$MF_{RGrade6}$	Grade 6	0.04–0.07
$MF_{RGrade7}$	Grade 7	0.03–0.05

Table 2.5. Encoded values of input and output variables

Parameters	Actual value of data 1	Fuzzy terms	Encoded values
pressure	0.5	low	1
wheel speed	2800	high	3
feed rate	0.06	medium	2
compound loading	1	low	1
number of passes	8	medium	2
type of compound	RC6	RC6	1
type of wheel	6X8	6X8	1
country of origin	China	China	1
roughness	0.04	Grade 4	4

In this stage, data 1 and fuzzy rule 1 from chromosome 1 will be used to illustrate how the processing stage works.

Step 1: *Random Generation of Chromosomes*

The first generation of chromosomes are randomly generated within the range specified in Table 2.3 for fuzzy rules and Table 2.4 for membership functions.

The encoded format of a chromosome would be as follows:

= [1, 1, 1, 1, 1, 1, 1, 1, 1, 6, 1, 1, 1, 2, 2, 2, 2, 2, 1, 2, 7, 1, 1, 2, 1, 2, 2, 1, 3, 3, 1, 7,
1, 1, 1, 1, 3, 3, 3, 3, 3, 2, 6, 1, 1, 1, 2, 2, 1, 3, 3, 1, 3, 7, 1, 1, 3, 2, 1, 2, 3, 1, 3, 1, 7,
1, 1, 3, 2, 2, 1, 2, 1, 3, 2, 7, 1, 1, 3, 1, 3, 1, 1, 2, 3, 2, 7, 1, 1, 3, 1, 3, 1, 1, 2, 3, 3, 7,
1, 1, 3, 1, 2, 3, 3, 1, 2, 1, 7, 1, 2100.0, 2100.0, 2100.0, 2800.0, 2100.0, 2800.0,
2800.0, 2800.0, 100, 100, 100, 104, 100, 129, 129, 200, 199, 200, 200, 200, 4, 4,
4, 6, 4, 5, 5, 7, 7, 8, 8, 8, 70, 70, 70, 72, 70, 95, 95, 113, 116, 130, 130, 130, 8, 8,
8, 10, 8, 9, 9, 12, 11, 12, 12, 12, 0.04, 0.05, 0.05, 0.07, 0.03, 0.04, 0.04, 0.05]

Step 2: *Data Fuzzification and Encoding*

For each chromosome, data are converted into fuzzy terms through fuzzification using the random-generated membership functions stored in that chromosome and encoded based on Table 2.3. The result of Step 2 is shown in Table 2.4.

Step 3: *Comparison of the Fuzzified Data and Fuzzy Rules*

The objective of this step is to determine which fuzzy rules are valid through the comparison of the condition parts of the encoded data 1 determined in Step 1 with the "IF" part of fuzzy rule 1. The comparison is shown in Table 2.6. The result showed that fuzzy rule 1 matched with encoded data 1. Their output parts will be compared in the next step. Otherwise, the fitness value of fuzzy rule 1 will be zero and data 1 will be compared with the next fuzzy rules in chromosome 1.

Table 2.6. Comparison of encoded data 1 with fuzzy rule 1

	Encoded data 1	Encoded fuzzy rule 1
pressure	1	1
wheel speed	3	3
feed rate	2	2
compound loading	1	1
number of passes	2	2
type of compound	1	1
type of wheel	1	1
country of origin	1	1

Step 4: *Comparison of the Output Part of the Fuzzy Rule and the Dataset*

The output part of the fuzzy rule 1 is then compared with the encoded data 1, which is shown in Table 2.7.

Table 2.7. Comparison of output parts

	Roughness grade
Fuzzified data 1	4
Fuzzy rule 1	5

Step 5: *Computation of the Accuracy of the Fuzzy Rule*

The accuracy of fuzzy rules, denoted by $A(R_i)$, is shown as follows:

$$A(R_i) = 1 - \frac{Dr}{13} \qquad (2.1)$$

where $A(R_i)$ is the accuracy of fuzzy rule i and Dr is the difference in roughness grade, *e.g.* the accuracy of the fuzzy rule 1 after the comparison in Step 3 is:

$$A(R_1) = 1 - \frac{(5-4)}{13} = 0.9166$$

Step 3 and Step 4 are repeated for the other fuzzy rules.

Step 6: *Determination of the Amount of Data Covered by Chromosomes*

Valid fuzzy rules are summarised and the summarised rule is used to determine the amount of data covered by chromosomes. The percentage of coverage of the chromosomes, denoted by C_i is determined by the following formula:

$$C_i = \frac{\text{number of data matched with summarised rule by chromosome } i}{\text{number of data}} \qquad (2.2)$$

Assume that the total number of data is 20 and the number of data matched with the summarised rule is 9 in chromosome 1. The percentage of coverage of chromosome 1 is:

$$C_1 = \frac{9}{20} \times 100\% = 45\%$$

Step 7: *Determining Fitness Values of Chromosomes*

Fuzzy accuracy and the data coverage are also taken into consideration in the design of fitness functions. Since the accuracy of rules affects the summarised rule, higher weight will be assigned in the fitness evaluation. The fitness function, denoted by $F(i)$, is defined in the following:

$$F(i) = (\text{average of accuracy of fuzzy rules}) \times 0.7 + (\text{data coverage}) \times 0.3 \qquad (2.3)$$

Assume that the average accuracy of fuzzy rules in chromosome 1 is 70% and the data coverage of the chromosome 1 is 50%. Therefore, the fitness of chromosome 1 is shown in the following:

$$F(1) = 70 \times 0.7 + 50 \times 0.3 = 64$$

and the maximum fitness is 100.

Steps 3 to 7 are repeated for the fitness evaluation of the other chromosomes. Step 8 will be performed until all the fitness values of chromosomes are determined.

Step 8: *Checking the Termination Criteria*

Termination criteria are used to determine whether the evolution process should be terminated or not. In this case study, the number of generations is used as a termination criterion. If the current number of generations does not exceed the termination criterion, Step 9 will be needed. Otherwise, Step 10 will be performed to decode the best chromosome among the population.

Step 9: *Crossover and Mutation*

Chromosomes are selected for crossover and mutation based on the fitness values so that a new generation of chromosomes can be formed in this step. After Step 9, Steps 3 to 7 will be repeated for this generation.

Step 10: *Chromosome Decoding*

If the current number of generations equals the number specified in the termination criterion, the best chromosome will be selected among the population and decoded. The decoded chromosome represents the optimal fuzzy rules and membership functions.

2.6 Results and Discussions

The Jgap package, which is written in Java, has been used to develop the program based on the methodology. Since the number of generations affects the total computation time, the number of generations is set to 50, 100, 150, and 200. The graph of fitness values against total number of generations is plotted, which is shown in Figure 2.12.

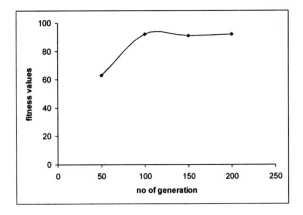

Figure 2.12. Fitness values versus the number of generations

According to Figure 2.12, fitness values for the generations after 100 are more or less the same. In the consideration of the computation time and accuracy, the termination criterion is 100. Two generalised rules were generated. They are verified through the comparison of the generalised rule and data. It was found that 78% of the data can be classified by the rules.

The performance of the system can be improved by two factors. The first factor is the amount of data. Since the number of variables involved in a rough polishing process is very large, generalised rules will be more concise and precise if the system can get more data. The second factor is the number of rules for each chromosome. There is a trade-off between the number of rules, the data coverage and the computation time. If the number of rules increases, the percentage of data coverage will be increased but computation time will be longer and *vice versa*. In the future development, it is necessary to design a methodology to determine the optimal number of rules for each chromosome. The optimal number of rules is the compromise between the data coverage and the computation time.

2.7 Conclusions

This chapter introduced a Web-based framework for provision of a collaborative system of polishing product and process analysis, and knowledge-base development. It is not unusual to see that there are many polishing companies following human-dependent work in many manufacturing activities, where the processes are actually based on personal knowledge and the experience of technicians. A Web-based portal system (WBPS) is developed to deliver a Web application system to conduct online polishing process planning, with a framework model to support knowledge sharing, retrieval, adaptation, and optimisation. The WBPS is constituted of four modules: Manual Planner, Repository, Business Case and AutoPlanner, these modules perform different planning and sharing functions to achieve product development and library construction. These special properties have been integrated and adopted in various complex systems for polishing problem solving. In particular, the optimisation mechanism of WBPS is such a sophisticated system that could potentially be used to design and represent polishing solutions as an applicable project paradigm when new projects are launched.

This research work aims at developing a Web-based generic system to provide effective instantaneous knowledge coordination in streamlining the polishing product and process, and in presenting the enterprises with a single point of access to their knowledge assets. WBPS is able to respond and behave adaptively in a dynamic environment to tackle polishing problems. Polishing enterprises can structure their precious polishing knowledge through effective and user-friendly Web system interfaces, and process the intellectual polishing solution accordingly. Enterprise management can easily target and acquire the correct polishing information where this information is stored in Repository. It allows users to share valuable information among individuals, project groups, departments, companies and even the whole industry on a real-time basis, depending on the necessity of manufacturing progress. Therefore, our Web-based portal system represents a generic distributed control framework, where the need for an optimal process planning can be truly fulfilled by the Intelligent Decision Support Mechanism of WBPS within a short period of time. All these are believed to award the companies using WBPS a competitive advantage. With implementation of this online polishing application system, not only the polishing enterprises can be benefited to facilitate a better utilisation of existing knowledge and new product development, but also it enables strong information-retrieval capabilities and key information exploration.

Regarding the knowledge-base development, a general framework for the construction of a knowledge base was developed and it has been implemented in the polishing process. Results showed that the classification result was satisfactory. With a proper knowledge base, it is expected that proper parameter settings could be extracted with given the required surface finishes. In addition, parameter settings could be kept permanently, they could be accessed at anytime and therefore the quality of surface finish would be consistent even though skillful polishing masters retire. Furthermore, through the Web-based system, technical opinions could also be delivered online to manufacturers quickly so that production lead-time could be shortened to increase competitiveness.

Acknowledgments

We would like to express our gratitude to Mr K.C. Cheng, Mr Eddie Tam and Mr W.H. Wong of PME Group Limited for their invaluable technical and financial assistance. We would also like to thank the Department of Industrial and Manufacturing Systems Engineering of The University of Hong Kong for providing Web resources for completion of our research. It would have been impossible to complete our Web project and this chapter without their dedication and support.

References

[2.1]   James, W. and Joe, F., 2000, "Understanding portals," *Information Management Journal*, **34**(3), pp. 19.

[2.2]   Rado, K. and Emily, H., 2001, "A model for enterprise portal management," *Journal of Knowledge Management*, **5**(1), pp. 86.

[2.3]   Yong Jin, K., Abhijit, C. and Raghav, H.R., 2002, "A knowledge management perspective to evaluation of enterprise information portals," *Knowledge and Process Management*, **9**(2), pp. 57–71.

[2.4]   Brian, M., 2004, "Enterprise portal adoption trends in 2003," *KM World*, **13**(2), pp. 16.

[2.5]   Setnes, M., Babuska, R., Kaymak, U. and Van Nauta Lemke, H.R., 1998, "Similarity measures in fuzzy rule base simplification," *IEEE Transactions on Systems, Man, and Cybernetics*, **28**(3), pp. 376–386.

[2.6]   Herrera, F., Lozano, M. and Verdegay, J.L., 1995, "Generating fuzzy rules from examples using genetic algorithms" In *Proceedings of 5th International Conference on Information Processing and Management of Uncertainty in Knowledge-Based Systems*.

[2.7]   Gonzblez, A., and Perez, R., 1999, "SLAVE: A genetic learning system based on an iterative approach," *IEEE Transactions on Fuzzy Systems*, **7**(2), pp. 176–191.

[2.8]   Park, D., Kandel, A. and Langholz, G., 1994, "Genetic-based new fuzzy reasoning models with application to fuzzy control," *IEEE transactions on Systems, Man, and*

Cybernetics, **24**(1), pp. 39–47.

[2.9]   Bonissione, P.P., Khedkar, P.S. and Chen, Y., 1996, "Genetic algorithms for automated tuning of fuzzy controllers: a transportation application," In *Proceedings of the 1996 International Conference on Fuzzy Systems*, pp. 674–680.

[2.10] Karr, C.L. and Genty, E.J., 1993, "Fuzzy control of pH using genetic algorithms," *IEEE Transactions on Fuzzy Systems*, **1**(1), pp. 46–52.

[2.11] Samuel, L.E., 2003, *Metallographic Polishing by Mechanical Methods*, Material Park, OH: ASM International.

[2.12] Davis, R.J. (ed.), 1994, Stainless steels, *ASM International Ohio*, pp. 423–429.

[2.13] Dickman, Al., 2002, "Polishing and buffing," *Metal Finishing*, **100**, pp. 31–50.

[2.14] Reyers, J., 2002, "Surface conditioning abrasives," *Metal Finishing*, **100**, pp. 66–70.

[2.15] Drucker, P.F., 1993, "Post-capitalist society," *Harper Business*, New York.

[2.16] Drucker, P.F., 1994, "The age of social transformation," *The Atlantic Monthly*, **274**(5), pp. 53–80.

[2.17] O'Brien, J.A., 2001, *Introduction to Information Systems: Essentials for the International E-Business Enterprise*, McGraw Hill, Boston, MA, USA, **1**, pp. 23.

[2.18] Heidi, C., 2003, "Enterprise knowledge portals," *American Management Association*, pp. 28.

[2.19] Bock, G.E., 2001, "Enterprise portals promise to put an end to corporate Intranet chaos," *Enterprise Application Webtop*, **440**(2), pp. 132–133.

[2.20] Kendler, P.B., 2000, "Portals customise information access," *Insurance & Technology*, **25**(10), pp. 47–51.

[2.21] Eckel, R., 2000, "A road-map to identify the portal for your company," *DM Direct Journal*, **14**(7), pp. 11–15.

[2.22] Moskowitz, R., 2001, "Campus portals come to higher education," *Matrix*, 7, pp. 54–6.

[2.23] White, M., 2000, "Corporate portal: realizing their promises, avoiding costly failure," *Business Information Review*, **17**(12), pp. 71–81.

[2.24] Aamodt, A. and Plaza, E., 1994, "Case-based reasoning: foundational issues, methodologies variations, and system approaches," *AI Communications*, **7**(1), pp. 39–59.

[2.25] Denai, M.A., Palis, F., Zeghbib, A., 2004, "ANFIS based modelling and control of nonlinear systems: a tutorial," In *2004 IEEE International Conference on Systems, Man and Cybernetics*, **4**, pp. 3433–3438.

[2.26] Cordón, O., Herrera, F., Magdalena, L., 2001, *Genetic Fuzzy Systems: Evolutionary Tuning and Learning of Fuzzy Knowledge Bases*, World Scientific, Singapore, pp. 1–16.

[2.27] Masao, M., 2001, *Fuzzy Logic for Beginners*, World Scientific, Singapore, pp. 17.

[2.28] Zadeh, L.A., 1996, *Fuzzy Sets, Fuzzy Logic, and Fuzzy Systems: Selected Papers*, World Scientific, Singapore, pp. 19–34.

[2.29] Zadeh, L.A., 1975, "The concept of a linguistic variable and its application to approximate reasoning – Part I, II and III," *Information Sciences*, 8–9, pp. 199–249, pp. 301–357, pp. 43–80.

[2.30] Michalewicz, M., Hinterding, R. and Michalewicz, M., 1997, "Evolutionary algorithms," In *Fuzzy Evolutionary Computation*, Pedrycz, W. (ed.), Kluwer Academic Publishers, Norwell, MA, USA, pp. 5.

[2.31] Gen, M., 1997, *Genetic Algorithms and Engineering Design*, John Wiley.

[2.32] Ishibuchi, H., Nozaki, K., Yamamoto, N. and Tanaka, H., 1995, "Selecting fuzzy if-then rules for classification problems using genetic algorithms," *IEEE Transactions on Fuzzy Systems*, **3**(3), pp. 260–270.

[2.33] Cordón, O., Herrera, F. and Lozano, M., 1997, *Fuzzy Evolutionary Computation*, Kluwer Academic Publishers, Norwell, MA, USA, pp. 46.

[2.34] Wang, C.H., Hong, T.P. and Tseng, S.S., 1998, "Integrating fuzzy knowledge by genetic algorithms," *IEEE Transactions on Evolutionary Computation*, **2**(4), pp. 138–149.

[2.35] Hong, T.P. and Lee, C.Y., 1996, "Induction of fuzzy rules and membership functions from training examples," *Fuzzy Sets and Systems*, **84**, pp. 33–47.

[2.36] Heichel, D. and Spengler, F., 2002, "Designing an efficient ceramic finishing process," *Ceramic Industry*, **152**(13), ABI/INFORM Global, pp. 19.

[2.37] Roal, J.M., Koong, K.S., Liu, L.C. and Yu, C.S., 2003, "An identification and classification of enterprise portal functions and features," *Industrial Management + Data Systems*, **102**(7), pp. 390.

[2.38] Brown, J.S. and Duguid, P., 1998, "Organizing knowledge," *California Management Review*, **40**(3), pp. 90–111.

[2.39] Andy, W., 2003, "Portals and knowledge management," *Malaysian Business*, **11**, pp. 11.

[2.40] Chad, H., 2001, "The intelligent enterprise," *InfoWorld*, **23**(6), pp. 45.

[2.41] Lee, M.A. and Takagi, H., 1993, "Integrating design stages of fuzzy systems using genetic algorithms," In *Proceedings of IEEE International Conference on Fuzzy Systems*, pp. 612–617.

[2.42] Leung, R.W.K., Lau, H.C.W. and Kwong, C.K., 2003, "An expert system to support the optimisation of ion plating process: an OLAP-based fuzzy-cum-GA approach," *Expert systems with applications*, **25**, pp. 313–330.

3

Integration of Rule-based Process Selection with Virtual Machining for Distributed Manufacturing Planning

Dusan N. Sormaz, Jaikumar Arumugam and Chandrasekhar Ganduri

Department of Industrial and Manufacturing Systems Engineering
Ohio University, Athens, OH 45701, USA
Emails: sormaz@ohio.edu, ja348693@ohio.edu, cg152603@ohio.edu

Abstract
Efficient utilisation of both design geometry information and process information is at the heart of modern CAPP systems. The need to consider them both in development and implementation of process planning algorithms and optimisation is the focus of this chapter. The chapter describes a framework for integration of process information in the form of rule-based process selection with geometry data in the form of virtual machining. A rule-based process selection module considers the part geometry and tolerances, and identifies a set of machining processes, tools, and/or machines needed to produce the feature with the required quality. Each selected process candidate is sent to a virtual machining module, which uses the process data and part geometry to generate a virtual machining model and simulates the process in a virtual 3D world enabling the users to verify the process parameters, identify any undesired part/tool collisions, and create an intermediate workpiece geometry required for fixture design. The virtual machining model also enables ranking of alternative candidates for the same feature and verification of the sequence of processes required for a single feature. Procedures and algorithms for both process selection and virtual machining and approaches for their integration are demonstrated on several complex mechanical parts.

3.1 Introduction

This chapter describes development of an integrative distributed process planning system IMPlanner. The integrative system consists of several independent or semi-independent modules, each performing a particular task in product development (some examples of modules are feature modelling, feature recognition, tolerance modelling, process selection, machine selection, cost estimation, sequencing, and cell formation). In this chapter the focus is on a rule-based process selection module and a virtual machining module, and methods for their integration. The data generated by the process selection module is the rudimentary data required by the virtual machining module. These modules may exchange data in several ways: through file interface (one module writes a file and the other reads it), database interface (both modules access the same database), object interface (the modules exchange objects using CORBA), XML interface (the modules exchange tagged text streams), or inter-process communication (remote method invocation, RMI).

This chapter is organised into several sections describing algorithms and procedures developed in modules and implemented in IMPlanner prototype. Section 3.2 describes the overall architecture of an IMPlanner prototype with the emphasis on the process selection and virtual machining modules. Section 3.3 describes the procedures for knowledge-based process selection and their implementation in two prototypes: Lisp-based 3I-PP and a newer version, a Java-based, rule-based process selector (RBPS) within IMPlanner. The knowledge representation, rule development and integration with a rule-based engine Jess are explained. Section 3.4 is devoted to a description of the virtual machining module. Virtual machining of selected process is achieved by using a Java 3D tool for development of animated, shaded virtual 3D models. The virtual machining procedure, which consists of developing geometric, kinematic and animation models, is described. Section 3.5 emphasises integrative properties of the approach by describing several different integration mechanisms between process selection and virtual machining. Section 3.6 provides a few step-by-step examples that demonstrate developed procedures and the prototype. Section 3.7 provides comparison of this work with related research. Section 3.8 provides the conclusions, and the chapter ends with a list of references.

3.2 IMPlanner Architecture

IMPlanner is a manufacturing planning prototype system [3.1] developed at Ohio University with the purpose of providing an integrative framework for development of process planning procedures and their integration into overall manufacturing planning. Data flow within the IMPlanner system for process selection and virtual machining is shown in Figure 3.1. The dark shaded modules represent outside software tools. Two modules shown by a sharp rectangle (with light grey shadow) in the middle represent a process plan representation model. The rounded rectangles without grey shadow represent modules that are being described in this chapter, namely RBPS (rule-based process selection) and IVM (IMPlanner virtual machining). Sharp rectangles on arrows represent interfaces utilised between different modules. It is important to note here that, due to software development tools selected for this system (namely, Java, JNI, and XML), all of these modules (existing and those under development) may exist on different computers with various operating systems, and on geographically distant locations.

Process plan representation is a key component of the system. It includes two related segments, a machining feature model and a process plan model, which interact in process selection. These models (as in other data models) are built using object-oriented modelling. Characteristics of object modelling such as data encapsulation, inheritance and method overloading are utilised. The feature model provides a link to the geometry model of the CAD system and contains references to the part geometry. The process plan model consists of a hierarchical semantic network that includes various entities required for representing process plans.

Existing systems for design (CAD), and process planning (CAPP) and databases are being linked into the system in order to utilise existing design, planning and data-management expertise. For example, any system need related to geometric computations on solid model may be satisfied by a CAD system (the current prototype is built for Unigraphics); tool, machine and cutting-parameter databases can be provided by a variety of tools (MS Access, Oracle, MySQL, XML, *etc.*).

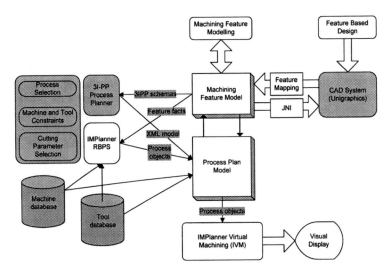

Figure 3.1. Architecture of the IMPlanner system

A high-level scenario for integration of process selection and virtual machining is shown by arrows in Figure 3.1. Part and stock designs are given as CAD models developed using a Unigraphics package. The part model is developed using feature-based design and utilised Unigraphics built-in form features. The stock model is given as a solid with planar faces such that the part solid model is a 3D subset of the stock solid model. The prototype currently does not check for these conditions, because stock design is outside the scope of the system. Both CAD models are imported into IMPlanner, and the machining feature model for the part is built by mapping form features into corresponding manufacturing features. The machining feature model is sent to either the 3I-PP system or RBPS module using appropriate data conversion. For 3I-PP, custom-defined schemas are created and sent using inter-process communication (IPC) and sockets. For RBPS, module facts are asserted using Jess API. Both of the process selection systems consider each feature individually, select appropriate manufacturing processes, verify capabilities and send them back to IMPlanner to build a process plan model. Virtual machining receives each manufacturing process object and generates the corresponding virtual world model using Java 3D. A virtual world model consists of: (a) a stock, tool, and feature geometric model; (b) a kinematic synchronisation model between workpiece and tool, and (c) an animation model that defines behaviours and transformation of geometric models in the virtual world to perform virtual machining.

This scenario can be executed in distributed or integrated applications, with the capability of displaying and verifying each of the steps and displaying each individual object in both the machining feature model and the process plan model upon request of a user.

3.3 Knowledge-based Process Selection

Process selection is that part of the process planning task in which a set of alternative processes is selected for a given feature [3.2]. The process selection module performs the following tasks for each feature: a) verification that the process candidate can satisfy the dimensional-tolerance and surface-finish requirements for the feature in full or partially, b) selection of multiple processes for a given feature when it is necessary to satisfy the design requirements, *i.e.* when a single process cannot produce the feature completely, c) selection of machining processes for a composite feature, d) specification of machine and tool constraints for processing of the feature by the machining process, and e) computation or estimation of machining time and cost for the process candidates.

In our approach we have developed two alternative process selection modules. The first module is part of the 3I-PP system [3.3], which utilised an object-oriented representation of process knowledge in terms of frames and schemas, and executed the process selection module as the first phase of integrative process plan development. The module is capable of selecting multiple processes required for individual features, both alternative processes (of which only one needs to be selected, for example, end milling or side milling of slots) and sequential processes (in which all processes have to be utilised, for example rough and finish milling of slots and pockets, or drilling and reaming for holes).

The new procedure for process selection has been developed as RBPS module in IMPlanner prototype using the idea of knowledge-based process selection from 3I-PP. However, this procedure has been implemented as a rule-based system in order to utilise separation of knowledge and inference and to allow incremental system development. While the basic procedure has remained the same, the implementation of the rule-based system involved several issues: knowledge representation, process selection rules, and the execution of the process selection procedure.

This section describes knowledge representation for the RBPS module, rules that utilise that knowledge in selecting processes for individual features, knowledge database, and the rule execution procedure that integrates Jess into IMPlanner.

3.3.1 Knowledge Representation

Process selection knowledge consists of three kinds of information (knowledge): 1) knowledge of applicable machining processes for different feature types, 2) knowledge about capabilities of individual manufacturing processes with respect to tolerance and surface finish, and 3) precedence requirements for different manufacturing processes.

Knowledge of applicable machining processes for various feature types relates to the understanding of process kinematics and physics and their relationships to the geometry of machining features. This knowledge is described in various manufacturing handbooks [3.4][3.5]. It may be of a general nature, as in those handbooks, or it may be specialised for each individual factory or manufacturing equipment. For example, it is a well-known fact that various drilling, boring and turning operations may be used to produce holes, and that prismatic features (meaning planar faces) are usually produced by milling operations. This knowledge is represented in a computer by utilising object-oriented programming and building a semantic net of manufacturing related entities [3.6]. The extent to which this semantic net is implemented defines the scope of the domain of the process selection module. The IMPlanner prototype contains the knowledge for machining of prismatic parts, though the principal procedure may be extended to other processes. The machining features included in the knowledge base are shown in Figure 3.2 and the machining process hierarchy is shown in Figure 3.3.

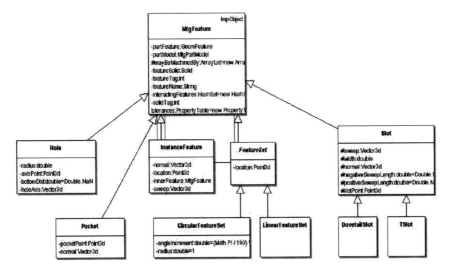

Figure 3.2. Hierarchy of features covered by IMPlanner

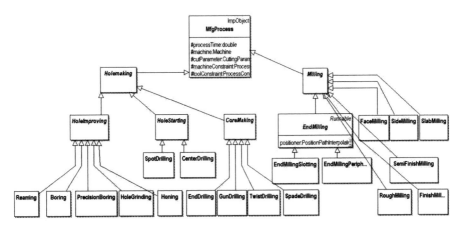

Figure 3.3. Machining process hierarchy in IMPlanner

Process capability data used in this project are based on publicly available data (from handbooks) collected in the USC/NIST project [3.7] for hole-making and milling capabilities. The standard-practice data related to tolerance and surfacefinish capabilities have been collected and implemented into the IMPlanner system. The small portion of the data used in the process selection module is shown in Table 3.1 (from [3.8]). It may be noted that some values are constant, while for some of them, the actual process capability depends on the feature dimensions, and the value is given by the formula (*e.g.* negative tolerance for twist drilling), which poses an additional requirement in the implementation.

Precedence requirements for manufacturing processes are based on the following data of a feature: stock, dimensions, specified tolerances and relations between them (*e.g.* true position versus straightness for holes). These requirements are mentioned as process selection rules in [3.9]. The process precedence data have been implemented in the form of a graph, shown in Figure 3.4. The graph is traversed from the initially selected process for a feature if all specified tolerances are not satisfied by the process. The traversal is recursive until tolerances are satisfied, or the RBPS module decides that a feature can not be made using stored knowledge.

Table 3.1. Process capability data

Parameter	Twist-drill	Reaming	Boring	Face milling	Peripheral milling	End-milling
Smallest T.D.	0.0625	0.0625	0.375	2.75	4	3/16
Largest T.D.	2	4	10	20	18	4
Negative tol.	$0.007 * D^{0.5}$	0.0004	0.0003			
Positive tol.	$0.007 * D^{0.5} + 0.003$	0.0004	0.0003	0.002–0.001		0.004
Straightness	$0.005 * (L/D)^3 + 0.002$	0.0001	0.0005			
Roundness	0.004	0.0005	0.0005			
Parallelism	$0.001 * (L/D)^3 + 0.003$	0.01	0.001	0.001		0.0015
Depth limit	12D	16D	9D			
True position	0.008	0.01	0.0001			
Surface finish	100	16	8	50–30	50–30	60–50
Perpendicularity		0.01	0.001		0.001	
Flatness				0.001	0.001	
Angularity				0.001		

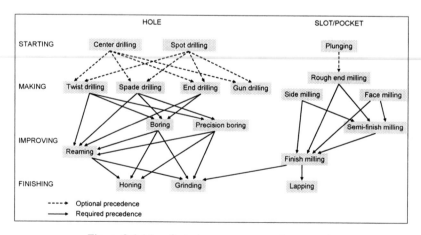

Figure 3.4. Manufacturing process precedence graph

The process capability and precedence knowledge have been implemented into an XML file with process classes implemented inside the IMPlanner package. For example, *edu.ohiou.implanner.processes.TwistDrilling* is a Java class that represents the twist-drilling process. The segment of the knowledge-base XML file is shown in Figure 3.5. It is clear that process capability data that require individual feature dimensions (diameter and depth) are stored as formulas for direct use by inference engine. The figure shows process knowledge for twist drilling (Table 3.1 and Figure 3.4) and a few data items read as follows: twist drilling precedes boring, precision boring and reaming operations; the largest hole that can be made by twist drilling is 2 in (50.8 mm); parallelism tolerance that twist drilling can achieve depends on the hole dimension and is given by the formula $(0.003 + 0.001*(l/d)^3)$, and so on.

```
<?xml version="1.0" encoding="UTF-8" ?>
- <ProcessRelation>
- <edu.ohiou.implanner.processes.TwistDrilling>
  <precedes>edu.ohiou.implanner.processes.Boring</precedes>
  <precedes>edu.ohiou.implanner.processes.PrecisionBoring</precedes>
  <precedes>edu.ohiou.implanner.processes.Reaming</precedes>
  <Parameter smallestToolDiameter="0.0625" />
  <Parameter largestToolDiameter="2.0" />
  <Parameter negativeTolerance="(* 0.007 (sqrt ?dia ))" />
  <Parameter positiveTolerance="(+ (*0.007 (sqrt ?dia)) 0.003)" />
  <Parameter straightness="(+ (*0.005 (** (/ ?depth ?dia) 3)) 0.002)" />
  <Parameter roundness="0.004" />
  <Parameter parallelism="(+ (* (** (/ ?depth ?dia) 3) 0.001) 0.003)" />
  <Parameter perpendicularity="(+ (* (** (/ ?depth ?dia) 3) 0.001) 0.003)" />
  <Parameter Depth="(* 12 ?dia)" />
  <Parameter truePosition="0.008" />
  <Parameter surfaceFinish="100.0" />
  </edu.ohiou.implanner.processes.TwistDrilling>
+ <edu.ohiou.implanner.processes.EndDrilling>
+ <edu.ohiou.implanner.processes.Boring>
```

Figure 3.5. Process capability knowledge base in XML format

Information from the knowledge-base XML file is parsed using a Java SAX parser into memory representation as the process precedence graph and as attributes for each machining process.

3.3.2 Process Selection Rules

Utilisation of the process selection knowledge is often given in the form of rules. For example, for manufacturing of holes it is necessary to ([3.9], p. 230):

"1. Check the core making processes (twist drilling, spade drilling, and end milling); matching the surface requirements with the process capabilities for each process. Do not use end milling if the surface ends in a taper.

10. Check boring for conditions not met by the core making processes. If this succeeds, list this as the process after core process in an alternate series and exit."

Therefore, it appears very natural to use the same paradigm (rule-based systems) to implement the process selection reasoning into CAPP system. We have used rule-based production system approach [3.10] for developing and implementing process selection module RBPS. Rule-based programming is an AI approach for problem solving that separates reasoning from knowledge and data (Figure 3.6). Four major components of a rule-based system are: knowledge base (production memory[1]), data base (working memory), inference engine (algorithm), and knowledge acquisition.

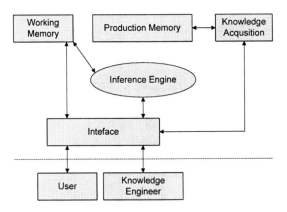

Figure 3.6. Rule-based systems architecture

Production memory consists of a set of rules that are completely independent of each other; rule execution is controlled by data, not by program flow. Rules represent localised pieces of knowledge as shown in the above example. A rule is fired (executed) when all condition patterns match and at that point the right side of the rule (action) is executed, which usually results in modification of the working memory to allow other rules execution. Working memory is a set of facts that are subject of reasoning. Facts are used as patterns to satisfy rule conditions, and rule execution modifies working memory (rules may add, delete or modify facts). The inference engine is the component that controls the rule-based system execution. It is done through a *match-select-execute* loop, where the *match* step performs pattern matching from working memory into the LHS of rules in order to determine which rules can be fired, the *select* step orders rules that can be fired according to the control strategy and select the one to be executed, and the *execute* step performs an action from the RHS of the selected rule. After the selected rule is fired, the execution steps are repeated and the loop runs as long as there are rules to be fired.

The production memory or rule base for process planning is usually developed separately for each phase in process planning. These groups of rules are the following:

---

[1] Terminology used in explaining production systems is usual in AI research.

1. Rules for relationships between features define precedence relations that determine which feature should be machined after which. These rules consider spatial, quality and tolerance relations between features and restrict their manufacturing order.
2. Rules for selection of processes define machining processes that may be used for manufacturing the feature, and create the corresponding operations.
3. Rules for selection of machines and tools use previously defined operations for the feature and knowledge-base that associate processes to machines and tools to select the appropriate machine and tool for a given operation.
4. Rules for selection of a resting face choose the appropriate orientation and corresponding faces of a part as the basis for fixture design by checking the machine toolhead orientation.
5. Rules for process sequencing define the order of machining based on feature precedence.

This chapter addresses the rules in groups 2 and 3, while group 1 belongs to feature recognition and modelling, and groups 4 and 5 belong to setup planning and sequencing, with all three of them being outside the scope of this chapter. The process selection knowledge described in Section 3.3.1 has been coded into a rule set using Jess language [3.11]. An example of rules developed to capture the process selection procedure is shown in Figure 3.7. This rule can be interpreted to execute the following reasoning: if there exists a fact of template *feature* in working memory with name *?f1* that is of type *slot* with status *input*, such that its quality is less than or equal to *30*, then modify its (*?f*) status to *process* and assert two new facts of template *operation*, one for the *end-milling-slotting* process and another for the *side-milling* process, link them with feature *?f1*, make them mutually exclusive (options *or*) and require another processing due to high quality (cut is set to *double*).

```
(defrule slot2.fine
        ?f <- (feature (name ?f1) (type slot) (status input) (quality ?quality&: (<= ?quality 30)))
        =>
        (modify ?f (status process))
        (assert (operation (feature ?f1) (process end-milling-slotting) (cut double) (options or)))
        (assert (operation (feature ?f1) (process side-milling) (cut double) (options or)))
)
```

Figure 3.7. Rule for selection of slot manufacturing processes

Implementation of such rules for all manufacturing process selection situations from groups 2 and 3 mentioned above results in the rule base for that task. We have implemented such a rule base (with a total of 32 rules for various feature/process/tool/machine combinations) in the initial development of the system in order to verify all processing and reasoning requirements. However, such a rule base carries the knowledge about process capabilities in itself. For example, a value of 30 µin for capability for slot quality is hard-coded in the above rule. Any change in knowledge, for example in cases where manufacturing improves (resulting in better capability), would require recoding of this rule. In order to accommodate further separation of knowledge and reasoning, the final rule-based prototype was extended such that rules provide a general reasoning mechanism, and that knowledge is read from XML into working memory, and some reasoning is performed by supporting functions.

In the extended prototype, rules for the process selection have been grouped into the following groups:

1. Rules for process consideration, which, based on feature dimensions and tolerances, decide which processes to consider:
   - If the hole has cast diameter, consider hole improving directly.
   - If the hole requires tight true position tolerance, apply *SpotDrilling* first.
   - If the slot is closed, consider only end milling; if it is open, consider both end and side milling.
2. Rules for capability testing, which test each individual tolerance requirement and decide if there is a complete match, a partial match or no match:
   - If the match is partial, accept the process and create an in-process feature (with unsatisfied tolerances) with processes from precedence.
   - If the match is complete, accept the process but do not expand the hole feature into an in-process feature.
   - If there is no match, still create an in-process feature (for this case, logic may depend on the process, but further processing is not performed in the current prototype).
3. Rules for tool, machine and cutting parameter selection, which select the machine and tool appropriate for the process based on feature dimensions, and select cutting parameters from the database.

These improved rules have also been implemented in Jess language with an example shown in Figure 3.8. This rule for process complete match may be read in plain English language as: "For hole *?jh*, process capability *?pc* for process *?processName*, and process/feature relation with status of 1 (COMPLETE MATCH), modify the process/feature relation status to 5 (SOLVED), create a new process instance of the *?processName* class on feature *?jh*, add this instance to the part *?part*, and delete (retract) the in-process feature *?nf*". The corresponding rules for the above-described situations have been implemented.

```
(defrule SelectHMProcessCompleteMatch
    ?jh <- (Hole (mayBeMachinedBy ?mbmb)(processes ?processList)
            (OBJECT ?o) (featureName ?fName) (partModel ?part))
    ?pc <- (ProcessCapability (name ?processName) (OBJECT ?pcObj))
    ?nf <- (Hole (OBJECT ?nfObj))
    ?do <- (ProcessFeatureRelation (oldFeature ?jh) (newFeature ?nf) (processCap ?pc) (status 1))
=>
    (modify ?do (status 5))
    (bind ?processInstance (createProcessInstance ?jh ?processName))
    (addProcessToPart ?part ?processInstance)
    (retract ?nf)
)
```

Figure 3.8. Rule for process *complete match* in extended RBPS

The procedure for process selection in Jess is illustrated on a single slot feature. A rule-based engine requires data for slot and part in the form of facts given in

Figure 3.9. Two facts are asserted, one for the part data giving the part material (carbon steel) and its stock dimensions (16×12×8 in), and another for the feature giving its type (slot), normal (z-pos), dimensions (width = 1.0, depth = 1.0 and length = 7.5), quality (50) and status (input).

```
(assert (part (name scfdemo04) (material CarbonSteel) (batch-size 1000.0)
        (x-dim 16) (y-dim 12) (z-dim 8)))
(assert (feature (name rs12) (type slot) (normal z-pos) (dim1 1.0) (dim2 1.0) (dim3 7.5)
        (status input) (quality 50)))
```

Figure 3.9. Fact assertion for a slot feature

The feature status "input" triggers the rule execution (shown in Figure 3.10) that demonstrates that *slot1.rough* rule was executed for the feature creating two alternative processes (*side-milling* and *end-milling-slotting*), and reasoning for both processes is completed by selecting tool, machine, cutting parameters and calculating the processing time, as seen in the list of facts after execution.

```
Jess> (run)
FIRE 1 MAIN::slot1.rough f-32
FIRE 2 MAIN::rule-mach f-32, f-31, f-34, f-1, f-0
FIRE 3 MAIN::tool-rule3 f-32, f-34, f-11
FIRE 4 MAIN::calc-mach-time-side-milling f-31, f-34, f-0, f-32, f-11, f-26
FIRE 5 MAIN::calc-mach-cost f-34, f-11, f-0
FIRE 6 MAIN::emsm-complete f-32, f-34, f-11
FIRE 7 MAIN::rule-mach f-32, f-31, f-33, f-4, f-0
FIRE 8 MAIN::tool-rule5 f-32, f-33, f-7
FIRE 9 MAIN::calc-mach-time-end-milling-slotting-slot f-31, f-33, f-0, f-32, f-7, f-28
FIRE 10 MAIN::calc-mach-cost f-33, f-7, f-0
FIRE 11 MAIN::ems-complete f-32, f-33, f-7
11
Jess> (facts)
f-31  (MAIN::part (material CarbonSteel) (x-dim 16) (y-dim 12) (z-dim 8) (batch-size 1000.0))
f-32  (MAIN::feature (name rs12) (type slot) (quality 50) (tolerance nil) (normal z-pos)
        (slope 0) (dim1 1.0) (dim2 1.0) (dim3 7.5) (diam03 0) (status process))
f-33  (MAIN::operation (feature rs12) (process end-milling-slotting) (cut single)
        (options or) (machine ML100) (tool T510) (op-time 2.81) (cost 3.66))
f-34  (MAIN::operation (feature rs12) (process side-milling) (cut single) (options or)
        (machine ML100) (tool T630) (op-time 0.72) (cost 0.94))
```

Figure 3.10. Rule execution of process selection for slot

3.3.3 Knowledge Base/Database

Our database consists of three segments: (a) a machine database that includes information for machines available for process planning (the attributes needed for process planning are provided for each machine); (b) a tool database that includes available tools for manufacturing (we extended tools from the previous research prototype RTCAPP [3.12] by introducing more tools made of different tool materials to cover a variety of scenarios); and (c) a cutting-parameter database that includes suggested speed and feed data for the processes under consideration and combinations of part and tool materials. A small database has been implemented in XML format, but the system can also be connected to other databases.

3.3.4 Integration of Rule Execution Engine into IMPlanner

Interactions between the rule inference engine, knowledge base, and feature model in IMPlanner, as well as execution of the process selection procedure are shown in Figure 3.11. The execution starts by importing features from the CAD model and creating corresponding Java objects in memory space. After that, process capability knowledge base is loaded into the inference engine. The next step is loading of rule base into the inference engine and creation of a Rete network. Then comes the creation of facts from feature Java objects. After that, the inference engine is run and the cycle explained earlier in Section 3.3.2 is executed as long as there are rules to be fired. During this procedure, process facts and their corresponding process plan model objects are created, which is the result of the inference. During the inference, the necessary in-process features are created in order to govern the needs for multiple processes on a single design feature. The execution of this inference engine will be shown in a complex example in Section 3.6.

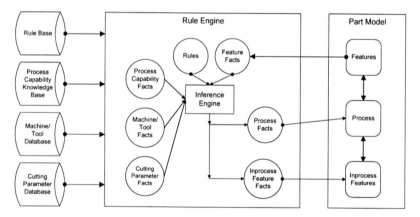

Figure 3.11. Execution diagram for process selection procedure

3.4 Virtual Machining of Milling Operations

Virtual machining is a method for verification of machining operations by providing a visual feedback to manufacturing engineers to enable a better understanding of relationships between various data items for the selected operation, for example possible intersection between tool and stock in non-cutting region, gauging of tool and fixtures, and so on. Our focus in virtual machining is on the geometric relations between the stock, tool and feature and it results in their synchronisation to create a virtual machining process and generate intermediate geometry. A virtual machining model is developed and implemented through three steps (Figure 3.12): (a) geometric model, (b) kinematic model, and (c) animation model. These models are explained in the rest of this section.

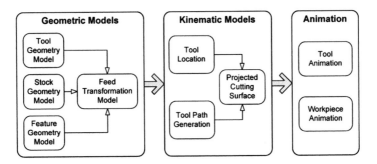

Figure 3.12. Three models for virtual machining

3.4.1 Geometric Model

The geometric model provides a geometric representation of the tool, the workpiece and the feature. All these components are represented as geometric objects in 3D Euclidean space using a simplified boundary representation of solid objects to represent coordinate points, lines and surfaces in 3-dimensional space.

The workpiece (or stock) is represented using a polyhedron with planar faces. The stock model can be imported from a CAD model, or it is generated from a parameterised box. The current implementation covers stocks with planar faces, and curved surfaces are not considered due to the extensive geometric computations involved in representing these surfaces. The topological information of the points in the model maintains the closed solid volume [3.13] for the stock. The point set and topological relations between individual points completely describe the stock shape and they constitute its geometric model. The solid representation is based on polygon-based boundary model data structures [3.14] that is implemented using the coordinate point data array and the topology through an index array, the utilities available in Java 3D [3.15].

Feature models are based on feature parameters. The feature may have a fixed number of parameters (dimensions, position), or a variable number of parameters (like number of curves in its profile). Features are represented in parametric form, which is based on research results in feature modelling and feature recognition, and from this parametric form a geometric model is computed. The set of features machinable by end-milling processes may be represented with the following types: slots, grooves, pockets, profiles.

Current implementation of virtual machining module includes slots and pockets. For each feature type a set of parameters that completely describe the feature has been defined. The reader is directed to [3.16][3.17] for complete specifications. As an illustration, Figure 3.13 shows the definition of an open blind pocket. The open blind pocket has these parameters: 1) a profile (usually floor profile), 2) a normal n, 3) local reference point P_0, 4) corner radius r, and 5) bottom distance bd. Figure 3.13(a) shows the open blind pocket, with its normal in the positive z-direction $n=(0,0,1)$, and profile A-B-C-D-E. The pocket's geometric model is computed from its parametric definition and stock's solid model by extending pocket faces and lines to intersect with the stock faces. Application of the algorithm described in [3.17] that generates the pocket's delta volume from stock is shown in Figure 3.13(b).

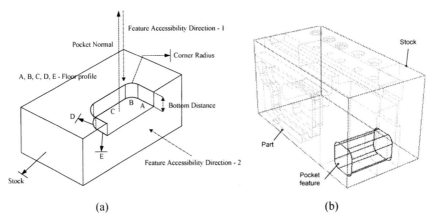

(a)   (b)

Figure 3.13. Open blind pocket: (a) pocket parameters, (b) geometry model

The tool geometric model is created to represent a tool as a hollow cylinder that moves in 3D space and interacts with the stock to perform machining. The tool geometry is aligned with the feature geometry such that the tool swept volume generates the feature faces. This model is shown in Figure 3.14. The cutting-edge surface is represented by a hollow cylinder that generates a rectangular surface after it is projected on the part's face along the feed vector. The combination of this projection surface with the feed vector direction, which intersects with the delta volume during transformation, generates the swept volume.

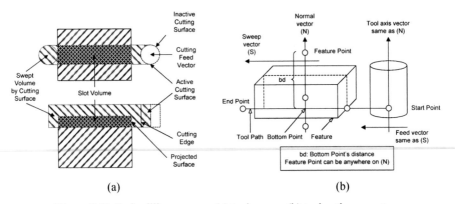

(a)   (b)

Figure 3.14. End-milling process: (a) tool sweep, (b) tool-path parameters

3.4.2 Kinematic Model

The kinematic model is a schematic representation of the interaction between the tool and workpiece, as mentioned in the previous section. It encapsulates the changing geometry of the workpiece due to the relative motion between the tool and stock. The kinematic model is developed in three steps: feature and stock face classification and segmentation, tool-path generation, and tool-path classification.

In order to perform cutting, the tool needs to be properly positioned with respect to the stock, and its initial location should be computed. The tool location calculation, which is based on the feature and stock parameters, links the tool and stock geometry models to generate the feature model. The line segment formed by these limiting points (start and end points) is called the tool path and forms the tool movement profile during machining. Figure 3.14(b) shows an example of a slot that is specified by a point on its central plane (P), normal vector ($\vec{N}$), sweep vector ($\vec{S}$) and the bottom distance from point P (bd), and corresponding tool-path parameters.

During motion along this tool path, the tool modifies some of the workpiece surfaces and generates an intermediate geometry. Therefore, it is necessary to classify stock faces as static and dynamic and identify the way feature faces are generated. Face classification is illustrated in Figure 3.15. It is clear that the stock top face is a dynamic face, which can be segmented into three parts: two sides that are not modified, and the slot top face, which is removed by machining. Also, all slot faces are generated during the milling process. Similar analysis is performed for pocket milling with the difference being that each tool path needs to be considered separately.

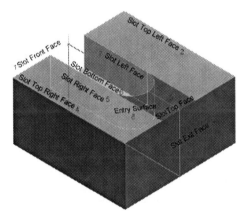

Figure 3.15. Face classification for slot milling

For machining a pocket, a list of tool-path segments needs to be computed both within the workpiece and while engaging the workpiece to extrapolate the motion of the tool. For pockets, end milling requires generation of tool-path segments. We have applied the well-known zig-zag method using offset procedures from [3.18] for generation of tool paths from the pocket geometry.

The animation of the workpiece along the tool path depends on the type of tool-path motion. Several cases need to be considered. For example, if the tool path segment starts outside the stock and ends outside the stock, the tool-path movement would result in the generation of a rectangular open slot. However, for milling of an open pocket, the first tool path should start outside and end inside. For two of these cases, changes of intermediate geometry happen in different ways, because for a pocket the tool must not leave the workpiece. In general, any tool-path segment may be classified based on whether the tool's initial and final locations are inside or

76   D. N. Sormaz, J. Arumugam and C. Ganduri

outside of the stock. This leads to four cases, as illustrated in Figure 3.16: startoutside_endinside ("AB"), startoutside_endoutside ("EF"), startinside_endinside ("BC") and startinside_endoutside ("CD"). These four tool-path-segment cases may further be classified based on the orientation of the workpiece entry or exit face and the tool's orientation. Illustration of this classification is shown in Figure 3.16(b). The trapezoid represents the stock for machining, the full arrow represents the sweep vector, and the hollow arrow represents the tool axis. The angle between the tool axis and the entry and exit face normal defines engagement and disengagement topology. When the tool axis is parallel to a stock face (front face in the figure), tool engagement/disengagement happens along a line of contact; when the axis is not parallel (the case of the back face), tool engagement/disengagement happens at a single point that needs to be calculated. These cases are classified as parallel start/end, bottom start/end, and top start/end. Each of these cases requires different calculation of the engagement/disengagement topology, with a detailed explanation given in [3.16][3.19].

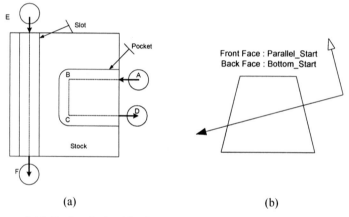

(a)   (b)

Figure 3.16. Tool-path classifications: (a) in-out classes, (b) orientation classes

3.4.3 Animation Model

Animation of a single tool path consists of four major steps: (1) displaying the tool motion according to the cutting parameters (feed and speed), (2) displaying the feature generation when the tool is engaging with the workpiece, (3) displaying the moving feature surfaces as the tool progresses, and (4) displaying the opening of the feature when machining is being completed. In Step (1), the tool geometry, cylinder, is translated along the tool-path's trajectory giving a notion of its interaction with the workpiece geometry. This movement needs to be synchronised with the modifying workpiece geometry at that time instant. The workpiece geometry modifications are carried out by combination of the three steps (2), (3) and (4).

In order to animate workpiece modification, time instants when the topology of the workpiece geometry changes need to be computed. We explored the tool-path time interval and decided to linearly interpolate the geometry between the two instants of changing topology. As a result, Figure 3.17 shows the relation between tool motion and corresponding time instants. Various subintervals "a", "b", "c" … "i", computed from the tool's initial location for the top start approach, represent the fraction of the total machining time from the beginning of the process. The fractions are "a": the time until the tool touches the workpiece; "b": the time needed to create a half-circle on the top face; "c": the time until the tool's bottom cutting edge touches the entry face; "d": the time needed to create a half-circle at the bottom of the feature; "i": the time after all faces are created and the tool is completely engaged; "e": the time interval for opening the top back face; "f": the time before the feature bottom is open; "g": the time interval for opening the bottom back face; and "h": the time interval for moving away from the workpiece. For each of the identified time instants, the corresponding geometry of the workpiece is computed and triangulation is performed using Java 3D utilities. Interpolation of geometry changes between these time instants (or for each time interval "a" to "h") is performed using linear morphing of the shape in Java 3D. The morphing geometry models are included into a virtual world scene graph to perform animation.

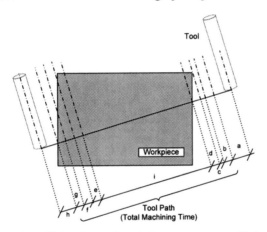

Figure 3.17. Total machining time and workpiece geometry modifying time instants

The illustration of geometry and topology changes is shown in Figure 3.18, which illustrates shrinking of the back face starting at time interval "e" (in Figure 3.17). Figure 3.18 shows the following cases: (a) the back face before interval "e"; (b) the back face change during interval "e"; (c) and (d) two instants during interval "f"; (e) and (f) the back face changes in interval "g"; and (g) the back face in interval "h", when the tool has finished cutting and it is not touching the workpiece.

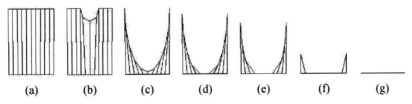

Figure 3.18. Shrinking slot back face at various stages

3.4.4 Virtual Machining Scene Graph

The animation model of stock, tool and feature (workpiece) are schematically represented in a tree-like diagram in the virtual world that is called a scene graph (SG) [3.15]. This section provides a description of the scene graph for virtual machining and its component subgraphs for the stock, tool and workpiece, which together generate the animation model of machining in the virtual world. The *EndMilling* class implements the top level SG, which includes other sub-SG for visualising the machining components in 3D space. The other sub-SG includes tool SG, stock SG, and workpiece SG, as explained in the following sections.

(1) *End-milling Scene Graph*
A high-level structure of the end-milling scene graph is shown in Figure 3.19. This SG is appended to the virtual universe for machining, which provides the view platform and lighting for the scene. The end-milling process provides nodes required for virtual machining of a feature. The ObjRoot branch group (BG) acts as the parent node with AnimRoot BG and Light BG as its child nodes. The Light BG has lights as leaf nodes, and it lightens the scene using a set of ambient lights, directional lights, point lights, and spotlights. These lights are positioned in space to impart effective brightness and shine in combination with material and appearance of shape 3D object.

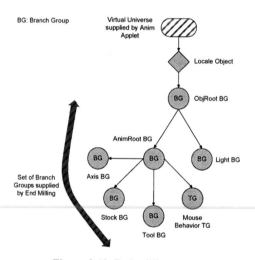

Figure 3.19. End-milling scene graph

The AnimRoot BG is the main BG that handles the process animation. The Axis BG, a child node of AnimRoot BG, sets up a coordinate system in the universe for visually displaying spatial orientation and location to the user. The AnimRoot BG contains nodes of tool and workpiece. The stock BG represents the stock geometry, which is replaced with the workpiece BG during the animation phase. The tool BG, which holds a cylindrical primitive shape representing the cutting tool, provides the tool transformation model for displaying feed motion. The AnimRoot BG also holds a utility transform group, named MouseBehavior TG. This group facilitates user interaction with virtual animation through a mouse device.

(2) *Tool Scene Graph*

The tool BG (see Figure 3.20(a)) represents the tool that is modelled using a solid shape located in space as defined by the tool path and the machining feature. The position and orientation of this cylinder are changed as required by the stock and feature. This change is achieved using a sequence of transformations (rotate by 90 degrees TG) applied on the cylinder's default location, as shown in Figure 3.20(a). Transformation groups are added for rotation and translation of the cylinder, which represent the cutting speed and feed motions. These two motions are defined by using a Rotational Interpolator (RI) and a Position Path Interpolator (PPI). The RI simulates the tool's rotation, and its angular speed is computed from the cutting speed. The PPI provides translation movement based on tool-path computation. The tool path supplies the starting and ending point locations and, using these points, the PPI computes the position on the path and corresponding position indices along which the tool traverses.

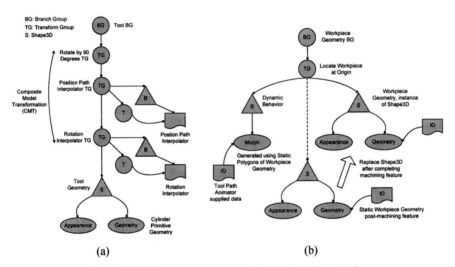

Figure 3.20. Component scene graphs: (a) tool, (b) workpiece

(3) *Stock Scene Graph*

The stock shape geometry model is represented in a 3D scene with the data structure stored in an Indexed Geometry Array [IGA]. This complete geometry along with its appearance is set to a Shape3D, which is part of the process SG. The stock scene graph is replaced by the workpiece scene graph during virtual machining.

(4) *Workpiece Scene Graph*

The workpiece geometry BG handles the workpiece dynamic geometry to execute virtual machining. This BG contains pre-machining and post-machining workpiece geometry. The transition between the two geometries is carried out using the animated feature generation, which morphs the pre-machining geometry to the post-machining geometry. The workpiece SG shown in Figure 3.20(b) provides an outline of this functionality. As the user starts the animation for virtual machining, the BG of the stock geometry is replaced with the BG pre-machining feature static workpiece geometry and, at the same moment, the morphing in dynamic behaviour is also initiated. An alpha object, which decides the timeframe for the animation is started at this instant. The same alpha object is used in the tool BG and, therefore, coordinates the translation motion of the tool with the changes in workpiece shown using intermediate geometry. This coordinating action is performed inside the dynamic behaviour node by smoothly transforming an intermediate geometry into the next geometry, assigned to that time instant. When the workpiece is completely transformed into its final geometry, a static workpiece geometry BG is replaced with the previous BG, having dynamic behaviour.

3.5 Integration Approaches

This section describes several approaches for integration of process selection and virtual machining modules. The paradigm for object visualisation is described first. Three alternative integration models: a) a distributed approach, b) an integrated application approach, and c) an XML-based Web distributed application are described afterwards.

3.5.1 Object Visualisation Paradigm

User interactions of the IMPlanner system are modelled through Java applets and applications [3.20]. This approach allows a variety of deployment scenarios: the modules of the system may be delivered to users in advance for local installation or they may be served in real time from a system server through a browser interface. Due to several disparate functions of the system, the design maintained the separation of system objects implementation and visual component implementation, and their integration was accomplished through interfaces (interface in Java jargon is a contract between several classes about methods and their signatures in order to guarantee the desired behaviour). The objectives of this approach are: a) unified model for 2D and 3D visualisation of any entity on the process plan model, and b) multiple views of each entity defined by the entity status or upon the request of the user, c) separation of model data from the model view (so-called model-view-controller or MVC approach).

   This approach is shown in Figure 3.21. *ImpObject* class as the top-level class for the process plan model objects contracts with *GUIApplet* class as the top-level class for the visual components *Viewable* interface, which specifies the desired behaviour. This behaviour may informally be described as the following. *GUIApplet* class expects that *ImpObject* will provide the necessary visual components and their content for display, while *ImpObject* class relies on *GUIApplet* class for actual display on the computer screen.

   This model has been extended to provide for several levels of display: display of 2D geometry (*Drawable2D*, *Draw2DApplet* and *Draw2DPanel*), display of 3D geometry (*DrawableWF*, *DrawWFPanel* and *DrawWFApplet*), and display in the virtual world (*Drawable3D*, *AnimPanel*, and *AnimApplet*). The planning object may decide (based on user request or based on the system status) to select a certain level of detail and modify it during the application session.

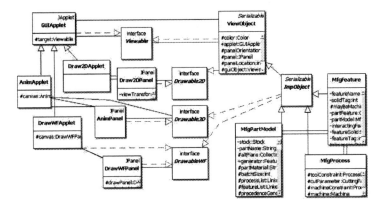

Figure 3.21. Interactions between *ImpObject* and visual components

As an example of the virtual world, the contract between the classes *AnimPanel* and *ImpObject* is specified through the *Drawable3D* interface as shown in Figure 3.22. *AnimPanel* is used to display a model and to control the view point, lighting, colour, *etc*. It relies on *ImpObject* for provision of the model and data to be displayed in the form of scene graph. The figure illustrates that *AnimPanel* implements methods *setView()*, *setLight()*, and *paintComponent()*, while *ImpObject* (and its subclasses) implement *getShapeList()* and *createSceneGrpah()* methods. With all of these methods declared in the *Drawable3D* interface, the actual display is achieved through cooperation of both *AnimPanel* and *ImpObject* instances.

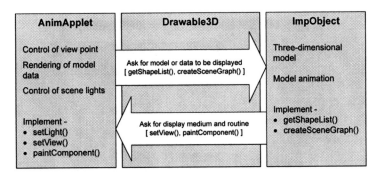

Figure 3.22. Interaction between *AnimPanel* and *ImpObject* through *Drawable3D* interface

3.5.2 Distributed Approach

The distributed approach for integration of 3I-PP process selection module and IMPlanner's virtual machining module (IVM) is achieved using TCP/IP sockets (see

Figure 3.23). 3I-PP plays the rule of process selection server and IVM acts as a client. Upon receiving a request from IVM, 3I-PP generates corresponding data using the process selection module and sends the summary of results as an XML document (or stream). The XML document is parsed at the client side using a DOM parser and the summary is shown as a list of features or processes in the IVM's GUI. The user may select any feature or process to obtain its details. This initiates another request to the 3I-PP server and, at this time, the XML stream with the complete data is retrieved by the IVM. The document is parsed by a SAX parser, and the required process plan model objects are created and displayed as a tree model. Selection of any manufacturing process in the tree triggers its virtual machining in *AnimApplet*.

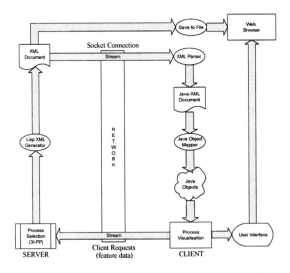

Figure 3.23. Data exchange between 3I-PP and IMPlanner

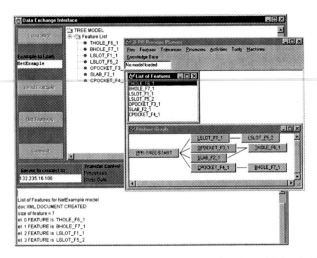

Figure 3.24. Distributed process selection in 3I-PP and virtual machining in IMPlanner

The integration of both systems consists in displaying both applications visually as shown in Figure 3.24. The IMPlanner GUI, called "Data Exchange Interface" is used for interaction with 3I-PP. This GUI directs the execution of the described procedure. The following tasks have been performed in the snapshot shown in the figure: (a) connect to the 3I-PP server, (b) load 3I-PP (3I-PP interface is shown in the front), (c) load an example (with seven features). The same set of seven features is shown in both modules. From this point, a user selects a feature and directs the request to 3I-PP to execute the process selection routine for that feature and to retrieve a summary. After that, the tree will be expanded and the process may be selected to execute virtual machining (run *AnimApplet*) for this process. The IVM interface may also be shown as an applet in a Web browser.

3.5.3 Integrated Application

The modular approach in model building and visualisation described in Section 3.5.1 enables rapid development of a tightly integrated application that demonstrates integration of rule-based process selection with virtual machining. The algorithms and procedures described earlier are integrated into the application/applet (see the figures in Section 3.6) that allows the user to perform the procedures described earlier. The application consists of:

Toolbar, which provides the user with buttons to perform a sequence of actions (open model CAD file, open stock CAD file, open/save XML file, load Jess engine, create facts, run engine) on a selected part CAD model to complete the process selection.

Tree model, which provides the user with a hierarchical model of the part with options to select any node in the tree and review/inspect its content.

Information tabbed panel, which provides the user with various views of the part model and its components; by selecting individual tabs, the user is able to inspect feature/process data (in *Data* tab), verify the geometry of the part, feature, or process (in *Wireframe* tab), create a virtual machining model of the selected manufacturing process (in *Animation* tab), or monitor rule engine execution (in *Jess Output* tab).

The application uses a unified interface to different views for process planning objects defined in Section 3.5.1 and provides a rich user-friendly environment for process planning. The application execution will be described in Section 3.6.

3.5.4 XML-based Web Distributed Application

The distributed application can also be achieved using pure XML data exchange, which provides the benefit of neutral data representation in all phases of process planning. As mentioned in the previous section, IMPlanner provides an XML representation of the hierarchical process plan model that includes parts, features and processes.

Each class in the process plan model implements an XML writer and an SAX parser that allow reading and writing XML streams. The necessary data for each of them is stored in XML format that is modelled in a fashion similar to STEP. At each stage of the described procedure, the resulting data can be saved into an XML file. This enables implementation of different applications (or applets) that can be run on separate computers, and even within Web browsers, and passing a partial process plan model between them.

The benefit of such an approach is in utilising underlying tools and applications on individual computers. For example, the feature-mapping task requires that a Unigraphics CAD package be installed and requires a licence for it, the process selection task requires the license and installation of a Jess engine, while virtual machining requires Java 3D as part of the installed Java Runtime Environment (JRE). Also, in practice, problems often arise when incompatible versions of software are run. XML-based data exchange avoids all these troubles by allowing each step of the process planning task run in an agent-based fashion and providing a unified data exchange mechanism, which carries both the data and its interpretation. The cases of utilising IMPlanner with XML data exchange and Web-based applets are described in [3.21] and [3.22].

3.6 Case Study

In this section, we demonstrate the application of the developed prototype system on a benchmark part "Scfdemo04" from the NIST design repository [3.23]. The part from the repository was recreated as an UG model using form features (see Figure 3.25), and fed into the integrated application, which generates a machining feature model containing a set of machining features with feature parameters. A feature-mapping routine from [3.17] that creates accessible, valid machining features corresponding to the design features is used for this purpose. The part has 18 features – 9 rectangular slots, 4 rectangular pockets, 2 countersunk holes and 3 counterbored holes (shown in Figure 3.26).

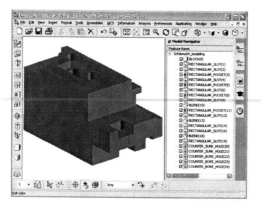

Figure 3.25. Unigraphics model of benchmark part, Scfdemo04

The feature parameters of any mapped feature can be inspected in the *Data* tab, as shown in Figure 3.27. Similarly, the geometry of any machining feature along with the part and stock geometry can be viewed in the *Wireframe* tab (see Figure

3.28). The prototype also allows for the input of feature and part information from an XML file (in prescribed format) as an alternative to UG part files. Parsers that read and write out information have been implemented.

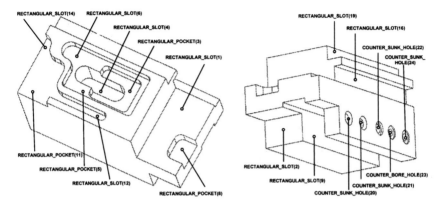

Figure 3.26. Geometry and feature of ScfDemo04 part

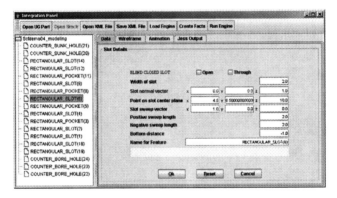

Figure 3.27. Details of a mapped rectangular slot

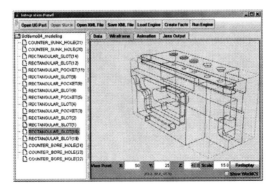

Figure 3.28. Geometry of a mapped rectangular slot

The retrieved feature information is directly fed, in the form of facts, into a rule-based process planner with a Jess inference engine, which on execution fires applicable rules to generate appropriate processes for features (see Figure 3.29). The details about the generated process are shown in the data tab corresponding to the process (see Figure 3.30). The next step is the generation of the tool path for the process that can produce the selected feature, as shown in Figure 3.31.

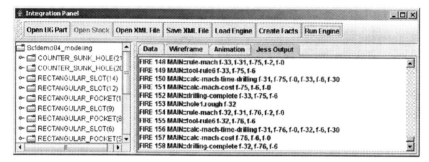

Figure 3.29. Execution of rules for process selection of all features

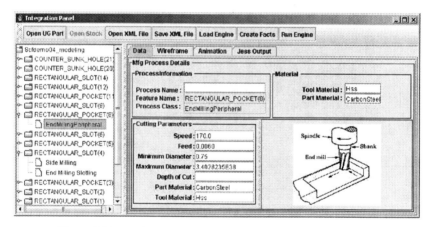

Figure 3.30. Details of the generated end-milling peripheral process

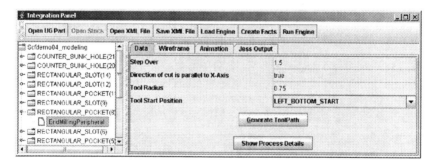

Figure 3.31. Generation of tool path for end-milling peripheral process

The final step is the virtual machining of the generated process in the animation tab. This serves as a verification tool for testing the suitability of the selected process for the corresponding feature (see Figure 3.32).

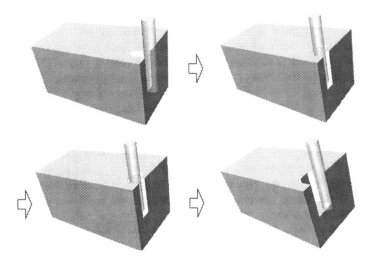

Figure 3.32. Animation of pocket milling

3.7 Related Research

The topic of knowledge-based CAPP has been present in research for many years. Early works on rule-based process selection include [3.24][3.25][3.12]. All of these systems have worked on a symbolic representation of part and features and some form of rule-based reasoning (implemented in Lisp and its derivatives), usually in custom-built languages for rule representation and parsing. Knowledge and data were also represented in the form of Lisp lists utilising Lisp's flexibility for representing various data. The value of those systems is in realising that for usable process plans it is necessary to provide the link to the geometry of a part for feature dimensions, and for tool-path generation.

From those early systems, work has been extended to involve CAM systems with process planning. As a result, the PART system [3.26] has been developed as an extension of tool-path generation with built-in knowledge reasoning based on relational database storage. Along the same lines, CAD/CAM packages have been used for extending their CAM features to include process planning steps, for example Unigraphics CAM NX3 [3.27] includes a knowledge-based module (in KnowledgeFusion) for automatic selection of hole-making operations. Another important approach was integration with feature-recognition systems, as in 3I-PP [3.3].

In the world of process visualisation, early efforts were made by commercial CAM vendors. Most CAM packages utilised tool-path animation as a means for NC-code verification. Early tools were based on wireframe animation without considering possible tool/part solid interactions. The first shaded images were based on pixel manipulation in order to provide an appearance of solid machining. Computer numerically controlled (CNC) simulation software Vericut [3.28] has gained widespread acceptance in the manufacturing sector, owing to its advanced, user-friendly features. Using this software, 3D solid simulation of CNC machines, detection of collisions, over-travel and errors can be done interactively. However, these visualisation approaches were tied directly into expensive CAD/CAM systems.

The invention of the Internet, and, particularly, VRML language [3.29], provided a media for machining animation. Several research reports [3.30]–[3.33] recognise virtual manufacturing (VM) as an integrated and synthetic manufacturing environment exercised to enhance all levels of decision and control. Various concepts related to VM have been implemented, each of which was built using different methodologies, modelling techniques and representations. Reference [3.34] describes a VM system used for CNC milling machines built using Java and Virtual Reality Modelling Language (VRML). The system converts G-code into machine movements and animates the tool and workpiece. The big drawback of the system is that NC code has to be generated before virtual machining, which makes it inefficient; for example if the animation model detects problems, all the effort in generating NC code has to be repeated.

Several efforts have been devoted to developing a better virtual model of the physical machining process. For example, a virtual lathe [3.35] is developed with capability to show, besides geometry, sounds of machining and actual sparks and chips resulting from turning. However, the complexity of this undertaking is enormous, so the reported results are only the first step in this direction. Another drawback of these efforts is in using VRML, a data representation language, which leaves the steps in the process planning disconnected from each other. Our approach, on the contrary, uses Java 3D, which is a procedural language and allows easier integration with other tasks of process planning and process selection.

3.8 Conclusions

In this chapter we have described procedures for rule-based process selection and virtual machining. The rule-based process selection procedure is based on the efficient representation of manufacturing planning knowledge and development of the set of rules to utilise it in selecting appropriate alternatives for machining operations. The Java-based engine Jess has been used to implement rules and integrate them into the IMPlanner prototype. The virtual machining model is based on the development of a detailed geometric model of the machining operation that includes stock, feature and tool and their space and time relations.

Implementation of the model using Java 3D enables virtual machining of features with realistic 3D shaded animation images. Integration of two models has been proposed in several ways. Integration with the 3I-PP system has been tested in the early phases of the development. The new approach, using distributed process planning has been demonstrated using XML data exchange, and a rapid development of a tightly integrated application is shown in a case study.

Development of the model and its implementation has been carried out using the contemporary set of tools (Java, Jess, Java 3D, and XML) to demonstrate portability of application, ease of application integration, and potential for distributed and Web-based processing.

Acknowledgements

The authors would like to acknowledge the work of members of the IMPlanner project that contributed to the results shown in this chapter: Prashant Borse, Sachin Jain, Deepak Pisipati, and Ajit Wadatkar. Portions of the IMPlanner project were funded by the Stocker Endowment Fund at Russ College of Engineering and Technology, Ohio Univeristy, Athens, OH and by Delphi Advanced Manufacturing Research Group, Shelby, OH whose support is highly appreciated.

References

[3.1]   Sormaz, D.N., Arumugam, J. and Rajaraman, S., 2004, "Integrative process plan model and representation for intelligent distributed manufacturing planning," *International Journal of Production Research*, **42**(17), pp. 3397–3417.

[3.2]   Chang, T.C., 1991, *Expert Process Planning for Manufacturing*, Addison-Wesley, Menlo Park, CA.

[3.3]   Khoshnevis, B., Sormaz, D. and Park, J., 1999, "An integrated process planning system using feature reasoning and space search-based optimization," *IIE Transactions*, **31**(7), pp. 597–616.

[3.4]   Kalpakjian, S., 1992, *Manufacturing Engineering and Technology* (2nd edn.), Addison-Wesley, Reading, MA.

[3.5]   Drozda, T.J. and Wick, C. (eds.), 1987, *Tool and Manufacturing Engineers Handbook: A Reference Book for Manufacturing Engineers, Managers, and Technicians, Vol. 1 Machining*, (4th edn.), Society of Manufacturing Engineers, Dearborn, MI.

[3.6]   Sormaz, D.N. and Khoshnevis, B., 1997, "Process planning knowledge representation using an object-oriented data model," *International Journal of Computer Integrated Manufacturing*, **10**(1–4), p. 92–104.

[3.7]   Khoshnevis, B., Tan, W. and Sormaz, D.N., 1993, "A process selection rule base for hole-making," *Research Report*, Factory Automation Systems Division, National Institute of Standards and Technology, Gaithersburg, MD.

[3.8]   Khoshnevis, B. and Tan, W., 1995, "Automated process planning for hole-making," *Manufacturing Review*, **8**(2), pp. 106–113.

[3.9]   Wang, H.-P. and Li, J.-K., 1991, *Computer Aided Process Planning*, Advances in Industrial Engineering 13, Elsevier, Amsterdam, The Netherlands.

[3.10]   Giarratano, J.C. and Riley, G., 1994, *Expert Systems: Principles and Programming*, 2nd edn., PWS Pub. Co., Boston, MA.

[3.11]   Friedman-Hill, E., 2003, *Jess In Action*, Manning Publications, Greenwich, CT.

[3.12]   Park, J.Y. and Khoshnevis, B., 1993, "A real time computer aided process planning system as a support tool for economic product design," *Journal of Manufacturing Systems*, **12**(2), pp. 181–193.

[3.13]   Requicha, A.A.G., 1980, "Representations for rigid solids: theory, methods and systems," *ACM Computing Surveys*, **12**(4), pp. 437–464.

[3.14]   Shah, J.J. and Mantyla, M., 1995, *Parametric and Feature-based CAD/CAM*, John Wiley and Sons, New York, NY.

[3.15] Sowizral, H., Rushforth, K. and Deering, M., 2000, *Java 3D API Specification, Second Edition*, Addison-Wesley Publication, Boston, MA.

[3.16] Borse, P.A., 2003, *Visualization of a Slot Milling Process for Verification and Validation of a Process Plan on the Internet*, MSc Thesis, Ohio University.

[3.17] Arumugam, J., 2004, *Analysis of Feature Interactions and Generation of Feature Precedence Network for Automated Process Planning*, MSc Thesis, Ohio University.

[3.18] Hansen, A. and Arbab, F., 1992, "An algorithm for generating NC tool paths for arbitrarily shaped pockets with islands," *ACM Transactions on Graphics*, **11**(2), pp. 152–182.

[3.19] Pisipati, D., 2004, *Virtual Manufacturing of Pockets Using End Milling with Multiple Tool Paths*, MSc Thesis, Ohio University.

[3.20] Horstman, C.S. and Cornell, G., 2000, *Core Java, Volume I*, The Sun Microsystems Press, Palo Alto, CA.

[3.21] Sormaz, D.N., Arumugam, J. and Neerukonda, N., 2004, "XML-based product and process data representation for distributed process planning," In *Proceedings of the 1st International Conference on Electrical/Electromechanical Computer Aided Design and Engineering*, Durham, UK.

[3.22] Sormaz, D.N., Pisipati, D.V. and Borse, P.A., 2006, "Virtual manufacturing of milling operations with multiple tool paths," *International Journal of Manufacturing Technology and Management*, **9**(3–4), pp. 237–264.

[3.23] Regli, W.C. and Gaines, D.M., 1997, "A repository for design, process planning and assembly," *Computer Aided Design*, **29**(12), pp. 895–905. (available online at http://edge.mcs.drexel.edu/repository/frameset.html)

[3.24] Descotte, Y. and Latombe, J.-C., 1984, "GARI: an expert system for process planning," In *Solid Modelling by Computers, from Theory to Applications*, Pickett, M.S. and Boyse, J.W. (eds.), Plenum Press, New York, NY, pp. 329–46.

[3.25] Berenji, H.R. and Khoshnevis, B., 1986, "Use of artificial intelligence in automated process planning," *Computers in Mechanical Engineering*, **5**(2), pp. 47–55.

[3.26] van Houten, F.J.A.M., 1991, "PART: a computer aided process planning system," *PhD dissertation*, University of Twente, The Netherlands.

[3.27] Unigraphics NX, 2005, http://www.ugs.com/products/nx/machining/, UGS Solutions.

[3.28] Vericut, http://www.cgtech.com/, CGTech.

[3.29] Web3D Consortium, http://www.web3d.org/.

[3.30] Brink, J. (ed.), 1994, *Virtual Manufacturing User Workshop*, Final Report, (available at http://www.isr.umd.edu/Labs/CIM/vm/lai1/final6.html) Lawrence Associates, Inc., Dayton, Ohio.

[3.31] Iwata, K. and Onosato, M., 1996, "Virtual manufacturing systems for manufacturing education," In *Proceedings of the 1996 International Conference on Education in Manufacturing*, March 13–15, 1996, San Diego, CA.

[3.32] Iwata, K., Onosato, M., Teramoto, K. and Osaki, S., 1997, "Virtual manufacturing systems as advanced information infrastructure for integrating manufacturing resources and activities," *Annals of the CIRP*, **46**(1), pp. 335–338.

[3.33] Yang, H. and Xue, D., 2003, "Recent research on developing web-based manufacturing systems: a review," *International Journal of Production Research*, **41**(15), pp. 3601–3629.

[3.34] Ong, S.K., Jiang, L. and Nee, A.Y.C., 2002, "An Internet-based virtual CNC milling system," *International Journal of Advanced Manufacturing Technology*, **20**, pp. 20–30.

[3.35] University of Michigan, 2000, *Revival of the Virtual Lathe*, Virtual Reality Laboratory, http://www-vrl.umich.edu/sel_prj/lathe/index.html.

4

CyberCut: A Coordinated Pipeline of Design, Process Planning and Manufacture

V. Sundararajan[1] and Paul Wright[2]

[1] Department of Mechanical Engineering
University of California at Riverside, Riverside, CA 92521, USA
Email: vsundar@engr.ucr.edu

[2] Department of Mechanical Engineering
University of California at Berkeley, Berkeley, CA 94720-1740, USA
Email: pwright@me.berkeley.edu

Abstract

CyberCut is a coordinated "pipeline" of software that can be used to design a part itself, design a mould for the part, and then machine an aluminium mould from the design. WebCAD is the "front end" to CyberCut and invites a mechanical designer to use specific design tools, strongly linked to downstream manufacture. Alternatively, we also allow the designer to use a CAD system of their choice and use the "feature recognition" described in this chapter to analyse the shape for downstream planning and machining. An ACIS graphics kernel facilitates the feature recognition step from the CAD file. Automated tool-path planners and automated tool selection procedures then generate computer numerical controlled (CNC) machining code. The algorithms that we have developed eliminate manual CNC programming, thus reducing significantly the mould cutting time. We also describe an optimal tool sequencing method by finding the shortest path in a single-source, single-sink, directed acyclic graph.

4.1 Introduction

A designer who desires a physical component has several ways to interact with the manufacturers who provide rapid prototyping (RP) or conventional manufacturing (MP) facilities. The nature of the interaction and the requirement for RP processes and conventional manufacturing processes is quite different. RP processes usually build the part in layers; hence process planning consists of a few well-defined steps [4.1]. Traditional processes, such as machining, on the other hand, require elaborate process planning as the motion and orientation of the tool and workpiece can be much more complicated. An investigation into the tasks involved in process planning may be found in [4.2][4.3], while issues involved in the integration of CAD and CAM are discussed in [4.4][4.5]. CyberCut uses the machining process –

specifically the 3-axis milling process as a rapid prototyping process to produce complete functional prototypes of parts.

The interaction between designers and manufacturers may be broadly classified into three categories (Figure 4.1):

1. Conventional, "over-the-wall" systems
2. Manufacturing-dependent CAD systems
3. Bidirectionally coupled CAD/CAM systems

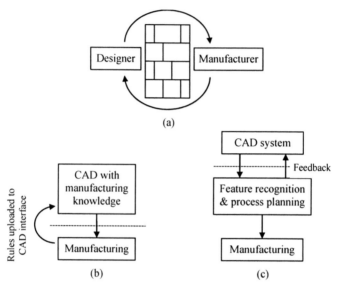

Figure 4.1. Architectures for designer–manufacturer communication: (a) over-the-wall manufacture; (b) manufacturing dependent CAD systems; and (c) bidirectionally coupled CAD/CAM systems

4.2 Conventional Approach

In CAD/CAM environments typical of the last three decades, there exists a loose interaction between the designer and the manufacturer (Figure 4.1(a)). During the design phase, the manufacturer is seldom consulted and the design is driven by requirements such as function, aesthetics and ergonomics. Occasionally, the design of the component is superficially checked for assembly problems by detecting interference with other components. Sometimes, crude tool availability checks such as tool diameter and depth are incorporated. However, the designer seldom has knowledge about fixturing, tool access problems and other manufacturing issues. The design is completed and sent to the process planner. Manufacturing problems typically manifest themselves at this stage and the design has to be modified iteratively, thus leading to a longer time to manufacture. The disadvantages of this approach forces one to consider other alternatives. In particular, a closer coupling between design and the manufacturing throughout the design phase is desired. Two such architectures to enforce the coupling are suggested below.

4.2.1 Manufacturing-dependent CAD Systems

The CAD system in this approach is geared towards the requirements of a manufacturing process. Only parts that can be produced by the downstream process can be designed [4.6]. The rules and restrictions for the design come from the manufacturing process and the designer is constrained by these rules during the creation of the part [4.7] (Figure 4.1(b)). This mode of operation is common in VLSI circuit design or MEMS (micro-electromechanical systems). Systems that have been constructed for these domains based in this principle are MOSIS (metal oxide semiconductor implementation service), MUMPS (multi-user MEMS process), and LIGA. A machining service that employs the manufacturing dependent CAD philosophy is described in [4.8][4.9]. The guarantee of manufacture of the parts or circuits obeying the rules makes these systems attractive for designers who want simple parts or circuits. Similarly, the manufacturers can easily vouch for the manufacturability of the design without much strain, since they can adhere to the common practices that they follow on the shop floor.

The principal disadvantages of these systems are the limitations that they impose on the designer. Manufacturers have a tendency to use rather conservative rules in order to assure manufacturability. Thus, it is common for parts to get rejected by the rules although they can be made with a little additional effort on the part of the manufacturer and would increase the value of the design at a small extra cost. Additional problems arise in the mechanical domain. Abstraction of manufacturing information is often a very difficult problem, since the range of components that may be produced by conventional means is huge. When one reflects on the components of a car, airplane or any machine, one is confronted with an immense diversity of size, form and intricacy, all of which have been manufactured using some conventional process. Even if the constraint of rapid manufacturing is imposed, a bewildering choice of manufacturing techniques presents itself. Assuming that some abstraction has been obtained, the presentation of this information to the designer creates a difficulty and the manufacturers are frequently unable to express their full capability. The designer, in turn, cannot exploit the full potential of the available manufacturing processes. Manufacturing constraints and rules are usually enforced using features that are a collection of geometrical or other entities that have significance for manufacturing. However, the design process often uses features that are different from the manufacturing features, making the design phase artificial and difficult [4.4]. The problem becomes acute if the component is to be used in an assembly. In this case, the chief characteristics of the interacting component are decided simultaneously, and one part is used to create the other for ease of design and to ensure the critical characteristics are consistent. Manufacturing dependent CAD breaks up this natural design process by demanding that the design be done in terms of manufacturing features. Thus, these systems may successfully be used when the desired part is within or at the boundary of the manufacturing rules and the interaction of the part with other components of the assembly is minimal. When these systems reject the part due to rule violations, a third alternative, described below, may be found to be more attractive.

4.2.2 Bidirectionally Coupled CAD Systems

From the above discussion, it can be seen that both the traditional over-the-wall approach and the constrained CAD approach have disadvantages that impede the design process. A happy compromise may be attained by allowing the designer to use conventional techniques for the design phase, but enforcing checking with a process planner for manufacturability from time to time. An architecture is shown in Figure 4.1(c).

It is clear that too frequent a check for manufacturability would slow down the design phase; hence the process planning phase has to be automated. Recall that the conventional approach where the process planning is done after the design is completed does not mandate automation and can also be done by a human process planner (perhaps even better). However, automation in the bidirectional architecture of Figure 4.1(c) is imperative.

Also, since the design phase uses information in a representation that is different from the format of manufacturing information, a translation between the two representations is required. The translator is popularly known as "feature recognition". The recognition presents the process planner with features that are used for planning. Feedback from the process planner can be communicated to the designer by mapping the manufacturing features back to the design. This approach allows considerable flexibility for both the designer and the manufacturer. It achieves a coupling of CAD and CAM through a cooperative mode of operation. The design phase is kept decoupled from the process planning phase in terms of representation and manipulation of component data.

This approach, though highly attractive, is fraught with difficulties arising from the feasibility of the automation of feature recognition and process planning. Both are considered to be difficult problems and have been areas of intense research for the last two decades, as can be seen from [4.4][4.10]–[4.12]. Even most systems based on this approach use large number of heuristics and computationally expensive algorithms that limit their use.

Early systems using feature recognition borrowed much of the terminology from the manual process planning domain and algorithms were sought to recognise these features. The process planning system was supposed to assist and not replace humans. This approach, however, tended to force the feature recognition algorithm into making fine distinctions between different types of features, which then needed special routines for each type that needed to be identified. Recent systems have moved towards greater automation of the entire process, eliminating many arbitrary distinctions between features. For instance, the older systems maintained distinctions between slots, shoulders, rectangular profile pockets and pockets with arbitrary contour. By contrast, newer systems such as the one in Ref. [4.13] and also the one described here unify these concepts into a single feature called a pocket. Distinction is maintained by flagging "open edges" for shoulder edges and leaving the tool-path planner to deal with these special edges, as will be explained later.

In summary, the conventional approach of over-the-wall manufacturing is clearly an undesirable scenario. The second approach of incorporating manufacturing into the design has the advantage of guaranteeing manufacturability, but suffers from the drawback of over constraining the designer. The third approach of having the design checked from time-to-time has the advantage of flexibility and ease of use for the designer, but requires sophisticated algorithms.

4.3 The CyberCut System

CyberCut is an integrated CAD/CAM system that targets automation of the design-to-manufacture process for rapid prototyping of components. A typical fully integrated CAD/CAM system would entail conceptual and detailed design, design validation, process selection and planning and the manufacture of the components. Several additional issues such as job scheduling, inventory management, production and quality control also fall under the purview of a manufacturing system. CyberCut restricts itself to the domain of design for manufacturability, design validation, automated process planning and manufacturing. CyberCut, being a pilot design-to-manufacturing system, has been engineered with certain restrictions. The following are the assumptions made:

1. Parts will be made with 3-axis CNC milling. This imposes limits on the geometry of the parts, the materials, batch size, and tolerance.

2. The materials for which the system is available are plastic, aluminium and steel. Other materials can be incorporated, provided information about their strength, machining characteristics, and suggested machining parameters are available.

3. Standard fixturing techniques such as toe-clamps and vices will be used to hold the part. Since there is no provision for special-purpose fixturing aimed at rapid location and clamping, the batch size is limited. A fixturing technique called RFPE [4.14] is available for holding complex parts, at the expense of decreased accuracy and longer process time.

4. Access directions are restricted to six orthogonal directions, corresponding to the three mutually perpendicular coordinate axes – both positive and negative directions. This implies that parts having features that require holding on inclined plates cannot be made. This is a temporary restriction to provide a stepping stone for more sophisticated fixturing and planning algorithms.

5. Cutting tools used are flat-end and ball-end mills, standard drills and reamers, standard taper tools and corner-rounding tools. Thus, undercuts are not permitted in the part.

6. The boundary of the part must be 2-manifold and connected. Disconnected boundaries imply that there is a hollow region in the part that cannot be machined. Manifoldness is needed to ensure that the precision and repeatability of the axes of the machine tool do not cause unexpected results such as thin walls.

4.3.1 Overview of the CyberCut System

For practical utilisations, the CyberCut system allows the part to be described in two ways:

1. The part can be designed in a commercial CAD system and can be described in the STEP format. The part is then processed by a feature recogniser and converted to manufacturing features.

2. The part can be designed directly in terms of subtractive features, *i.e.* design-by-features methodology. CyberCut provides a specialised Web-based interface for this approach called WebCAD.

Figure 4.2 shows the architecture of the CyberCut system.

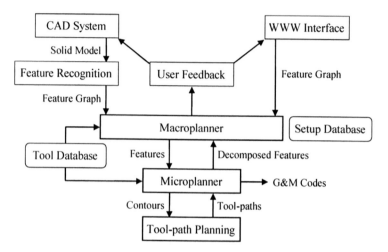

Figure 4.2. Architecture of the CyberCut system

The feature recogniser and the Internet-based design-by-features interface generate a feature graph (discussed in the next section) containing a description of the features of the part. The feature graph is sent to the *Macroplanner*. The *Macroplanner* uses the setup database and a feature validator to assign features to setups and to sequence the setups. Each feature is then sent to the *Microplanner* for detailed planning of the machining. The *Microplanner* uses the tool database to determine the optimal sequence of tools and the cutting parameters for the tools. It decomposes features into regions that can be machined by the various tools and assigns appropriate cutting parameters to them. These decomposed features are then sent to the *Tool-path Planner* to generate G&M codes for the machine tool.

4.3.2 Definition of Features

CyberCut, being a feature-based system, requires a consistent definition of features. The feature definition captures the geometry and attributes of individual features as well as the spatial relationships between features. CyberCut does not have a predefined set of features. Rather, features are described by specifying a 2D planar contour, an access direction corresponding to the normal to the planar contour, a radius for sharp corners, a depth attribute for 2.5D features or a bottom surface for freeform features, and other qualifying attributes such as "through". If the feature opens to the side of the stock (as in a step, slot or shoulder), the corresponding edges of the features that adjoin the stock boundary are marked "open". Consider the part shown in Figure 4.3(a). Figure 4.3(b) shows the features from the +Z direction. Features 1 and 2 correspond to the shoulders on the right-hand side of the part. The contour of the feature is the top face of the cuboidal shape of Feature 1 or Feature 2, while the height of the cuboid indicates its depth. The thick edges in the figure are the open edges of the contour and denote that the feature shares a boundary with the stock face of the part. The presence of two adjacent open edges at right angles to each other indicates that this feature is a shoulder. This interpretation is optional and may be done in the process planner as necessary.

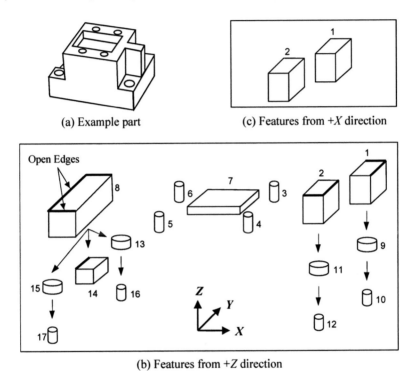

(a) Example part   (c) Features from +X direction

(b) Features from +Z direction

Figure 4.3. Partial feature tree for *example part*

In addition to the above feature descriptions, the definition captures two types of spatial relationships:

1. *Parent–child relationships between features that can be accessed from the same direction.* Feature A is the parent of feature B, a) if the contour of feature B is completely contained within the contour of feature A, b) if the contour of feature B is "deeper" in the part than the contour of Feature A, *i.e.* if a ray were cast from outside the part in a direction opposite the access direction, then contour A is hit before contour B. In Figure 4.3(b) the arrow from the shoulder Feature 1 to the hole Feature 9 shows the contour of the

hole is entirely contained with the contour of the Feature 1. This signals a constraint that Feature 9 can be removed only after Feature 1 is removed. However, this is a soft constraint as the hole may be drilled before the shoulder is machined.

2. *Alternative representations*: Feature A is an alternative representation of feature B if they have different access directions and the intersection of their volumes is not null. For example, a shoulder, as it is usually defined, can be accessed from more than one direction. According to the feature definitions of CyberCut, a feature has only one access direction. The various features associated with the shoulder are linked as alternative representations. Figure 4.3(c) shows the same shoulders (*i.e.* Features 1 and 9) in Figure 4.3(b) represented from the +X direction as Features 1 and 2. Note the location of the open edges. However, since the features from the two access directions represent the same removal volume, the features are linked in the data structure. The linking is performed in the feature recogniser and is available for use by the process planner.

The features for an individual access direction for a tree of features as shown in Figure 4.3(b) while the entire set of features from all access directions form a graph because of the alternate representations (Figure 4.4).

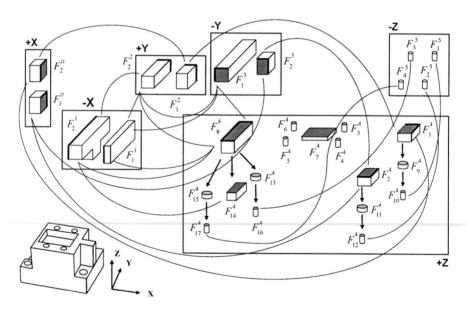

Figure 4.4. Complete feature graph for *example part*

4.4 Architecture

In this section, we describe the key components and algorithms used in the process planner.

4.4.1 WebCAD

WebCAD is an Internet-based design interface that allows designers to describe the part by subtracting features from an initial rectangular stock. The interface allows the basic viewing and editing operations expected of CAD interfaces. WebCAD is process-aware, *i.e.* WebCAD incorporates manufacturing rules into the interface. Rather than allowing designers to create arbitrary geometries, the software attempts to ensure that the user is guided towards the creation of a machinable design. The result is a part geometry that has already been checked for manufacturability before any attempt is made to create a detailed manufacturing plan. The user begins by specifying the dimensions of the rectangular stock. The user is presented with a palette of features consisting of holes and pockets. If the designer chooses to drill a hole, the WebCAD system insists that the user defines a depth and selects a standard drill bit that is capable of drilling to that depth. Similarly, for non-drilled features, the designer selects a depth and a milling cutter that is long enough for the depth. The user then draws a contour on any of the surfaces of the original stock currently. Only circular arcs and line segments are available. The circular arc is restricted to have a radius larger than the radius of the selected tool. When the user completes the design of the pocket, it is immediately checked for narrow channels and minimum distance from the edges of the stock. These design methods and rules constrain design choices, helping a designer compose manufacturable parts. The user can then save the part and submit it to the process planner on the server.

4.4.2 Feature Recogniser

The feature recognition algorithms are described in detail in [4.15]. The input to the system is a boundary representation of the component. The algorithm assumes that the component is located in a cuboid (the stock) that is the bounding box of the part. The features are removal volumes. These can be obtained by subtracting the part from its box and by decomposing according to the definitions above. A starting access direction is assumed. The faces of the part that are encountered first from the access direction (as in a ray test) are subtracted from the corresponding face of the stock. This gives disconnected sheets, each of which is turned into a feature. These sheets are then descended recursively into the part to detect changes in cross sections that then give new features and so on. The hierarchical information obtained during the recursion is recorded in a tree. Each node of the tree is a feature and the edges of the tree denote a parent–child relation such that if the contours of all the features are projected onto a plane with its normal along the sweep direction, the child contours are completely contained within the contours of the respective parents. This process is repeated for each of the 6 orthogonal directions to give the complete feature graph. Extensions of this algorithm for freeform features are described in [4.16].

The steps in the feature recognition process for the part in Figure 4.3(a) are shown in Figure 4.5. The stock is the rectangular bounding box of the part (shown in dotted lines). In order to recognise features from the $+Z$ direction, the $+Z$ face of the part is extracted and all its edges are marked "open" (indicated by the thick lines). The z faces of the part (*i.e.* planar faces with the z-axis as the normal) are extracted and sorted in decreasing z-coordinate. The faces with the largest z-coordinate are subtracted from the stock face. The results of the subtraction are the contours 1 to 8. Note that the open edges from the stock have been propagated to the Features 1, 2 and 8 (the numbering of the features is arbitrary and can be randomly assigned). Each contour is then descended into the part. For example, if contour 2 is chosen, the faces of the part that lie directly below it (*i.e.* the next highest z-direction) are subtracted from the contour. The result is contour 9. The distance between the top-level z-face and the z-face that resulted in contour 9 is the depth of Feature 2. The process continues as shown in the figure. As contour 9 is further descended, it exits the part without any intersections with other faces, thus signalling that the corresponding Feature 9 is through. The process is repeated for each of the other contours to yield the tree of features shown in Figure 4.3(b).

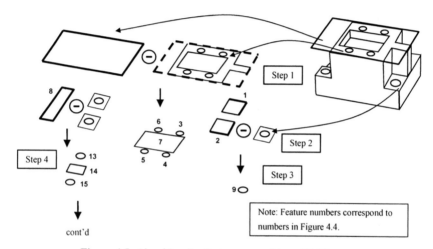

Figure 4.5. Algorithm for feature recognition of 2.5D parts

4.4.3 Feature Validation

The feature graph generated by the feature recogniser or by the WebCAD interface is first validated to ensure their essential feasibility. The *Feature Validator* first compares the diameter and the length of the tools in its database to the depth and the size requirements of the features and returns a list of tools that can be used. It then computes the area that can be accessed by the tool. For the feature to be valid, there must be at least one tool that can access the entire area of the feature. The tools and the features are then checked for collision with tool holders to ensure that there is no collision during machining. This is just a preliminary check to eliminate obvious problems. Detailed checks for collisions can only be performed when the current state of the component and the fixtures being used for the setup are known. These checks are postponed until the microplanning stage. If tools are not available for the features, or every tool configuration leads to collisions during machining, an error report is generated and is passed on to the user. For the part in Figure 4.3(a), all the features from the +Z-direction and the −Z-direction can be machined. However, features from other directions are not machinable. For example, Feature F10 from the −X-direction and Feature F01 from the +X-direction both have sharp corners that cannot be achieved using milling.

4.4.4 Macroplanner and Setup Planner

The *Macroplanner* is responsible for selecting fixtures, assigning features to setups and coordinating the activities of the microplanner, tool-path planners and NC-code generators. The input to the macroplanner is the feature graph that has been validated earlier. Since each feature can potentially have many alternative representations as described earlier, a single part can have numerous feature-based models (FBMs), each of which corresponds to a certain way of machining the part. These different ways of machining the part in turn correspond to the various ways in which the setups can be sequenced and the fixture configured for these setups. A detailed description of performing this in an automated fashion and as a part of the software being described in this chapter can be found in [4.17]. The steps in the basic algorithm are briefly described below, though improved algorithms have also been presented in the above work for toe-clamps and parallel-sided fixturing vices. The first step is to construct all possible feature-based models (FBMs). Here, only those representations of features that have been validated by the *Feature Validator* are used. For each FBM, the number of different access directions over all the features is determined. The number of different access directions in an FBM gives a lower bound for the minimum number of setups (and hence the minimum time) that would be needed to machine it. The FBMs are then sorted in ascending order of this preliminary minimum setup time. This procedure helps reduce considerably the number of FBMs that have to be considered. Next, the precedence constraints between the various features of each FBM are identified based upon the feature hierarchy and machining rules. Based on these precedence constraints, an FBM is split into various sets of features, each of which is designated as a "simply fixturable setup" (SFS). Now, the fixturing constraints are applied to each SFS, which in general splits the SFS into the actual setups that the part will be machined in. Finally, for each FBM, these setups are ordered in a linear fashion to generate the setup plan. The setup times and sequences for each of these FBMs are then computed and the FBM with the least setup time is chosen. Algorithms for fixture planning are described in detail in [4.17][4.18].

Figure 4.6 shows the vise configurations for the part in the figure. Note that all the features can be manufactured from the +z-direction in one setup. Figures 4.6(a)–(c) show different vise configurations if the geometry of the part were changed. Figures 4.6(d)–(f) show toe-clamp configurations for the same part.

4.4.5 Microplanner

After setup planning, each setup is sent to the *Microplanner*, which selects the optimal set of tools out of the available tools in the database for the given feature geometry. The microplanner first obtains a list of feasible tools (*i.e.* tools that can machine the area without collisions) and computes the accessible regions for each of these tools. It is clear that the larger tool can be used to remove material in a shorter amount of time, but will not be able to reach all regions of the contour geometry. A

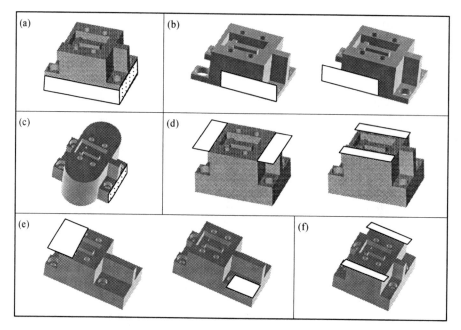

Figure 4.6. Fixture plans for various geometries of the part: (a) two configurations for vise – all features machined from one setup; (b) two setups required due to insufficient area and deep features; (c) non-planar faces allow vise clamps only on shaded region; (d) two setups required for toe-clamps; (e) previously machined faces used for toe-clamps; and (f) single setup is sufficient for toe-clamps

smaller tool will take longer to machine, but will be able to reach all regions of the feature. The problem is to find an optimal combination of the larger and smaller tools that will minimise the time taken to machine the entire feature geometry. Figures 4.7(b)–(d) show the regions of a pocket (Figure 4.7(a)) that are machined using three different tools. Each of these regions is called a decomposed feature for the corresponding tool. For a single feature, the problem of selecting an optimal sequence of tools can be posed as the problem of finding the shortest path in a single-source, single-sink directed acyclic graph [4.19]. The edges represent the cost of machining, while the nodes represent the state of the stock after the particular tool has completed machining. In order to generate the machining costs, this module interacts with the tool-path planning module, which calculates the actual tool paths and machining cost. The machining costs are stored in the edges of the graph and the minimum cost path is computed. At this stage, tool holder collision checks are also performed to ensure collision-free tool paths.

Freeform surface microplanning involves two major operations – rough machining and finish machining. For rough machining, the freeform surface is decomposed into slices of specified thickness [4.20]. The thickness of the slice controls the amount of finishing that needs to be performed later and the time taken for rough (and finish) machining. The freeform pocket is first decomposed into slices of equal thickness. The tools and their accessible areas for each of the slices are computed. Some of these slices can be merged depending upon the maximum allowable depth of cut of the tools that are to be used. Thus, the tool is allowed to plunge to its full depth capacity to machine the accessible area of the deepest slice. The unmachined areas remaining in the "higher" slices are then removed in turn to complete the rough machining. In addition, a similar graph-based strategy to the 2.5D microplanner is adopted to compute the optimum set of tools that will machine the slices. The final machining is done on a tool-by-tool basis rather on a per slice basis, *i.e.* all the regions that can be removed by a particular tool are machined first before proceeding to the next tool. Thus, the tool paths for a particular tool have to be connected optimally to minimise air-travel time. This will be discussed in the next section.

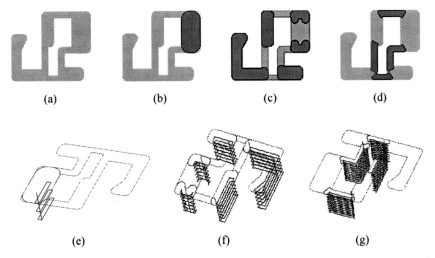

(a)   (b)   (c)   (d)

(e)   (f)   (g)

Figure 4.7. Microplanning and tool-path planning of example feature: (a) original feature; (b) decomposed feature for 0.5" tool; (c) decomposed feature for 0.25" tool with areas left over by the 0.5" tool; (d) decomposed feature for 0.125" tool; (e) tool path for 0.5" tool; (f) tool path for 0.25" tool; and (g) tool path for 0.125" tool

Different strategies for finish machining have been explored in the CyberCut system. One approach [4.21] is similar to that used for microplanning where multiple tools are used to machine the surface. The surface is decomposed into sub-surfaces based on four independent criteria (this leads to four possible decompositions): (1) Based on the feature contour similar to the approach for microplanning; (2) Based on bottom surface – large tools cannot access all regions of the surface due to self-intersections and gouging. The surface can be decomposed into different sub-surfaces that would correspond to intersection-free and gouge-free regions for the various tools; (3) Based on flat and steep regions – flat regions can be easily machined by designing the tool paths for the 2D contour of the feature and then projecting onto the offset surface of the design surface, whereas steep regions can be machined by using a slicing approach; and (4) Based on directionally flat and steep regions – given a direction and a critical slope angle, the surface can be decomposed into sub-surfaces that are directionally flat or steep. The first two criteria are concerned with tool selection, whereas the last two are concerned with the tool-path strategy. The different decompositions are combined in order to give minimum cost and machining time.

Another approach that CyberCut uses is the inverse tool offset and zigzag tool-path strategy [4.22]. This method does not use multiple tools as it is observed that switching tools during finish machining leaves marks on the surface due to tool-offset measurements at the machine tool and other alignment problems. The surface is first discretised by sampling the design surface at a resolution that depends upon the curvature and slope of the surface with respect to the access direction. Next, the offset surface is computed using the inverse tool-offset method, *i.e.* by taking the inverse of the tool in the access direction with the centre of the inverted tool moving on the design surface. For each discretised point, the inverse tool-offset algorithm gives a tool centre location so that the tool is tangent to the surface. These points of the offset surface, however, do not ensure that the tolerances or scallop requirements are met. Points are interpolated to the points of the offset surface both in the feed-forward and the side-step direction to meet these requirements.

4.4.6 Tool-path Planner

The tool-path planning module has two functions. First, it computes the contour parallel tool paths for a given contour. The tool paths for a region consist of a number of concentric regions called tool-path cells. Figure 4.8(c) shows the tool-path cells for Feature 3 of the sample part in Figure 4.8(a). These tool paths are required by the microplanning module to compute the machining cost in the graph algorithm as explained in the previous section. The other function of this module is to generate the connected tool paths for each tool in the setup. A tool may remove regions from different features, which have constraints due to the feature hierarchy, in the setup. A sample part and its feature hierarchy are shown in Figure 4.8. Only one tool is used here and the regions removed by this tool must obey the feature hierarchy. Due to the parent–child constraints in the feature hierarchy (Figure 4.8(b)), the tool-path cells exhibit the same precedence constraints, which are shown in Figure 4.8(d). All these constraints must be satisfied while connecting up the tool-path cells for a given tool. For each tool, the feature tree is queried and only those features that use that tool are considered and their tool-path cells obtained. After doing this for all relevant features, a tree of tool-path cells is obtained, which must be connected optimally while satisfying the parent–child feature constraints. This problem reduces to the asymmetric travelling-salesman problem with precedence constraints [4.23]. This problem is NP-hard and thus a heuristic algorithm is used to find the optimal or near-optimal solution for most practical-sized problems. After this step, the final connected tool paths for each tool are obtained and they are used to generate the G&M code for the NC machine.

4.5 Implementation and Results

The algorithms are implemented in C++ using the ACIS® Solid Modelling Kernel. Figure 4.9 shows the detailed software sequence diagram.

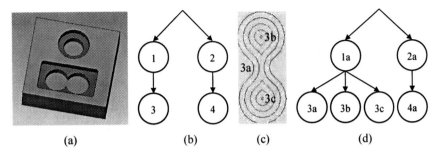

Figure 4.8. Sample part and feature hierarchy for air-time minimisation: (a) example part; (b) feature hierarchy; (c) tool-path cells; and (d) tool-path cell hierarchy

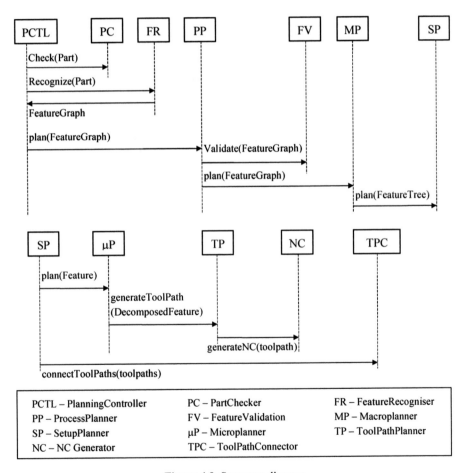

PCTL – PlanningController   PC – PartChecker   FR – FeatureRecogniser
PP – ProcessPlanner   FV – FeatureValidation   MP – Macroplanner
SP – SetupPlanner   μP – Microplanner   TP – ToolPathPlanner
NC – NC Generator   TPC – ToolPathConnector

Figure 4.9. Sequence diagram

Figure 4.10 shows some example parts that were planned and machined using the CyberCut process planning system. Figure 4.10(a) shows a part that has only freeform surface features. Figure 4.10(b) shows two mould halves, both of which have only 2.5D features, whereas Figure 4.10(c) shows a mould half that has a mixture of freeform and 2.5D features.

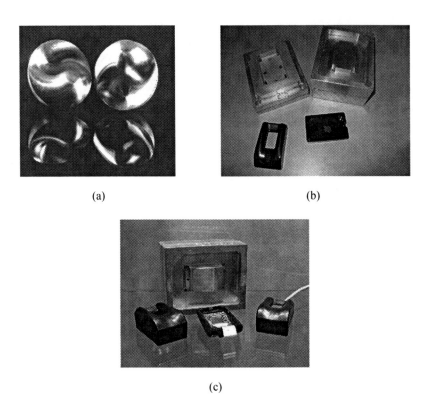

(a)   (b)

(c)

Figure 4.10. Examples of parts machined using the CyberCut system: (a) machined yinyang part with freeform surfaces; (b) machined mould halves with 2.5D features; and (c) machined mould that has a combination of freeform and 2.5D features

4.6 Conclusions

CyberCut is an integrated process planning system that automatically generates G&M codes and machinist instructions from CAD models. The system restricts itself to parts that can be made on a generic 3-axis milling machine. Subject to this constraint, the system performs feature recognition of the CAD model to generate features, generates setups and fixtures, selects an optimal combination of tools for the removal of feature volumes, generates tool paths and connects them to reduce redundant air-time motion. The system has been used to machine parts that range from simple brackets to complex moulds with sculptured surfaces. With CyberCut, a user gets feedback about the manufacturability of the part within seconds. If the part is not manufacturable, the user is presented with a list of problematic features and the reason for their failure.

Although CyberCut captures an important subset of machined parts, especially in the prototyping domain, a substantial amount of research and development is still needed to realise the vision of a comprehensive bidirectionally coupled system. It should be noted that CyberCut's feature-recognition module can automatically extract features from six orthogonal access directions. Features from directions other than these directions need to be extracted manually. Also, since the feature-recognition module uses a "visibility"-based approach, as seen from the parent–child hierarchy of features, certain classes of features such as T-slots, dove-tails and other undercut features cannot be recognised. Also, the system needs to include other manufacturing processes, specifically turning. The planning module also does not use tolerance models – a limitation for the manufacture of precision components. Despite these limitations, CyberCut provides a complete CAD to G&M code service that can provide a template for the research and development of these extended systems.

References

[4.1]   Kai, C.C. and Fai, L.K., *Rapid Prototyping: Principles & Applications in Manufacturing*, Wiley, New York, NY, USA.

[4.2]   Chang, T.-C. and Wysk, R., 1985, *An Introduction to Automated Process Planning Systems*, Prentice-Hall, Englewood Cliffs, NJ, USA.

[4.3]   Feng, S.C., 2000, "Manufacturing planning and execution software interfaces," *Journal of Manufacturing Systems*, 19(1), pp. 1–17.

[4.4]   Shah, J.J. and Mäntylä, M., 1995, *Parametric and Feature-based CAD/CAM – Concepts, Techniques and Applications*, John-Wiley & Sons, Inc., New York, NY.

[4.5]   Lin, E., Minis, I., Nau, D.S. and Regli, W.C., 1995, *Contribution to Virtual Manufacturing Background Research*, Final Report Prepared for the Manufacturing Technology Directorate, Air Force Wright Laboratory.

[4.6]   Sarma, S.E., Schofield, S., Stori, J., MacFarlane, J. and Wright, P., 1996, "Rapid product realization from detail design," *Computer Aided Design*, 28(5), pp. 383–392.

[4.7]   Inouye, R. and Wright, P.K., 1999, "Design rules and technology guides for web-based manufacturing," In *Proceedings of the 1999 ASME Design Engineering Technical Conference*, September 12–15, 1999, Las Vegas, NV, USA.

[4.8]   Kim, J.H., Wang, F., Sequin, C.H. and Wright, P.K., 1999, "Design for machining over the Internet," In *Proceedings of the 1999 ASME Design Engineering Technical Conference*, Las Vegas, NV, USA.

[4.9]   Ahn, S., Sundararajan, V., Smith, C., Kannan, B., D'Souza, R., Sun, G., Mohole, A., Wright, P., Kim, J., McMains, S., Smith, J. and Sequin, C., 2001, "CyberCut: An Internet-based CAD/CAM system," *ASME Journal of Computing and Information Science in Engineering*, 1(1), pp. 52–58.

[4.10] Shah, J., Shen, Y. and Shirur, A., 1994, "Determination of machining volumes from extensible sets of design features," In *Advances in Feature Based Manufacturing*, Shah, J., Mäntylä, M. and Nau, D. (eds.), Elsevier Science Publishers, Amsterdam, pp. 127–157.

[4.11] Han, J., Pratt, M. and Regli, W., 2000, "Manufacturing feature recognition from solid models: a status report," *IEEE Transactions on Robotics and Automation*, 16(6), pp. 782–796.

[4.12] Shah, J., Anderson, D., Kim, Y.S. and Joshi, S., 2001, "A discourse on geometric feature recognition from CAD models," *ASME Journal of Computing and Information*

Science in Engineering, **1**(1), pp. 41–51.

[4.13] Regli, W.C., 1995, *Geometric Algorithms for Recognition of Features from Solid Models,* Ph.D. Thesis, University of Maryland, USA.

[4.14] Sarma, S. and Wright, P.K., 1997, "Reference free part encapsulation: a new universal fixturing concept," *Journal of Manufacturing Systems,* **16**(1), pp. 35–47.

[4.15] Sundararajan, V. and Wright, P., 2000, "Identification of multiple feature representations by volume decomposition for 2.5D components," *ASME Journal of Manufacturing Science and Engineering,* **122**(1), pp. 280–290.

[4.16] Sundararajan, V. and Wright, P., 2004, "Volumetric feature recognition for machining components with freeform surfaces," *Computer-Aided Design,* **36**(1), pp. 11–25.

[4.17] Kannan, B. and Wright, P.K., 2004, "Efficient algorithms for automated process planning of 2.5D machined parts considering fixturing constraints," *International Journal of Computer Integrated Manufacturing,* **17**(1), pp. 16–28.

[4.18] Sundararajan, V. and Wright, P., 2002, "Feature based macroplanning including fixturing," *ASME Journal of Computing and Information Science in Engineering,* **2**(3), pp. 179–192.

[4.19] D'Souza, R., Wright, P.K. and Sequin, C., 2002, "Handling tool holder collisions in optimal tool sequence selection for 2.5-D pocket machining," *Journal of Computing and Information Science,* **2**, pp. 345–349.

[4.20] D'Souza, R., 2004, "Automated tool sequence selection for 3-axis machining of free-form pockets," *Computer Aided Design,* **36**(7), pp. 595–605.

[4.21] Sun, G., Sequin, C. and Wright, P.K., 2001, "Operation decomposition for freeform surface features in process planning," *Computer Aided Design,* **33**, pp. 621–636.

[4.22] Wright, P.K., Dornfeld, D., Sundararajan, V. and Mishra, D., 2004, "Tool path generation for finish machining of freeform surfaces in the CyberCut process planning pipeline," *Transactions of the North American Manufacturing Research Institution of SME,* **32**, pp. 159–166.

[4.23] Castelino, K., D'Souza, R. and Wright, P.K, 2003, "Tool-path optimization for minimizing airtime during machining," *Journal of Manufacturing Systems,* **22**(3), pp. 173–180.

5

Process Planning, Scheduling and Control for One-of-a-Kind Production

Paul Dean[1,2], Yiliu Tu[2] and Deyi Xue[2]

[1] Gienow Windows and Doors LLP, Calgary, AB T2C 2B6, Canada
Email: pdean@gienow.com

[2] Departments of Mechanical and Manufacturing Engineering
University of Calgary, Calgary, AB T2N 1N4, Canada
Emails: ytu@ucalgary.ca, dxue@ucalgary.ca

Abstract
This chapter describes a framework for computer-aided process planning, scheduling and control for flexible manufacturing systems (FMS) producing One-of-a-Kind Products (OKP). This constitutes a form of mass customisation (MC) using one-piece flow (mass production) to produce variations on a number of different kinds of products (customisation). The basic concept is to produce customised forms of a standard product in the same time and for the same cost as the standard product. There is a need to capture customer requirements that specify the variation of the standard product easily, quickly and accurately. Also, there is a need to be able to respond to changing demands in the market for new variations and options for standard products. This chapter includes a case study of a company that produces a significant number of products daily that require the use of computer-aided engineering software systems and CNC (computer numerical control) equipment. The planning and control of these FMS are complicated because of the non-standard nature of the products.

5.1 Introduction

Manufacturing companies today need to be innovative, both in their product design and processes. Over a period of time, demand for customised products has been increasing. In demanding this customisation, the customer also expects the product to be delivered in the same kind of time frame and at the same cost as mass-produced items. There is also an assumption of quality by the customers. In order to compete with large mass-producing companies, small-to-medium enterprises have to offer their customers something extra that usually translates into some unique aspect of the product either in appearance or functionality. Examples of such companies include Apple computers for their unique designs, and Dell for their quick delivery of personalised computers. Both companies offer something different that satisfies and meets their customers' expectations.

It is observed that customer expectations today do not include an emphasis on longevity in commodity products. Today's customers do not seem to want to pay the cost for long-lasting but prefer their disposable nature so that they can avail themselves of newer models with improved appearance and functionality. Product examples include such items as cell phones and motor cars. Together with environmental consciousness, this leads to additional requirements in design and in the process of improving the disposal and recycling of products.

One way to be successful in this type of manufacturing process is to develop products in such a manner that they can easily be customised and have opportunities for improvement. For example, the cell phone can be customised both in its appearance and its functionality (*e.g.* the cell-phone plan has many options such as voice mail, call forwarding, call display and others). Cell phone functionality has developed since its first introduction with the addition of text messaging, MP3 players and video displays. The case study in this chapter focuses on a manufacturer of customised windows and doors that uses an incremental product development method and a product production structure similar to that described by Tu *et al.* [5.1][5.2]. Gienow's integrated system was developed over the last ten years since 1995. It was instrumental in Gienow being able to reduce their production cycle from 10–12 days to 24 h.

Gienow's products can be customised in numerous ways including size, geometric shape, appearance options and functional options. Over the product's lifecycle, new innovations are introduced. In the past they have included extended glazing options, grill patterns and add-on options.

First, the base product is designed and tested in its basic form. In the case of windows and doors this includes the basic frame, other components and the glass. The product is tested to ensure that it meets technical specifications and the prescribed building codes according to the geographic location where it will be installed. The design of the base product is also based on market studies that determine appearance and functional requirements. Once the base product is designed and tested, the limits of the product variations are determined and established as design constraints. These constraints include such properties as size (minimum and maximum), shapes, combinations, and options that are inclusive and/or exclusive. Consequently, the various customised windows that meet the technical specification and building codes of the base product can be designed and mass produced through the variations of the base product within these predefined limits. This production mode is called one-of-a-kind production (OKP), which is a very efficient means to realize so-called mass customisation. It has much higher flexibility than the batch-production mode that normally uses standard modules to achieve limited customisation with a required minimum production batch size to compensate its production cost. Compared with a job shop, an OKP company has nearly the same flexibility but with much higher production efficiency.

In terms of the base product, the product improvement program and process can also be defined. This can include warranty and maintenance services programs, and the ability to add on new improvements after installation or complete replacement programs. This process should be a collaborative process that includes the customers and suppliers so that any improvements developed by suppliers are included as soon as they become available. Customer involvement ensures that new requirements and expectations are registered once they occur. This collaborative process leads to the design improvement process becoming shorter and more effective.

An OKP company normally demonstrates the following characteristics [5.3]:

1. "Once" successful approach on the product, *i.e.* no prototype or specimen will be made in OKP, and the batch size can be one.
2. Product is usually designed by modifying and combining existing products to avoid the risk of long lead time and high costs for developing a radically innovative product.
3. Frequent changes of product design, manufacturing processes, and production schedules due to changes in customer requirements, arrivals of customer orders or raw materials, and manufacturing process, as well as machine breakdowns.
4. Mixed-product production.
5. Frequent changes of production system to adapt customised product production requirements.
6. Optimally utilising technologies and resources to continually improve production efficiency and reduce the production costs.

Dedicated manufacturing systems, *e.g.* mass or batch production systems, are characterised by rigid equipment designed specifically for a product or a restricted family of products. Typically, processes are manual with limited use of CNC equipment. They have high production rates and are expensive to change. Included in these systems are dedicated machines and dedicated flow lines. The latter is used in mass production.

Flexible manufacturing systems [5.4] are defined by CECIMO (Commit Europeenne de Cooperation des Industries de la Machine Outil) as *automated manufacturing systems capable, with minimum human action, of producing any part type of a predefined family; these are generally adopted for the production at small to medium volumes, in variable lots sizes that differ also in their composition. The system flexibility is usually limited to the family of parts on which the system was conceived.* FMS are characterised by their use of CNC equipment and low volume or small batch sizes. Mass customisation uses FMS in dedicated flow lines in order to achieve some economies of scale at the same time as attaining some customisation. An OKP system is an FMS with a lot size of one. It is a customisation that produces a unique variation from one kind of product.

OKP companies are normally different from one-off and job-shop manufacturing companies mainly in two aspects:

1. OKP companies are product-oriented manufacturing companies, whereas one-off or job-shop manufacturing companies are capability-oriented manufacturing companies.
2. OKP companies often adopt automated and highly efficient manufacturing processes, *e.g.* product flow line processes, whereas one-off or job-shop companies normally adopt high flexibility but low-efficiency manufacturing process, *e.g.* universal and functional equipment and job-shop process.

Due to these differences, OKP companies are able to effectively compete with large mass-production manufacturing companies and gain market shares through customisation, fast delivery and low-cost production, to realise mass customisation. The EEC (European Economic Community) research program, "ESPRIT basic research action 3143 – Factory of Future (FOF) production theory", envisaged that OKP would become a novel manufacturing paradigm in this century along with a clear market trend toward customisation and responsiveness [5.5].

The characteristics of OKP result in great uncertainties and complicated product data flow, which leads to a longer product development cycle and higher production costs. Particularly along with the OKP, companies are moving to so-called virtual manufacturing systems (VMS) that logically consist of manufacturing resources belonging to different physical manufacturers. Uncertainties increase and product data flow becomes even more complicated as a result. Computer Aided Design (CAD) and production control systems currently available either overlook these problems or are not developed to meet OKP needs. This complex data flow requires good communication channels and electronic data interchange (EDI) methods. If a customised product is ordered then there is likely to be customised sub-components that are outsourced. These outsourcings need to be manufactured in a similar customised manner by the suppliers. Distribution of manufacturing further complicates data requirements and the whole supply-chain management.

The various processes of mass customisation have been identified by MacCarthy [5.6]. These processes are summarised in Table 5.1.

Table 5.1. Processes of mass customisation

Process	Definition: The process of
Order taking and coordination	Defining the order and product specification
Product development and design	Designing the product and its compliance with internal and external standards
Product validation and manufacturing engineering	Confirming manufacturability Producing the BOM, routing and processing instructions
Order fulfilment management	Managing the order fulfilment by determining when production can start Managing the value-added supply chain Scheduling, monitoring and controlling the process
Order fulfilment realisation	Executing the production and delivery Realising the supplier activities
Post order process	Installation Warranty claims Service

Central to these processes is the products' manufacturing engineering knowledge. Without accurate information about the bill of materials, routings, engineering constraints and data requirements, it will be impossible to manage a mass-customisation process for one-of-a-kind products. This information is required for all the processes from order taking to service. The method of developing this knowledge is beyond the scope of this chapter, but it is assumed to be available in order to manage the planning, scheduling and control processes.

5.2 Literature Review

In the case study, Gienow developed a make-to-order JIT production environment to market their product. This process provides customers with a low-cost and high-quality product when they need it. In their paper [5.7] on the JIT approach, Banerjee and Armouti made the statement that *"The make-to-order production environment requires the tightest production activity control, because commitments are already made to customers, orders booked and inability to meet deadlines may have more severe repercussions than other environments."* In recognising the importance of this, the application at Gienow includes a control mechanism at the item and sub-component levels. This is possible because the system is based on the detailed generic bill-of-materials and operations (GBOMO) [5.8]–[5.10] information, generated by the knowledge-based system.

As discussed by Sanoff and Poilevey in 1991 [5.11] and by Kumar *et al.* in 2004 [5.12], they emphasised the requirement for an integrated system architecture for manufacturing systems. In their final scheme [5.12], they identify the knowledge-base system (KBS) as being central to the design and integration with all functions. As stated, this was the approach taken in developing the application for Gienow, where engineering knowledge is captured in a database and is utilised by an expert system to provide all the necessary information for sub-functions to operate.

In 1993, Westbrook and Williamson [5.13] discussed the development of mass customisation in Japan applied to customised suits and customised bicycles. In reviewing their paper, it was mentioned that the National Bicycle Industrial Company was considering offering the bicycle in increments of 1 mm. In the case study, Gienow offers their product in increments of 1 mm for both width and height and all other geometric dimension of the product (radius and angles). In the conclusion of their paper, Westbrook and Williamson indicated the need to have close ties with their suppliers, an ability to manage workflow on the shop floor and implement a flexible manufacturing system using IT technology.

Piller [5.14] reflects on the state of mass customisation in 2004. A complete analysis of his paper in comparison with the situation at Gienow is beyond the scope of this chapter. However, one comment is worthwhile; although we agree with Piller that there are not many "mass customisation" companies, the case study at Gienow deserves recognition. In the 1980s the window and door industry sold its products from an inventory of standard sizes. In 1985 Gienow decided to re-engineer their manufacturing process to a make-to-order and JIT delivery system using FMS, offering to their customers a product "any size, any shape, and when you need it". This meant a significant investment in CNC equipment and computer technology. By the 1990s Gienow had 45% of their local market, zero finished goods inventory and several suppliers delivering sub-components on a JIT basis. By the early 2000s Gienow's major competitors had also adopted the same method of production.

Kumar [5.15] argues that mass customisation can give a company an advantage in all four (five) areas of competitiveness; price, quality, flexibility, delivery (and service). The paper is extensive in its discussion, but one point needs to be emphasised, that is, there are five areas of competitiveness, and "mass customisation" has an impact on service, as will be described in the case study later in this chapter.

The other point discussed in the two papers [5.14] and [5.15], is the central importance of customer co-design. This is beyond the scope of this chapter but the reader should be aware of the importance that Gienow places on this co-design. It should also be appreciated what this co-design means in practice. Both authors mention the term "within the solution space"; the Gienow system implements this concept by using a one-of-a-kind production system. By this it is meant that Gienow offers a number of products and allows their customer through a computer interface to specify the variations in a product they require. This co-design process creates a one-of-a-kind product that the customer is entirely satisfied with. Gienow then produces the variation product at the same cost as a stocked standard product. Gienow's customers assume that quality is built into products. For a period of time the system did create a competitive advantage on price and flexibility, but what became important to customers was the on-time delivery that was achieved with the system. In future, service will become the dominant factor and the requirement to know what customisation was done to each product will be important in keeping service costs down.

Vandaele and De Boeck, in their paper on advanced resource planning [5.16], discussed the interdependency between material and resource planning and the conflict between resource utilisation and customer satisfaction. They defined lead time as a measurement of demand versus capacity, where lead time is asymmetric with a bias to the right. As interruptions occur in processing, so lead times are extended and customer dissatisfaction increases. As utilisation decreases, lead time increases in a nonlinear manner. Factors affecting utilisation include demand (quantity and time), capacity (processing times, setup times, number of resources, shift patterns), and outsourcing capabilities. These factors can be influenced by lot sizes, sequencing and release mechanisms. Nonlinear effects of utilisation on lead time are further affected when the average parameters from demand and capacity are not deterministic; they have a stochastic nature. Variability in product mix, operations and management decisions also has an impact on the factors. This is the case in OKP where each product is one-of-a-kind, varying in materials, resource usage and operations. They proposed the use of an advanced planning system that would account for variability in these factors and would determine optimum lead times, resource usages, lot sizes and sequence of setups.

In contrast to this, the application developed at Gienow was based on running the OKP system with lot sizes of one (because of the one-of-a-kind nature of the product) and the use of CNC-controlled equipment that would effectively reduce setup times to zero. The application then planned, scheduled and controlled the production process based on releasing orders in groups, while creating a fixed production sequence. The net result was that orders were released according to available capacity and in accordance with customer requirements. Required capacity was determined in advance and resources allocated accordingly. As processes occurred, future schedules needed to adjust as the result of interruptions caused by material delays and machine downtime.

In designing both product and process for OKP where variations on a product provide customisation, it is important to decide whether to build up the variation from a base product or to scale down from a full option set [5.17]. In Gienow's case, all processes including data generation start from basic products that provide the minimum requirements of a window or a door. Options are then selected by the customer to meet both functional and technical requirements.

Fujimoto and Ahmed [5.18] discussed the need to consider both process and product in order to provide customised items at a reasonable cost using economies of scale. Variety is caused by differences in basic functions (thermal properties of windows), adaptability requirements (different size and shape of windows), optional functions (windows that open), and non-functional requirements (welded frame or screwed frame). Variety impacts the manufacturing process in a number of ways; high inventories, feeding complexities, excessive capital investment, change in assembly sequence and complexity in line balancing. Assuming that the FMS has been designed with both these considerations in mind, the processes of planning, scheduling and control are impacted and as a result are not the same as conventional processes.

In recognising the complexity of scheduling for an automated assembly system constructed in the various forms of flow lines and cell groups, Little and Hemmings [5.19] propose the use of a simulator that is "run-ahead" and provides an analysis of what can happen when production starts. In their conclusion, they indicate that the delivery of sub-assemblies and the organisation of components in the correct sequence is a major problem for these systems, particularly when the main orders are to be supplied in a JIT-type manner to the customers. This is addressed in the case study for this chapter.

Zhang *et al.* [5.20] propose the use of Petri nets for scheduling flexible systems. The authors identify some of the advantages as being a formal model to define the internal relationships of discrete-event processes, identification of constraints and monitoring the current state of the production system. Even though the focus is on products that do not have a predetermined sequence of assembly, their approach of using a time-based Petri net to determine the production time can be used to determine the production time of a predetermined sequence and determine if there are any constraints likely to occur in the proposed schedule.

Anderson [5.21] recognises that product mix has a reduction effect on available capacity. In his paper, he proposes that human management (discretionary capacity management) of product mix increases the reduction effect on capacity when there are interruptions in the schedule. The implication in the conclusion is that discretionary capacity management is detrimental to throughput, and conversely any computer-aided assistance with capacity planning for product-mix variation should have a positive effect.

With regard to the impact of interruptions on scheduling, Foley [5.22] recommends a number of alternatives – adjusting the schedule, recalculating the schedule or absorbing the interruption in the current schedule. In adjusting or recalculating the schedule, a hierarchical computer system is proposed that would determine which appropriate products should be produced next.

One interesting conclusion from the results of their experiment was: *"The implication is that, at the occurrence of an interruption at the high load level, simple myopic modifications to the predetermined schedule may end up leading to a much worse performance than continuing to rigidly follow it."* From this we conclude that since Gienow lines are more often loaded to high levels, when an interruption occurs, the current schedule should not be modified. Subsequent schedules would be modified to help absorb the interruption, according to how large the interruption is. This creates the requirement to provide information on the performance of each production line and to show the impact of any interruption on the lines. Because of the type of product mix for orders and the fact that Gienow is using a FMS, the total plant schedule has to be considered.

The reference to the theory of constraints [5.23] is important, because the impact of interruption at a bottleneck is far greater than at a point that is not a bottleneck. Therefore, to know where the potential bottlenecks are in a schedule would be useful information to management in the event of an interruption. A simulation should provide this information as mentioned before.

In discussing the impact of the interruption on a system that has a rigid schedule (as in the case of Gienow), the observation was made that the impact of the interruption would propagate quickly down the rest of the line. To some extent this is mitigated by designing the system as a one-piece flow. For some types of interruptions it will be possible to put the interrupted product aside and carry on with the rest of the schedule until the problem has been expedited. This has led to the development of some functions in the application to assist with the control of these types of interruptions.

According to Bielli and Dell'olmo [5.24] to solve a large discrete-event stochastic control problem it is necessary to integrate a number of strategies. In their paper on the IS-OPTIMUS system, Palacios et al. [5.25] focused their research on material-cutting optimisation, static scheduling and dynamic rescheduling after production conditions change. The system accounts for product mix and short lead times. In comparison, the majority of CNC cutting equipment at Gienow contains optimisation routines and therefore Gienow is able to concentrate on the development of a schedule, and the control and monitoring of the performance of the schedule. It also provides information that helps new schedules to consider the changing production conditions. In addition, the proposed simulator will also provide useful information to the scheduling process.

Matsui et al. [5.26] discussed the size of buffers, considering the differences between finite and infinite buffers. The basis of Gienow's production system is one-piece flow. However, since the cutting optimisation capability of the machines is used by Gienow, some of the buffer sizes are greater than 1. Depending on the cutting machine and the size of the material being moved, these buffers are set at 25, 50 or 100. The paper also compared fixed and dynamic routing and concluded that if machine loading is balanced then throughput with fixed routing is almost the same as with dynamic routing. It is possibly more economical because fixed routings are simpler to manage and control. In the case study, the production process uses fixed routings but they do have multiple paths because of the product mix caused by different options selected by the customers.

The impact of sequence changes caused by material shortages, outsourcing delays and machine downtimes was considered when developing the system. Dupon et al. [5.27] researched this issue and concluded that sequence changes would not materially affect production time, but would adversely affect lead time to the customer. As a consequence, consideration was given in the application to provide information that would help management expedite issues. Accordingly, lead time delays are kept to a minimum. Similar to the conclusion in the paper, management at

Gienow emphasises the application of the FIFO rule in expediting issues and this helps maintain the level of customer satisfaction.

In his paper, Miltenburg [5.28] recognised the requirement to solve the two problems of model sequencing and line balancing for a FMS running in JIT mode. JIT is a pull system and the customer sets the main sequence of production, thus setting the sequence required for other facilities. Miltenburg proposes a genetic algorithm to determine the sequence for optimum model production and line balancing. The problem at Gienow, with largely fixed processing work cells, emphasises the challenge of determining the optimum model sequence for the main lines, while levelling the production of the ancillary facilities. This will be discussed later in the chapter.

Much more has been written on these subjects. The following sections describe the results of similar research and the development of a production planning, scheduling and control application.

5.3 Process Planning

Process planning involves long-term and short-term planning. In conventional manufacturing systems, this usually involves the use of manufacturing requirements planning (MRP-II) and master production scheduling (MPS)-type systems. However, conventional systems tend to rely on the fact that large quantities of standard products with standard bills of materials (BOM) are being used to plan and schedule processes and resources. This is not the case in mass customisation, particularly when one-of-a-kind products are being manufactured.

Also, because of the customised nature of sub-components, the planning of any distributed manufactured parts requires additional attention. Operations cannot rely on a steady stream of standard parts from an outside source as each part is different and needs to be specified accurately before being ordered. Timing is important, as otherwise promised delivery dates are not going to be met and delivery times become extended and no longer compare with mass-produced products.

5.3.1 Long-term Process Planning

The essential part of long-term planning is the availability of consistent sales and production data that is easy to collect and maintain. Often, sales are measured in terms of revenue, while production is measured in terms of units. This is acceptable in conventional manufacturing systems because there is usually a constituent relationship between revenue and units for each product (or kind). In OKP, this relationship is no longer consistent or linear. By this we mean that one day 100 units can be produced and sold for $1000 and the next day 80 units can be produced and sold for $1200 and the following day 120 units produced and sold for $800. It is very dependent on the customer and the customisation or variation of the kind of products ordered.

Hence, it becomes important to measure both sales and production in a way that is meaningful and then to maintain this consistency of measurement. In this way, variations will give an indication of changing markets, customer demands, and efficiencies in production.

It is necessary to maintain production data at a detailed level. For example, if there is a warranty claim or a service request by a customer, then the precise specification of the product is required in order to process the claim or the service order. Otherwise there is a costly exercise needed to re-specify (measure) the original product in order to perform the replacement (part) or service. This detailed data is then stored for a period of time depending on the lifecycle of the product or warranty terms.

This production data can then be analysed in a number of ways, such as

- inventory usage by production line (if hybrid production method is being used) or by product kind (one-to-one relationship with production line);
- unit production by production line and time (preferable using standard time and actual time).

Long-term planning enables a company to plan for capacity expansion, more space, more equipment or more employees. It also enables a company to establish strategic alliances with suppliers of raw material and with distributed manufacturing facilities for customised sub-components. In Section 5.6, these issues are further discussed and a method of processing data is proposed to help with the planning process.

5.3.2 Short-term Process Planning

Short-term planning is required to meet the promised or expected delivery date of the customer. In conventional mass-production systems this means producing one or x thousands of the ordered products. Based on the standard (average) production rate, this will take a certain amount of time. However, when planning for mass customisation and particularly one-of-a-kind products, the standard (or average) rate does not work. Depending on the specification (size, shape and options), each product will take a different amount of time. This complicates the planning process, resource availability, material supply and sub-component delivery. Consider the following shop-floor layout from Gienow for one of their window products.

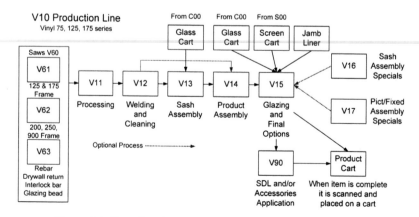

Figure 5.1. Shop-floor layout for a production line and work centres

Delivery-date Confirmation

As orders are entered into the system, the delivery date may be confirmed or may be tentative. If tentative, a process has to be followed to confirm the delivery date so that the planners know when to schedule the order. At Gienow, customers with tentative orders are phoned several days prior to the delivery date and asked to confirm. If the order is not confirmed, it is then tentatively rescheduled for a later delivery date.

Since Gienow uses a fleet of trucks to deliver their (large) products, there is a trucking schedule in place. This means as an order is processed; only dates available to the location of the delivery address are seen as potential delivery dates by the schedulers. During this process and as the various delivery dates come closer to "today", the scheduler's initial task is to load trucks to their potential capacity. When the time comes to confirm the schedule for the next day for any truck that is not full to near capacity, the scheduler will attempt to fill the truck by contacting customers that are known to have flexibility in their receiving dates.

Line Balancing

Once the initial schedule based on the trucking sequence has been determined, the loading of production lines is reviewed. If lines are under- or over-scheduled to their available capacity, the scheduler's task is to balance the load on each line. The balancing is achieved by either increasing or decreasing the available capacity by redistributing employee resources or by rescheduling orders to increase or decrease the load on the lines.

As stated earlier, it is essential to have the BOM and routing information for each customised product because this will enable the system to accumulate the standard time for each product on the production lines (by work centre if necessary). This also means that as the scheduler reschedules orders he/she can see the immediate effect on the loading of lines required to make all of the different products on the order. At Gienow, reports and graphical views are available to help the schedulers in this task.

Figure 5.2 shows an analysis of all orders according to their current scheduled date. The analysis processes detailed orders and shows how many units are to be produced by each production line daily. The analysis also shows how much (standard) time will be required to produce the products. Therefore, the planners are able to see how the production lines are loaded and to perform the balancing operation.

The graph in Figure 5.3 shows how each production line is performing and what the size of the forward load is by day. This helps the planner to see the possible orders to move. The program also has an algorithm that calculates the lead time of each production line.

Lead time is calculated based on the assumption that an order will enter the system on the next production day after being posted in the order-entry program and delivered to site that day plus the maximum lead time days. Under normal circumstances, the lead time is 10 days or 2 weeks. Each week has 5 production days, *i.e.* Monday to Friday.

Summary of Forward Load By Tentative Hours

Part 1

Board Date	SD0	SS0	V10	V50	V20	V30	V40	V70	V80	V90	W00	W30
Capacity	350	300	650	250	75	240	75	50	250	20	350	100
7/28/2006	425 7.8	417 2.4	711 0.3	247 0.0	60 0.0	263 0.3	58 0.0	61 0.0	348 2.5	51 0.0	304 176	100 0.8
7/31/2006	387 3.6	284 0.0	771 0.0	235 0.0	37 0.0	266 0.0	74 0.0	49 0.0	282 0.0	13 0.0	723 80	224 0.2
8/1/2006	415 2.3	314 0.0	762 0.0	194 0.0	43 0.0	249 0.0	17 0.0	31 0.0	209 0.0	0 0.0	171 353	34 1.1
8/2/2006	486 39.9	446 42.6	745 34.3	185 37.9	49 21.6	305 33.5	30 18.4	59 60.6	298 31.8	7 0.0	870 455	130 24.0
8/3/2006	421 36.2	429 43.0	766 31.2	186 19.5	63 21.4	389 39.4	48 30.7	93 61.8	365 22.5	9 69.5	601 335	231 22.3
8/4/2006	672 42.6	587 45.9	1095 28.0	270 29.5	106 38.9	454 42.2	83 31.7	101 48.7	570 36.0	47 0.6	804 427	202 24.5
8/8/2006	435 42.8	475 63.1	661 46.8	173 41.8	33 31.4	267 42.1	63 8.1	42 26.5	279 34.1	10 0.0	555 462	64 29.1
8/9/2006	556 30.4	671 47.3	642 36.1	203 32.9	37 42.4	331 29.2	63 12.6	44 37.4	383 29.7	3 0.0	837 729	192 30.0
8/10/2006	427 51.5	305 54.9	735 33.5	160 44.4	44 39.1	240 43.7	69 26.1	66 59.8	237 26.0	11 0.0	807 470	203 57.3
8/11/2006	439 34.0	436 47.8	666 34.3	220 33.6	65 52.9	274 33.6	77 42.8	94 68.2	271 27.3	37 0.0	658 89.5	179 91.6
8/14/2006	409 30.3	392 45.6	642 21.2	194 28.7	33 44.4	207 37.3	67 35.0	57 46.8	235 42.8	1 100.0	611 368	137 72.7
8/15/2006	346 45.4	454 45.6	573 31.5	214 44.7	39 33.3	243 41.3	52 42.9	43 72.0	215 38.7	45 0.0	1022 559	350 69.8
8/16/2006	214 55.4	267 57.3	325 39.4	174 27.7	56 5.2	137 49.3	79 43.0	16 44.0	148 27.1	10 0.0	700 943	126 56.7
8/17/2006	128 34.8	82 23.9	357 17.0	358 4.9	41 7.4	123 20.6	36 10.2	36 19.2	82 24.1	9 204	291 320	13 18.9
8/18/2006	289 51.1	328 48.6	681 36.8	149 24.7	62 17.9	149 53.0	24 15.3	16 81.3	235 31.8	30 0.0	78 0.0	61 35.0
8/21/2006	312 63.0	213 63.6	346 62.4	174 33.8	81 7.2	156 52.2	48 53.7	13 55.0	236 18.7	14 0.0	165 798	116 94.2
8/22/2006	442 83.9	320 81.9	521 66.0	202 54.2	97 33.4	293 71.8	58 61.9	63 51.1	290 73.3	19 58.4	338 963	48 93.0
8/23/2006	220 78.3	237 75.9	307 55.4	135 42.4	108 11.8	186 42.6	42 41.5	33 68.8	158 68.3	7 0.0	212 1000	99 87.9
8/24/2006	89 72.0	135 69.5	324 22.8	271 12.3	29 28.0	111 44.7	39 60.4	16 100.0	239 73.3	9	316 728	54 64.0
8/25/2006	243 24.1	214 56.0	276 36.8	76 7.4	9 17.0	88 47.0	18 40.5	9 0.0	167 21.5	31 0.0	120 51.9	47 70.2
8/28/2006	54 72.2	81 63.2	113 48.7	70 9.8	44 4.6	60 33.3	17 11.0	7 84.0	51 77.9	10 0.0	184 226	51 12.1
8/29/2006	159 46.5	295 54.8	183 55.0	70 18.1	48 29.7	84 58.2	24 19.2	2 71.9	120 23.5	1 0.0	103 67.4	55 44.0
8/30/2006	170 51.1	67 73.7	236 42.3	189 0.9	81 9.9	107 31.5	33 5.6	7 99.2	95 43.9	1 92.6	75 1000	17 69.4
8/31/2006	68 100.0	48 100.0	103 70.4	145 22.9	40 0.0	58 65.7	40 60.3	0	9 0.0	0	106 1000	14 100.0

MPC version 2006.001

Figure 5.2. Forward load by tentative hours

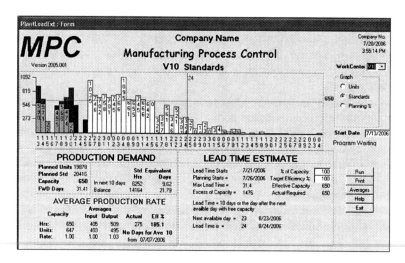

Figure 5.3. Production line graph

Under circumstances when a production line has more than 10 days worth of work, the lead time will be extended. This means that within the lead time there are several days that have more orders than capacity (excess orders). The lead time is calculated under the assumption that all orders in the system that lie within the lead time will be made before any new orders are processed.

The next available production day with available time is found by calculating the excess order time and allocating that time to production days that have free capacity starting at the first planning day. This is done because the next two or three production days are already fully allocated and fixed.

Once the next production day with free time is found, if this is less than 10 days from today, then the lead time is set to 10 days. If this is greater than 10, then the lead time is set to the day after the next available day, allowing for production and shipping to complete their tasks.

The performance of the lead time calculation can be affected by two parameters:

1. A portion of the maximum capacity, expressed as a percentage of maximum capacity or simply recorded as *capacity %*, can be used to adjust the *effective capacity* of the production line. If a board is being blanked or reduced within the 10-day lead time without making the day a non-production day in the calendar, then the *capacity %* can be changed. It can also be reduced to less than 100% in cases where allowance is being made for rush orders or service items.

2. Likewise, the *efficiency %* gives an indication of the amount of actual hours required to complete a specific amount of standard hours. For example, if the current efficiency of production line V10 is 120%, then 100 standard hours of work should be completed in $100/1.2=83.3$ h.

This lead time (for each production line and therefore each product type) is transferred to the knowledge base for the order-entry system to use online.

Sequencing

Each operation will be different and the method of sequencing will, as a result, be different. However, for one-piece flow of customised items it is necessary to have a sequence for each product on each line. In Gienow's case, the sequence is initially driven by the trucking plan. This plan dictates which truck is going to leave at what time of day. The location that the truck is going to will determine which orders it will carry. Therefore, the major sequence of orders is predetermined by this trucking sequence.

Once this is established, then the orders are first sorted into a sequence that starts with the trucking sequence and then followed by the order number (this ensures that the earliest orders get made first) and then the height of the product (in descending sequence as this helps with production efficiency[1] and storage on shipping carts – the tallest product is placed at the back). The orders are then organised according to their details and the different products allocated to their appropriate production lines. This then generates the sequence for each production line for this schedule.

Distributed Manufacturing – Outsourced Components

Part of the planning process involves the customised components that are outsourced. This outsourcing can be to a supplier or member company, either constitutes a part of the supply chain. Depending on the suppliers' capability to match the JIT (just-in-time) process, different methods of planning have to be used.

---

[1] Generally, products with the longest production time are scheduled first. Also, if a process using a CNC machine can be gradually changed instead of randomly changed, then the process is more efficient.

It would be ideal if all suppliers had the capability to deliver the sub-component just-in-time, but for SMEs (small and medium-sized enterprises) this is not always practical or possible owning to their limited buying power.

Hence, if the supplier does not have JIT capability, the sub-component has to be ordered according to the lead time. The longer the lead time, the less flexible the main production becomes. In a situation where the customer changes the order, and the sub-component has been ordered, there can be additional costs. In the case where the customer and/or scheduler reschedules and delays the production of the order, the sub-component will be delivered early and this can also lead to additional costs, potential damage, and reordering. All this leads to an increase in the cost of the customised product, which is contrary to the goal of mass customisation.

If the supplier does have JIT capability, then when the schedule for the main plant is confirmed, the customised sub-components for each supplier are extracted from the system and communicated to the suppliers using some form of electronic data interchange (EDI). At Gienow this function includes transmitting a computer file with the specification of customised components and details of what to print on a bar-coded label. These labels are then attached by the supplier to the product and delivered to the main plant within a specified period of time (currently less than 24 hours).

Irrespective of the supplier's capability, all sub-components are controlled by the receiving department. Each day, the receiving department is provided with a list of all sub-components required for that day's schedule. The receiving department's task is to deliver the sub-components to appropriate production lines. The bar-coded labels have information that identifies the production lines as well as the specific product. This product identification number is a compound number that also identifies the date of production and the sequence number. In the case study this is defined as "*MMDD-nnnn*", where *MM* is the month, *DD* is the day and *nnnn* is the production sequence number.

One of the computers available at Gienow shows all production lines and which sequence number each line has just completed. If necessary the receiving department can use this computer display to determine when the sub-components can be delivered to a line and whether the line has enough space to accept the delivery. If so, the receiving department can deliver the lines' components at the start of the schedule. For large quantities of sub-components, the receiving department is given a function similar to the one described next in internal components.

Internal Components

There are a number of issues with the production of internally made sub-components and their delivery to various production lines. First, there is a lead time to produce these components. In Gienow's case, the quantity is so large that it takes the full schedule of time to make them. Therefore, these sub-components have to be made and delivered in the sequence that the main product is being made. Since all sub-components can not be made at the same time, there will be multiple deliveries throughout the schedule. Gienow uses the cart method to deliver sub-components to the production lines. Therefore, these carts have to be identified and the items within them also identified and in the same sequence as the main production lines.

To achieve an even distribution of sub-components to multiple production lines in a JIT mode, it was necessary to develop a specific algorithm. At Gienow, there are two algorithms, one for a flow process and the other for a batch process.

Flow Process
- Determine lines involved, *e.g.* V10, V20, V30 and V40
- Determine total number of items to be made on each line
- Determine size of delivery batch for each line
- Determine the sequence of delivery to lines, *e.g.* V40, V20, V10 and V30
- Create sequence of sub-assembly production

Table 5.2. Flow method – batch size

Production line	# of items	Size of batch	Sequence
V10	50	5	3
V20	10	2	2
V30	50	5	4
V40	5	1	1

The sample data above would create the balanced sequence given in Table 5.3.

Table 5.3. Flow method – production sequence

V40 items	1	V20 items	2
V20 items	2	V10 items	5
V10 items	5	V30 items	5
V30 items	5	V40 items	1
V40 items	1	....	

V20 and V40 are the slower lines. By delivering the sub-assemblies to these lines early in the production cycle, ensures that waiting does not occur, while the other two lines maintain a steady production rate.

Batch Process
- Determine lines involved, *e.g.* V10, V20, V30 and V40
- Determine total number of items to be made on each line
- Determine size of cart delivered to each line
- Determine number of carts required for each line
- Determine the number of deliveries
- Determine how many carts are produced for each delivery
- Create sequence of sub-assembly production

The model allows for closer synchronisation of the sub-lines with main lines. This closeness is very dependent on the product and the manufacturing processes involved. In the case of Gienow, some of the sub-components require a certain amount of "curing" or "waiting" time before they can be used in the fabrication process. This creates an automatic buffer that means that synchronisation does not have to match exactly.

If closer synchronisation is required, then details of the main production-line schedules are available. This is achieved by using the actual scheduled production numbers for each line instead of assuming that there is an even production rate. This again is caused by each customised item requiring a different amount of time to make. These varying production rates are discussed in Section 5.4. An example of (batch) requirements and calculated production sequence for a sub-component is show in Tables 5.4 and 5.5.

Table 5.4. Batch method – requirements

Production line	Work centre	# of items	# of carts
A00	A05	40	2
V10	V13	587	24
V10	V15	687	28
V20	V23	75	3
V20	V25	286	12
V30	V35	100	4

Table 5.5. Batch method – production sequence

Production line	A00	V10	V10	V20	V20	V30	Total
Deliver to	A05	V13	V15	V23	V25	V35	
1	25	50	50	25	25	25	200
2	–	75	100	–	25	–	200
3	–	75	75	–	25	25	200
4	–	75	75	25	25	–	200
5	15	75	60	–	25	25	200
6	–	75	100	–	50	25	200
7	–	50	75	–	50	–	200
8	–	50	75	25	50	–	200
9	–	62	77	–	36	–	175
Total	40	587	687	75	286	100	1775

The algorithm is a modified version of the minimum-cost-assignment algorithm. It attempts to provide a smooth delivery of sub-components to each production line that matches its production rate. In Section 5.5, Adaptive Planning and Control, the issue of main lines getting out of synchronisation with each other is discussed. This happens when a production line either gets ahead of schedule or falls behind schedule due to machine downtime or lack of manpower or material resources, which happens in real-life situations.

5.4 Process Control

At this stage it is now necessary to implement the schedule and control the process. In the case study, a number of tools or computer aids were developed to help with controlling the process. There are fourteen production lines, three main sub-component lines and outsourced components. All of these have to be coordinated in such a manner that complete orders are finished in the same time period and trucks can be loaded and dispatched according to the trucking schedule.

As mentioned in the previous section, each of the production lines has its own schedule. These schedules include the products to be made and the sequence in which to make them. As each production line begins, it is usually necessary to get some raw material. Since each product is customised, the raw material for each product is going to be different. In Gienow's case, since most of the material is of a linear nature (that is varying lengths of profiles), the lengths are all different. At the beginning of the process, there is the cutting of these profiles. Since the system knows what products are going to be made and in what sequence, the system prepares the cutting list for each product. Depending on the type of cutting machine, the forms of these lists are different. Sometimes they are downloaded files that go directly to CNC machines that may have linear optimisation functions included in them to reduce the scraps. Alternatively, there may be lists for picking the standard lengths of material and process instructions to cut these to the custom size.

These custom-size lengths of material are then passed to the appropriate production line and the main processing begins. The product goes through a number of workstations as shown in Figure 5.1. Various processes for each product are determined by the routing data described in the introduction section. As the product goes through various work centres, additional parts and material will be required and in some cases this is controlled by computer lists or by Kanban systems. Also, the required sub-components will be delivered to the appropriate work centres when required. These sub-components are identified by the main product number *MMDD-nnnn* as defined earlier. They are also delivered in the sequence of the main production sequence as defined by their item number. To control this process, a number of computer software modules were developed at Gienow. The first is an overall scheduling module for each line, as shown in Figure 5.4.

In this example, when the quarter shifts (2 hours each) are set for the schedule on this production line, they are for 17, 20, 18, and 29 units, respectively. Again, this variable production time for each item is an important aspect of an FMS to control, and can only be achieved if accurate detailed BOM and routing are available. This computer module tells the supervisor a number of details about the state and performance of the production line. It gives an indication of how the line is performing according to the schedule, where problems may be occurring, what the production rate is and what to prepare for in the next schedule. At a more detailed level, the current items being manufactured can be viewed in real time.

Figure 5.5 shows the specific products being manufactured on the production line. The view as shown in Figure 5.5 is divided into two sections. The upper section shows products that have been put aside to wait for a sub-component of some kind. Such components can be from an external supplier that is late or they could be replacements for the components that were damaged or rejected for quality reasons.

This view provides the supervisor with information to improve control over the performance of the production line.

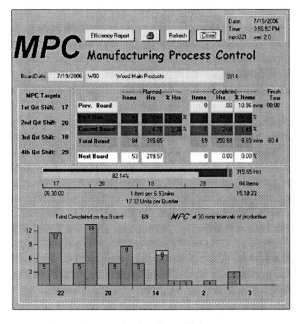

Figure 5.4. Production line efficiency report

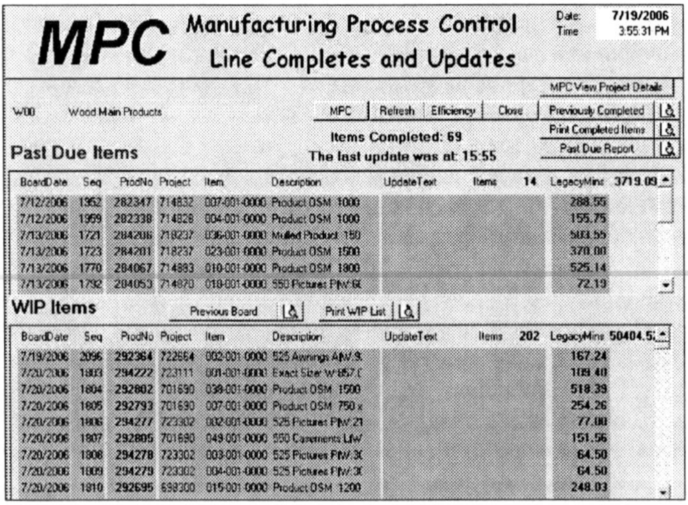

Figure 5.5. Details of production line schedule

In addition to this, a detailed view of any product can be shown by selecting the specific line item on the screen, as illustrated in Figure 5.6. This figure shows a graphic of the product, which visually confirms certain properties of the product to the production line. It also provides information on sub-components in case they need to be remade.

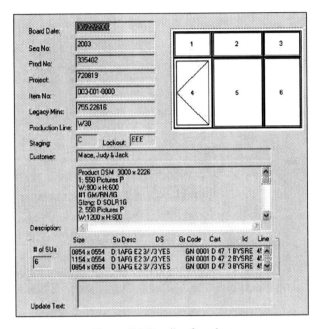

Figure 5.6. Details of product

At the end of each schedule, supervisors complete the daily efficiency report, as shown in Figure 5.7, by reviewing the report and confirming the number of staffs, total working hours and any changes to the work (repairs) or resources (transfer in/out staff). The information is then accumulated to a plant total for management to review. It is also used for historical purposes and for future use in planning, as described in Section 5.5 of this chapter.

The plant total report shown in Figure 5.8 provides daily performance information on each production line. It shows what was planned (including any outstanding items from previous schedules) and what was completed. With the information on how many staff worked on the schedule and for how long, the efficiency of each line is calculated. Management can tell from the check marks which lines have been updated before any decisions are made. The tools also give an indication of when incomplete schedules will be finished.

5.5 Adaptive Planning and Control

On a daily basis as the schedules for the coming days are being prepared, the current state of each production line has to be taken into consideration. If a line falls behind schedule or moves ahead of schedule, future schedules have to be adjusted to keep all the lines in synchronisation.

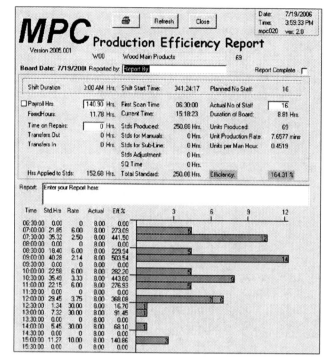

Figure 5.7. Production line – efficiency report

		Planned					Completed									
Line	Description	Past Due	1st 3:30	2nd 11:00	Total Units	Std Hrs	Units	Std Hrs	Actual Hrs	Eff %	Units/ ManHr	Mgr	Time of Update	Est. Finish	Actual Finish	Hrs to Finish
A00	Thermalock	2	60	0	62	47.59	62	58.59	52.54	111.51	1.18	☑	11:36	16:10	11:24	04:45
B00	Industrial Windows	0	0	0	0	0.00	0	11.13	9.92	112.20	0.00	☑	09:10	15:00	00:00	15:00
D00	Industrial Doors	47	50	0	97	195.32	67	109.05	88.64	123.02	0.76	☑	11:36	18:42	11:32	07:10
DH0	Wood Double Hung	0	0	0	0	0.00	0	0.00	6.00	0.00	0.00	☑	17:04	00:00	00:00	
MP0	Master Piece	0	0	0	0	0.00	0	0.00	0.00	0.00	0.00	☐	17:04	00:00	00:00	
PD0	Vetter Patio Door	0	0	0	0	0.00	0	0.00	0.00	0.00	0.00	☐	17:04	00:00	00:00	
PH0	Patio Doors	1	10	0	11	41.76	10	39.74	41.40	95.98	0.24	☑	04:56	23:59	14:50	09:09
SD0	Steel Doors	363	111	0	474	816.38	128	204.33	269.16	75.91	0.48	☑	04:56	23:59	22:48	01:11
SS0	Steel Doors Specials	127	29	0	156	788.33	40	213.32	175.49	121.56	0.23	☑	04:56	23:59	22:29	01:30
V10	Vinyl 75, 125, and 175	683	260	159	1,102	1,084.70	592	589.66	461.29	127.83	1.28	☑	04:57	23:59	23:44	00:15
V20	Vinyl Specials	51	19	0	70	113.44	24	30.92	50.00	61.84	0.48	☑	04:57	23:59	12:37	11:22
V30	Vinyl 200, 250	182	133	97	411	243.83	276	170.92	189.93	89.99	1.45	☑	04:58	23:59	23:54	00:05
V40	Vinyl Patio Doors 900 !	28	41	0	69	91.20	42	62.56	50.49	123.90	0.83	☑	14:22	20:55	14:02	06:52
V50	Vinyl Multi Lites	95	127	0	222	361.26	133	228.96	268.88	85.15	0.49	☑	04:58	23:59	22:45	01:14
V70	Vinyl Bending Process	0	4	0	4	27.28	4	33.44	48.70	68.66	0.08	☑	12:23	12:06	12:06	
V80	Vinyl Casement	11	132	0	143	172.14	145	208.94	247.20	84.52	0.59	☑	04:59	23:59	22:59	01:00
V90	Vinyl SDLs	7	40	0	47	20.45	21	5.71	7.65	74.68	2.75	☑	04:59	23:59	14:07	09:52
W00	Wood Main Products	65	99	0	164	540.97	72	182.95	287.02	63.74	0.25	☑	05:00	23:59	15:29	08:30
W30	Wood Specials	12	15	0	27	149.86	10	80.88	34.90	231.74	0.29	☑	05:00	23:59	14:42	09:17
WB0	Wood Basement Bucks	0	0	0	0	0.00	0	0.00	0.00	0.00	0.00	☐	17:04	00:00	00:00	
N00	Wood Clad Sliding Doc	9	10	0	19	92.50	8	34.34	49.90	68.82	0.16	☑	04:55	23:59	15:09	08:50
WS0	Wood Specials Bendin	0	0	0	0	0.00	0	0.00	0.00	0.00	0.00	☐	17:04	00:00	00:00	
		1,683	1,140	256	3,078	4,787.01	1,634	2,265.42	2,339.11	96.85	0.70					

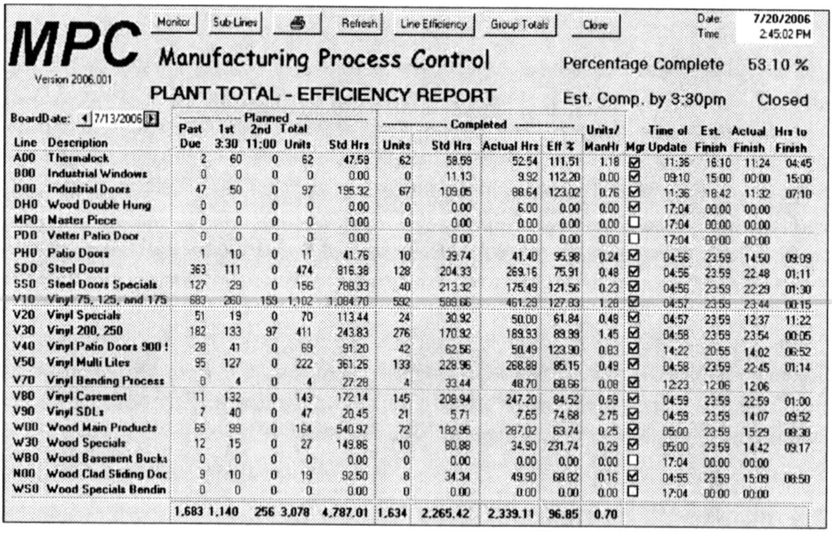

Figure 5.8. Plant total – efficiency report

This is primarily due to the policy of shipping complete orders, which means that all the products for an order have to be finished within the same time period. If an order is incomplete, then either the customer is dissatisfied with a partial order, or the truck dispatch is delayed, which causes further problems down the line. Each plant operation is different but as a general rule it is desirable to keep the amount of work-in-progress and back-orders to a minimum.

MPC
Version 2006.002

Manufacturing Process Control
Plant Load 5 Days from 8/21/2006

Page 1 of 1
Saturday, August 19, 2006

	Planned Capacity	Averages Units	Hours	Std Hrs	PerHr	Past Due	+/- Hours	Forward Load in Days 0821	0822	0823	0824	+/- Adj Std Hrs	+/- Hrs
A00 Thermalock	55	27	34	38.06	0.81	0	0.00	50	48	71	44	-103	36.53	-16.58
B00 Industrial Windows	27	27	15	18.58	1.76	0	0.00	0	0	0	6	27	2.79	18.58
D00 Industrial Doors	64	63	99	120.03	0.64	0	0.00	66	55	56	33	38	69.72	69.51
DH0 Wood Double Hung	32	2	16	8.48	0.11	0	0.00	0	0	0	0	2	0.00	8.48
PH0 Patio Doors	14	10	37	29.64	0.28	0	0.00	15	19	19	18	-31	40.02	-41.90
SD0 Steel Doors	212	159	244	288.82	0.65	0	0.00	149	132	186	192	-3	319.66	-107.01
SS0 Steel Doors Specials	59	38	193	188.64	0.20	0	0.00	46	50	51	81	-55	298.68	-244.30
V10 Vinyl 75, 125, and 175 S	633	514	495	524.81	1.04	0	0.00	579	573	490	624	-310	639.93	-337.14
V20 Vinyl Specials	46	27	43	40.38	0.62	0	0.00	33	34	43	39	-41	73.26	-107.96
V30 Vinyl 200, 250	378	278	192	173.91	1.45	0	0.00	321	345	315	401	-271	241.94	-156.82
V40 Vinyl Patio Doors 900 Se	50	26	34	37.32	0.75	0	0.00	24	21	21	31	7	50.50	6.91
V50 Vinyl Multi Lites	143	96	171	157.78	0.56	0	0.00	134	74	160	315	-100	208.65	-342.15
V70 Vinyl Bending Process	6	3	44	33.45	0.06	0	0.00	0	0	0	0	3	34.42	33.45
V80 Vinyl Casement	180	139	201	184.16	0.69	0	0.00	165	191	186	148	-137	205.58	-228.06
V90 Vinyl SDLs	29	19	10	12.57	1.94	0	0.00	21	34	15	34	-28	24.64	-32.32
W00 Wood Main Products	102	79	222	196.79	0.36	0	0.00	80	87	113	62	-26	333.73	-531.01
W30 Wood Specials	18	21	33	63.92	0.63	0	0.00	11	18	30	46	-21	256.07	-295.15
WB0 Wood Basement Bucks	10	17	17	0.00	1.00	0	0.00	0	0	0	0	17	0.00	0.00
	2,039	1545	2099	2097.35		0	0.00	1696	1779	1741	1854	-1033	2836.12	-2303.48
								0	200	443	723			

Calculations: Carry Forward, One Shift
Averages based on 10 days from 8/11/2006
Capability based on Averages

Figure 5.9. Plant load of 5 days

MPC
Version 2006.002

Manufacturing Process Control
Plant Forward Load in Units from 8/21/2006

Page 1 of 1
Saturday, August 19, 2006
7:54 PM

	Planned Capacity	Forward Load in Units 1st Wk	2nd Wk	3rd Wk	4th+	Total	Line AveEff%	AveUnits Comp'd	Ave Daily Std Comp'd	Estimated Days
A00 Thermalock	55	269.00	277.00	194.00	455.00	1195.00	113.59	27.20	38.06	43.93
B00 Industrial Windows	27	8.00	15.00	22.00	39.00	84.00	121.20	27.00	18.58	3.11
D00 Industrial Doors	64	326.00	343.00	242.00	1102.00	2013.00	121.55	63.20	120.03	31.85
DH0 Wood Double Hung	32	0.00	11.00	0.00	46.00	57.00	52.59	1.80	8.48	31.67
MP0 Master Piece	0	0.00	0.00	0.00	0.00	0.00	0.00	1.00	1.00	0.00
N00 Wood Clad Sliding Door	13	62.00	58.00	45.00	213.00	378.00	75.51	6.30	23.90	60.00
PH0 Patio Doors	14	97.00	92.00	44.00	223.00	456.00	81.01	10.20	29.64	44.71
SD0 Steel Doors	212	951.00	1493.00	820.00	3841.00	7105.00	110.23	158.90	268.82	44.71
SS0 Steel Doors Specials	59	355.00	587.00	280.00	1153.00	2375.00	97.98	37.70	188.64	63.00
V10 Vinyl 75, 125, and 175 S	633	3395.00	4218.00	2397.00	10073.00	20083.00	105.99	514.20	524.81	39.06
V20 Vinyl Specials	46	302.00	419.00	118.00	962.00	1801.00	94.55	26.60	40.38	67.71
V30 Vinyl 200, 250	378	2057.00	2926.00	1614.00	6899.00	13496.00	90.78	278.40	173.91	48.48
V40 Vinyl Patio Doors 900 Se	50	134.00	244.00	139.00	643.00	1160.00	108.31	25.80	37.32	44.96
V50 Vinyl Multi Lites	143	674.00	998.00	726.00	2510.00	4908.00	92.10	96.40	157.78	50.91
V70 Vinyl Bending Process	6	0.00	1.00	0.00	0.00	1.00	75.37	2.60	33.45	0.38
V80 Vinyl Casement	180	987.00	1485.00	833.00	3087.00	6372.00	91.46	138.70	184.16	45.94
V90 Vinyl SDLs	29	169.00	174.00	107.00	257.00	707.00	129.30	18.90	12.57	37.41
W00 Wood Main Products	153	726.00	1179.00	729.00	4966.50	7600.50	88.67	118.35	196.79	64.22
W30 Wood Specials	48	368.80	296.40	33.80	785.20	1474.20	193.81	54.34	63.92	27.13
WB0 Wood Basement Bucks	10	0.00	308.00	0.00	120.00	428.00	5.95	16.80	1.00	25.48
WS0 Wood Specials Bending	0	0.00	0.00	0.00	0.00	0.00	0.00	1.00	1.00	0.00
Units	2,152	10,850.80	15,124.40	8,343.80	37,374.70	71,693.70	Plant Eff%	1,625.39		
Days		5.34	7.78	4.37	20.39	37.88	99.70	2,130.68	2,124.25	44.11

Report Type: Units
Averages based on 10 days from 8/11/2006
Planned Capacity based on Current

Wood Units counted as Boxes
Values in Days
Efficiency based on 100%

$$\text{Total Days} = \frac{\text{Total FWD Load Hours}}{\text{Total Planned Capacity in Hrs}} \times \frac{100}{\text{Efficiency}}$$

$$\text{Estimated Days} = \frac{\text{Total FWD Load Units}}{\text{Ave. \#Units Completed per Day}} \times \frac{100}{\text{Efficiency}}$$

Figure 5.10. Plant forward load

The 5-day planning report, as shown in Figure 5.9, calculates the recommended adjustment to the schedule that is currently being planned. It makes a number of assumptions that may be incorrect. The planner has to use his/her judgement and experience to determine the exact amount of adjustment. For example, if some of the production lines are behind, the program would make the recommendation to reduce the future work schedule to these lines so that they can catch up the due dates. The system does not know that overtime is planned for the coming week to allow the lines to catch up the schedule. Therefore, future workloads may not need reducing.

A second report as illustrated in Figure 5.10, analyses the next four weeks of the forward load. This helps the planner in the short-term to determine whether it is necessary to start increasing or decreasing the available capacity of each production line. This can be achieved by transferring people between lines or hiring additional staff. Long-term planning of resources by using an adaptive planning method will be discussed in Section 5.6.

Additional research is ongoing to model and simulate an OKP system [5.7]. The simulation model uses a coloured Petri net and proposes the use of an adaptive control structure as shown in Figure 5.11.

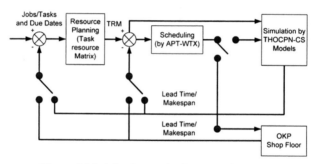

Figure 5.11. Adaptive production control structure

Constraints of the sort sequenced as defined in Section 5.3.2 may not produce the optimum production sequence. The details of a schedule can be processed through a *Make Span* program that can determine the allocation of resources to meet the production rate. In the case of the V10 line, this is set to 2 minutes, as shown in Figure 5.12.

The results of this demonstrates that in order to maintain an average production rate of one product every 2 minutes, it is necessary to allocate resources to the work centres V11 with 4 persons, V12 with 6 persons, V13 with 6 persons and V15 with 16 persons, respectively. However, this may not be the optimal allocation of resources. Therefore, a simulation of the production schedule is recommended to validate performance of the schedule. An algorithm was also developed by Li *et al.* [5.29] to produce the optimal sequence for batches of fifty products, maintaining the overall delivery priority to the customers. The algorithm APT was based on average processing times (APT) and selects the optimal sequence from the best of three sub-algorithms. Since Gienow is using the OKP method, schedules are constantly changing. Therefore, with the use of the modified coloured Petri net called temporised hierarchy object-oriented coloured Petri nets with changeable structure

(THOCPN-CS) [5.30][5.31], difference scenarios on the shop floor can be simulated. In conjunction with the adaptive control structure as shown in Figure 5.11, it is possible that an improved schedule may be developed.

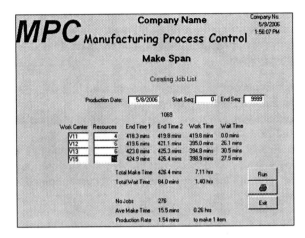

Figure 5.12. Makespan for production sequence

5.6 Long-term Resource Planning

An important aspect of process planning and control is the long-term planning of resources. For one-of-a-kind production (OKP), this is difficult to perform without good historical data (sales and production), good sales forecasting and a strategic plan. Sales forecasting is dependent very much on having good marketing information and a good understanding of customers' future demands. Once a good forecasting model has been developed it is then necessary to use that in conjunction with historical data to predict long-term requirements for resources and then plan accordingly.

Simple linear relationships can be used to predict the required resources for forecasted sales. However, if these relationships are analysed it is often found that they are not all linear in nature, particularly for customised products. This causes inaccuracies in prediction results by using these methods. A number of researchers have proposed the use of neural networks to compensate for the nonlinear nature of the relationships [5.32]. Razi and Athappilly [5.33] compared various techniques for developing predictive models and concluded that a model based on a neural network could produce acceptable results. However, they did make the observation that regression models are easier to construct and verify. Other research suggests that the use of adaptive techniques [5.34][5.35] would improve the accuracy of predictive models, even if they are based on linear relationships.

Research was carried out in Gienow to determine which method would be appropriate. Dean *et al.* [5.36] used a number of methods that include developing a model based on neural network technology. The conclusion was that maintaining historical data at the branch, market, and product group level and using the last three months of data as new training (learning) data would assist in predicting resource demand. The resources include both material and labour. The forecasting results from Gienow's study also agreed with previous conclusions that the use of adaptive techniques did improve the accuracy of a predictive model that was based on linear relationships.

Hence, a system was developed to store the historical production data at a detailed level for various functions (not all covered in the scope of this chapter), including its use in prediction of resources. For the prediction model, a number of relationships are established:

I_{ip} the average amount of each inventory item used to produce one product on a production line over the selected period of time;

T_p the average amount of resource time required to produce one product on a production line over the selected period of time;

R_{bmp} the revenue generated by a production line for each branch and market over the training period;

U_{bmg} the number of units made on a production for each branch and market over the training period;

TR_{mb} the total revenue for each branch and market over the training period;

FR_{mb} the total forecast revenue for each branch and market over the forecasting period;

A_i the amount of inventory of item i;

TFU_p the number of units to be produced on production line p over the forecasting period.

The total number of products to be made over the forecasting period on a production line is determined by

$$(TFU)_p = \sum_{b=1}^{n} \sum_{m=1}^{q} (FR)_{bmp} \frac{U_{bmp}}{(TR)_{bm}} \tag{5.1}$$

where b = bth branch, m = mth market, p = pth production line, n = number of branches, and q = number of markets.

The total amount of resource required over the forecasting period can then be predicted as

$$\sum_{p=1}^{n} (TFU)_p T_p \tag{5.2}$$

and the amount of inventory requiring an item over the forecasting period is

$$A_i = \sum_{p=1}^{n} (TFU)_p I_{ip} \tag{5.3}$$

where i = ith inventory item, p = pth production line and n = the number of production lines.

Depending on the number of branches, markets, production lines and inventory items, the amount of data required to perform these operations can become significant. However, once these data are in place with a method of maintaining it, the data become very useful in the planning process for one-of-a-kind production.

Having observed this model over one and half years of production as it was developed and modified, some of the observations made include:

- The markets determined a specific product mix. Each branch was active in a number of markets but not all. Thus, by modelling at the branch and market levels, changes in product mix were recognised by the model. Through experimentation, it was found in the case of Gienow that a three-month period was suitable to recognise a permanent change in the direction of a market or branch. This then affected the number of units predicted for each of the production lines since these were also included in modelling parameters.
- As the units for each of the production lines varied so did the resulting forecast for inventory items. At the same time, the fact that inventory relationships were constantly revised as new inventory items were introduced and other items became obsolete; these too were reflected in the model.
- Since the average production line rates are being recalculated as efficiencies increase or decrease, this will be reflected in the results of the model by requiring more or fewer resources to make the units on each production line. Individual production lines can change their efficiencies independent of other production lines.

The report as illustrated in Figure 5.13 shows resource requirements for each of the production lines. The first six months (January to June) are actual hours required by the production lines, while the balance (July to December) are the predicted hours for each production line based on the sales forecast figures.

Forecasting: July

Plant Resource Plan – Standard Hours

Page 1 of 1

Production Line		Jan	Feb	Mar	Apr	May	Jun	Jul	Aug	Sep	Oct	Nov	Dec	Total
Production Days in Month		20	19	23	19	22	22	20	22	20	21	22	16	
A00	Thermalock	729	706	1,092	865	1,039	912	468	668	668	668	601	634	9,030
B00	Industrial Windows	670	578	736	498	385	390	323	421	403	407	378	313	5,478
C00	Sealed Units	9,753	8,298	9,719	8,813	10,130	10,405	9,723	10,899	9,562	9,938	9,759	6,969	113,968
D00	Industrial Doors	2,006	2,007	1,745	1,197	1,728	2,499	943	1,345	1,343	1,344	1,210	1,074	18,432
DH0	Wood Double Hung	336	275	1,191	895	1,131	393	2,099	2,427	2,326	2,463	2,273	1,680	17,489
G00	Grills	9,651	9,429	12,114	11,380	14,187	14,916	13,412	16,034	13,190	13,708	13,461	9,614	149,998
MP0	Master Piece	0	0	0	0	0	0	0	0	0	0	0	0	0
N00	Wood Clad Sliding	612	389	595	465	793	813	378	431	399	422	397	290	5,843
PH0	Patio Doors	674	454	480	390	419	655	535	593	514	535	529	373	6,131
S00	Screens	2,575	2,467	2,878	2,528	3,124	3,027	2,875	3,223	2,828	2,939	2,386	2,061	33,411
SD0	Steel Doors	4,973	4,985	5,379	4,985	5,304	5,519	6,910	7,861	6,849	6,914	6,943	4,824	71,026
SS0	Steel Doors Specia	4,629	4,365	4,640	4,445	5,064	4,939	2,813	3,119	2,709	2,816	2,766	1,964	44,287
V10	Vinyl 75, 125, and	12,987	8,314	9,569	9,214	11,227	12,218	12,252	13,691	12,008	12,496	12,270	8,733	134,979
V20	Vinyl Specials	1,159	933	1,165	1,496	1,732	1,209	682	767	658	684	876	477	11,820
V30	Vinyl 200, 250	3,577	3,260	3,681	3,285	3,956	4,359	4,986	5,510	4,791	4,984	4,925	3,476	50,751
V40	Vinyl Patio Doors 0	1,028	1,172	1,328	903	1,247	1,219	1,302	1,462	1,279	1,334	1,306	928	14,696
V50	Vinyl Multi Lites	2,230	6,226	7,068	5,228	5,944	6,038	6,159	6,832	5,937	6,176	6,106	4,307	68,251
V70	Vinyl Bending Proc	723	680	795	622	838	903	594	662	591	606	595	421	7,998
V80	Vinyl Casement	3,624	3,892	4,959	3,826	4,859	4,707	5,790	6,421	5,574	5,796	5,735	4,044	58,926
V90	Vinyl SDLs	91	130	135	112	205	239	202	224	195	202	200	141	2,077
W00	Wood Main Produc	5,833	4,760	6,862	5,809	7,049	7,851	5,832	6,614	6,766	6,811	6,866	4,176	71,100
W30	Wood Specials	2,089	1,819	2,229	1,887	1,914	2,142	1,112	1,259	1,117	1,176	1,128	824	18,476
W80	Wood Basement B	250	412	0	272	343	312	207	229	199	207	206	144	2,800
W90	Wood Specials Be	0	0	0	0	0	0	0	0	0	0	0	0	0
Total		70,000	65,098	77,201	68,855	82,699	85,853	79,577	89,372	78,684	81,825	80,134	57,368	916,666
Revenue		8,307,489	7,882,789	9,233,754	7,896,968	9,677,809	10,578,112	9,427,456	10,567,448	9,271,411	9,635,477	9,461,768	6,357,614	108,697,895
Daily Rev		415,374	414,884	401,468	415,630	439,900	480,823	471,373	480,339	463,571	438,832	430,080	422,338	441,861
Daily Std		3,500	3,426	3,357	3,624	3,759	3,902	3,979	4,062	3,934	3,896	3,642	3,585	3,726

Figure 5.13. Resource plan

The predicted values seem reasonable when compared to the actual values. There is not a linear relationship with the total revenue figures as each branch and market show some signs of different changes in their product mix. Production lines are in fact showing an improvement in their efficiencies, which reflects a lower resource requirement.

5.7 Conclusions

This chapter describes a framework for a process planning, scheduling and control system that has been implemented in Gienow. With appropriate modification, this framework can be applied to other similar manufacturing companies that aim at achieving mass customisation by using a one-piece flow method or OKP mode. It emphasises the need for a central expert system that contains sufficient engineering knowledge to be able to generate the bill of materials and process routings as well as the extensive data system.

The framework describes the long-term and short-term planning requirements of OKP. It establishes the need for strategic plans to meet the long-term demands of the customers. Detailed information (as described in the case study) is required in order to determine the capacity increases required and the supply of materials and manufactured sub-components. In short-term planning the customers' immediate demands have to be met; this also establishes the need for detailed information on each product so that they can be scheduled in a one-piece flow process.

This chapter also presents the process controls that are required to plan and monitor the manufacturing process in real time. In Gienow's case, the task of some of the production lines is less than two minutes. The sequence of items for one-piece flow production is important and provides a means of controlling the manufacturing process including the remake process when production gets out of synchronisation.

Adaptive control methods are necessary so that the planning and scheduling processes take into consideration what has transpired recently. Changes in product mix and plant efficiencies are also important considerations. Also, modifying the schedule according to the current state of various production lines is important. Time frames will vary from operation to operation but the basic principles will still apply. The use of the adaptive method allows for the calculation of error factors in forecasting and production. These can be used to adjust the ongoing forecasting to arrive at more accurate results.

Finally, the computer-aided OKP planning and scheduling system as presented in this chapter is based on the availability of detailed bills of material and process routings for each customised product being manufactured. These supply all the information for planning and scheduling, and provide necessary information for the flexible manufacturing processes. The information includes the material list, processing instructions, CNC download files, and fabrication instructions.

Acknowledgment

The authors would like to thank Gienow Windows and Doors LLP, Calgary, Alberta, Canada for the information on their OKP process. The case study provides an industrial application of the computer aides designed by the primary author, described in this chapter.

References

[5.1]   Tu, Y.L., Chu, X.L. and Yang, W.Y., 2000, "Computer aided process planning in virtual one-of-a-kind production," *Computers in Industry*, **41**, pp. 99–110.

[5.2]   Tu, Y.L. and Xie, S.Q., 2001, "An information modelling framework to support sheet metal parts intelligent concurrent design and manufacturing," *International Journal of Advanced Manufacturing Technology*, **18**, pp. 873–883.

[5.3]   Tu, Y.L., Dean, P., Xue, D., Li, W. and Li, X., 2006, "One-of-a-kind product design and manufacture," In *Proceedings of the 7th International Conference on Frontiers of Design and Manufacturing*, Guangzhou, China, **3**, pp. 185–190.

[5.4]   Matta, A. and Semeraro, Q., 2005, "A framework for long term capacity decisions in AMSS," in *Design of Advanced Manufacturing Systems* 2005, Springer, The Netherlands, Chapter 1, pp. 1–35.

[5.5]   Rolstadås, A., 1991, "ESPRIT basic research action No. 3143 – FOF production theory," *Computers in Industry*, **16**, pp. 129–139.

[5.6]   MacCarthy, B., Brabazon, P.G. and Bramham, J., 2003, "Fundamental modes of operation for mass customisation," *International Journal of Production Economics*, **85**, pp. 289–304.

[5.7]   Banerjee, P. and Armouti, H.A., 1992, "JIT approach to integrating production order scheduling and production activity control," *Computer Integrated Manufacturing*, **5**(4), pp. 283–290.

[5.8]   Hegge, H.M.H. and Wortmann, J.C., 1991, "Generic bill-of-material: a new product model," *International Journal of Production Economics*, **23**, pp. 117–128.

[5.9]   Du, X., Jiao, J. and Tseng, M., 2002, "Graph grammar based product family modelling," *Concurrent Engineering: Research and Applications*, **10**(2), pp. 113–128.

[5.10] Huang, G.Q., Zhang, X.Y. and Liang, L., 2005, "Towards integrated optimal configuration of platform products, manufacturing processes, and supply chains," *Journal of Operations Management*, **23**, pp. 267–290.

[5.11] Sanoff, S.P. and Poilevey, D., 1991, "Integrated information processing for production scheduling and control," *Computer Integrated Manufacturing*, **4**(3), pp. 164–175.

[5.12] Kumar, K.D., Karunamoorthy, L., Roth, H. and Mirnalinee, T.T., 2004, "An infrastructure for integrated automation system implementation," *International Journal of Flexible Manufacturing Systems*, **16**, pp. 183–199.

[5.13] Westbrook, R. and Williamson, P., 1993, "Mass customisation: Japan's new frontier," *European Management Journal*, **11**(1), pp. 38–45.

[5.14] Piller, F.T., 2004, "Mass customisation: reflection on the state of the concept," *International Journal of Flexible Manufacturing Systems*, **16**, pp. 313–334.

[5.15] Kumar, A., 2004, "Mass customisation: metrics and modularity," *International Journal of Flexible Manufacturing Systems*, **16**, pp. 287–311.

[5.16] Vandaele, N. and De Boeck, L., 2003, "Advanced resource planning," *Robotics and Computer Integrated Manufacturing*, **19**, pp. 211–218.

[5.17] Levin, I.P., Schreiber, J., Lauriola, M. and Gaeth, G.J., "A tale of two Pizzas: building up from a basic versus scaling down from a fully-loaded product," *Marketing Letters*, **13**(4), pp. 335–344.

[5.18] Fujimoto, H. and Ahmed, A., 2003, "Assembly process design for managing manufacturing complexities because of product varieties," *International Journal of Flexible Manufacturing Systems*, **15**, pp. 283–307.

[5.19] Little, D. and Hemmings, A., 1994, "Automated assembly scheduling: a review," *Computer Integrated Manufacturing Systems*, **7**, pp. 51–61.

[5.20] Zhang, W., Freiheit, T. and Yang, H., 2005, "Dynamic scheduling in flexible assembly system based on timed Petri nets model," *Robotics and Computer Integrated Manufacturing*, **21**, pp. 550–558.

[5.21] Anderson, S., 2001, "Direct and indirect effects of product mix characteristics on capacity management decisions and operating performance," *International Journal of Flexible Manufacturing Systems*, **13**, pp. 241–265.

[5.22] Foley, W.J., 2002, "Impact of interruptions on schedule execution in flexible manufacturing systems," *International Journal of Flexible Manufacturing Systems*, **14**, pp. 319–344.

[5.23] Goldratt, E.M., 1999, *The Theory of Constraints*, North River Press, MA, USA.

[5.24] Bielli, M. and Dell'olmo, P., 1993, "Flexibility of scheduling tools for order production problems," *Computer Integrated Manufacturing*, **6**, pp. 64–70.

[5.25] Palacios, M.C., Alvarez, E., Alvarez, M. and Santamaria, J.M., 2006, "Lessons learned for building agile and flexible scheduling tool for turbulent environments in the extend enterprise," *Robotics and Computer Integrated Manufacturing*, **22**, pp. 485–492.

[5.26] Matsui, M., Uehara, S. and Ma, J., 2001, "Performance evaluation of flexible manufacturing systems with finite local buffers: fixed and dynamic routings," *International Journal of Flexible Manufacturing Systems*, **13**, pp. 405–424.

[5.27] Dupon, A., van Nieuwenhuyse, I. and Vandaele, N., 2002, "The impact of sequence changes on product lead time," *Robotics and Computer Integrated Manufacturing*, **18**, pp. 327–333.

[5.28] Miltenburg, J., 2002, "Balancing and scheduling mixed-model U-shaped production lines", *International Journal of Flexible Manufacturing Systems*, **14**, pp. 119–151.

[5.29] Li, W., Tu, Y.L. and Xue, D., 2006, "Adaptive production scheduling and control for one-of-a-kind production shop floor," *Transactions of NAMRI/SME*, **34**, pp. 159–166.

[5.30] Jiang, Z.B., Zuo, M.J., Fung, R.Y.K. and Tu, Y.L., 2000, "Temporized coloured Petri nets with changeable structure (CPN-CS) for performance modelling of dynamic production systems," *International Journal of Production Research*, **38**(8), pp. 1917–1945.

[5.31] Jensen, K., 1995, *Coloured Petri Nets: Basic Concepts, Analysis Methods and Practical Use*, Springer, New York.

[5.32] Carmo, J.L. and Rodrigues, A.J., 2004, "Adaptive forecasting of irregular demand processes," *Engineering Applications of Artificial Intelligence*, **17**(2), pp. 137–143.

[5.33] Razi, M.A. and Athappilly, K., 2005, "A comparative predictive analysis of neural networks (NNs), nonlinear regression and classification and regression tree (CART) models," *Expert Systems with Applications*, **29**(1), pp. 65–74.

[5.34] Alterman, R., 1988, "Adaptive planning," *Cognitive Science*, **12**(3), pp. 393–421.

[5.35] Lim, M.K. and Zhang, D.Z., 2004, "An integrated agent-based approach for responsive control of manufacturing resources," *Computers and Industrial Engineering*, **46**, pp. 221–232.

[5.36] Dean, P., Xue, D. and Tu, Y.L., 2006, "Prediction of manufacturing resource requirements for mass-customisation production," *Transactions of NAMRI/SME*, **34**, pp. 71–78.

6

Setup Planning and Tolerance Analysis

Yiming (Kevin) Rong

Worcester Polytechnic Institute, Worcester, MA 01609, USA
Email: rong@wpi.edu

Abstract
A computer-aided manufacturing planning (CAMP) system is introduced in this chapter. When product design information is identified, a production plan can be rapidly generated with comparison of alternatives. It is based on the concept of production and process similarity and the best-practice knowledge in the automotive industry. Therefore, it is the intention that the system is applied in the mass-customisation environment, *i.e.* achieving a mass-production economic goal with the flexibility of product design changes. It is a CAD integrated system with defining bill of process in three levels: feature, part, and machine levels. A tolerance-analysis-based automated setup planning strategy is developed to generate new production plans. The system is validated by several production cases.

6.1 Introduction

Computer-aided manufacturing planning (CAMP) acts as a bridge between computer-aided design (CAD) and computer-aided manufacturing (CAM), to help engineers convert product design specifications into manufacturing plans. The overall objectives of manufacturing planning are to ensure product quality, minimise production cost and maximise manufacturing efficiency, although different production systems may place special emphasis on different aspects of these objectives. For example, mass production emphasises more on throughput, while job shop and batch production focus more on production cost.

Today, in order to enhance an enterprise's ability to quickly respond to dynamic changes in the global marketplace, the concept of mass customisation has been introduced into industry [6.1]. It allows customised products to be made to suit special customer needs, while maintaining near-mass-production efficiency. Compared to conventional mass production, mass customisation allows for more product variety in which products are grouped into families. By the use of certain modularity principles, products are decomposed into modules. The reuse of certain modules in the new product may simplify the product design. On the other hand, the low cost in mass customisation is achieved primarily through the utilisation of manufacturing process capability as much as possible to produce a greater variety of products. Hence, flexible manufacturing resources are widely used to increase the process capability of an enterprise in mass customisation.

To help realise manufacturing planning for mass customisation, a CAD/CAM platform is needed and the major issues must be recognised and resolved. The main goal of CAMP for mass customisation is to help design feasible and optimal manufacturing plans quickly. Integrated information models are needed to prescribe the relationships between product design and manufacturing. Thus, changes in product design will prompt corresponding changes in manufacturing plans quickly and automatically. Since flexible manufacturing resources are used in mass customisation, such as multi-part fixtures, multi-axis machines and combination cutters, manufacturing planning systems must be designed to deal with these flexible manufacturing resources.

A framework of a comprehensive CAMP system is introduced. It is supported by an automated setup planning method and an information modelling technique, which are presented in the next two sections of this chapter. Fixture design is an important component of the system, but relatively independent in technique. It will be addressed separately. CAMP for rotational parts is not included in this chapter.

6.1.1 Current State-of-the-Art

CAMP can be divided into part information modelling, feature manufacturing strategy, manufacturing resource capability analysis, setup planning, and fixture design. Significant advances in research have been made during the past 3 decades.

Part information modelling. Part information includes geometry information and design specifications (tolerance, surface finish, *etc.*), which are defined in CAD models or neutral files (STEP, IGES, *etc.*). Feature technology is widely recognised as a useful tool for representing part information [6.2]. By the use of graph theory, part information can be represented by a feature-tolerance relationship graph (FTG), where parts are composed of features with design specifications described by relationships between these features [6.3]. The remaining challenge is to design a comprehensive part information modelling system that would allow new features to be added without programming effort.

Feature manufacturing strategy. Feature manufacturing method needs to be specified to link candidate manufacturing processes to features. It can be represented in two models. One is associating a list of candidate processes to a feature type. The other one is associating all product features that can be produced by a process type to the process type [6.4]. Both representations are needed in general to define the relationship between features and processes, including cutters and machine tools used in these processes. If a new feature type or process type is added, all the pre-defined feature manufacturing methods have to be updated. The maintenance effort involved in this updating procedure is huge and tedious.

Manufacturing resource capability analysis. Manufacturing resources include machine tools, cutting tools and fixtures [6.5]. In the mass-customisation environment, the challenge is how to evaluate the capabilities of candidate manufacturing resources and derive an optimal process design based on that. No practical solution is yet available to properly cover the enormous varieties of manufacturing resources in the marketplace.

Setup planning. The objective of setup planning is to determine the number of setups needed, the orientation of the workpiece and the machining process sequence in each setup. The existing research has been focused on the following aspects:

- *Setup constraint modelling.* Geometric, manufacturing and kinematic constraints have been considered in setup planning [6.6]. Geometric constraints, including feature orientation [6.7][6.8] and tolerance analysis [6.7][6.3] for both rotational and non-rotational parts, have been studied. Manufacturing constraints have been modelled by manufacturing knowledge such as operation precedence constraints [6.7][6.8] and best practices in industry [6.9].
- *Decision-making strategies.* Various techniques such as knowledge-based expert systems [6.10], neural networks [6.11], and graph-based analysis [6.3] have been developed to aid setup planning. The existing research, however, has only taken into account limited manufacturing resource capabilities [6.7][6.11]. None will generate setup plans for mass customisation when multi-axis CNC machines, flexible fixtures and complicated cutting tools are involved.
- *Inter-setup tolerance modelling.* Currently, graph-based representation has been recognised as an effective tool to describe the relationship between datum and machining surfaces [6.3][6.12][6.13]. By the use of graphs, it is easy to track the tolerance stackup relationships among manufacturing processes.
- *Information integration with other modules in manufacturing planning systems.* Setup planning has a close relationship with tolerance analysis, fixture design and manufacturing planning. An information exchange standard needs to be established to make the integration more reliable and flexible.

Fixture design. The objective of fixture design is to generate fixture configurations to hold workpieces firmly and accurately during manufacturing processes. Previous work has been focused on an automatic modular fixture design, using standard fixture components to construct different fixture configurations [6.8], dedicated fixture design with predefined fixture component types [6.14], variation of fixture design for part families [6.15], and fixture design verifications [6.16]. In the mass-production or mass-customisation environment, multi-part fixtures are commonly used to achieve optimal cycle time. Optimisation of multi-part fixture design for mass customisation, however, is not well developed. More research is needed for fixture base selection/design, part layout and orientation, and multi-part fixture configuration. Furthermore, manufacturing planning needs to determine an optimal process sequence for all machining features, and to determine the optimal tool path to machine these features as presented in the fixture, based on the process sequence. Currently, no research work is being done in this area.

In summary, the current state-of-the-art technology involves several major limitations that can be discussed on three levels.

1. On the feature level, features and their manufacturing methods are restricted to a predefined format. Adding new features and new processes requires reprogramming.

2. On the part level, setup planning lacks a mechanism to consider both product feature tolerance relationships and flexible manufacturing resource capabilities. Without such a mechanism, realistic setup plans cannot be generated.

3. On the machine level, no research has been conducted with multi-part fixture design and the corresponding global tool-path generation.

A comprehensive CAMP system for mass customisation will be presented. The major contributions of this research are that: (1) new features, processes and manufacturing resources can be added and utilised without extra programming work due to the use of a comprehensive feature, setup, and manufacturing information model, and (2) the best manufacturing practices for a part family are organised on the three distinct levels described above. The manufacturing planning system is therefore modular and expandable so that manufacturing plans for new parts can be generated easily based on existing plans in the part families.

6.2 Manufacturing Planning System

The architecture of the proposed computer-aided manufacturing planning system for mass customisation is developed as shown in Figure 6.1. First, the part information is represented by a feature tolerance relationship graph (FTG) [6.3]. Product features are the basic units in an FTG and can be extracted from CAD models. Feature manufacturing methods can be generated based on the available manufacturing resource capability. Setup planning is divided into two steps. In the first step, manufacturing features are divided into functional groups and a datum machining feature relationship graph (DMG) is generated [6.3]. The DMG is constructed based on a tolerance analysis and a manufacturing resource capability analysis. In the second step, the setups are generated with the incorporation of fixture planning. Conceptual fixture design is used to determine the optimal part layout on fixture bases and a manufacturing plan generation module is used to generate alternative solutions. In order to get an optimal setup plan, several criteria are used to evaluate these solutions. The manufacturing resource capability modelling is one of the critical functions in computer-aided manufacturing planning.

6.2.1 Feature-based Part Information Modelling

In the computer-aided manufacturing planning system for mass customisation, parts that have geometric similarity and can provide the same functions are grouped into part families based on industrial practice. Part information is composed of features and the relationships between features. In order to represent part family information, the definition of a feature is extended to combined features that are composed of primary surfaces, including flat surface, cylindrical surface and cone surface. Figure 6.2 shows the combined features in a simplified part model.

The combined features are defined with geometric entities that can be machined together in one or more manufacturing processes. According to the successful practice of part family and available manufacturing resources, each combined

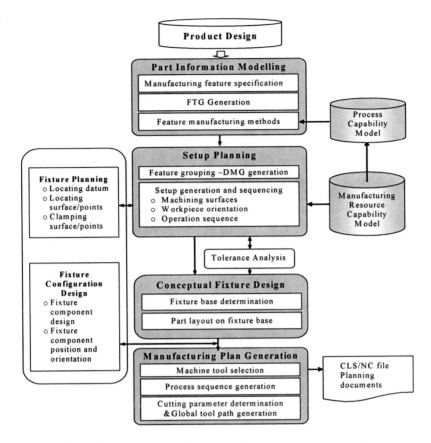

Figure 6.1. Architecture of computer-aided manufacturing planning

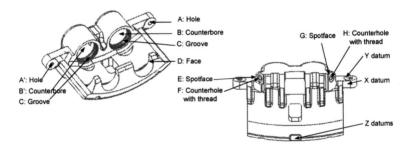

Figure 6.2. Notations of a simplified part model

feature corresponds to a sequence of predefined manufacturing processes in which combination cutters are used to reduce the manufacturing cycle time.

Figure 6.3 shows the information structure of combined manufacturing features. Surfaces are considered as the atomic primary features and represented by operational datasets. Then, an object-oriented programming technique can be applied for reasoning. Main surfaces (MS) are the surfaces that determine the feature type, main parameters, position, and orientation. Auxiliary surfaces (AS) are those surfaces that are attached to main surfaces. The feature information can be further linked to a local tool-path representation.

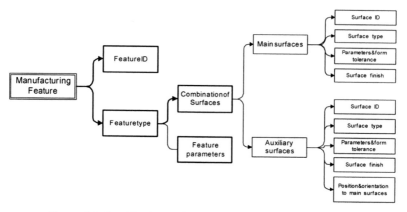

Figure 6.3. Combined manufacturing feature information structure

Figure 6.4 presents the definition of *hole* features in a sample part. The surface parameters are directly extracted from the part CAD model. And the feature parameters can be calculated based on surface parameters.

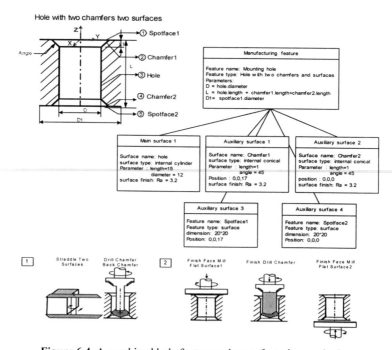

Figure 6.4. A combined hole feature and manufacturing methods

6.2.2 Feature Manufacturing Strategy

The goal is to generate geometric shapes, dimensions and tolerances for the features with selected manufacturing processes. In mass customization, combined features can be achieved by various means determined by available manufacturing resources.

The manufacturing resource capability would influence the selection of feature machining methods. Therefore, it is necessary to identify the relationships between manufacturing resources and features, as shown in Figure 6.5.

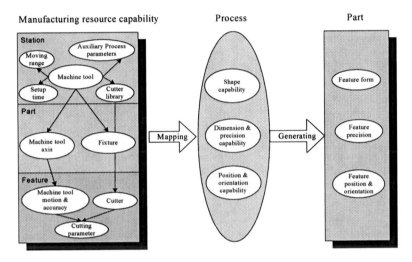

Figure 6.5. Relationship between manufacturing features and resources

From Figure 6.5, the manufacturing resource capability is described in 3 levels: station (machine tool), part, and feature levels. A process model describes the association between manufacturing features and manufacturing resource capability. In the feature level, cutter capability is the critical factor and the process model includes cutter, cutter motions and feature generation accuracy. Each manufacturing feature may have several alternative manufacturing processes. Each process may have specific requirements for cutter design and motions.

On the process model, the cutter type determines the basic motion types that are divided into primary and feed motions. Both of these motions can be represented mathematically. The cutter parameters and the feature parameters determine the machine motion parameters. Figure 6.6 shows several examples of machine motions. Figure 6.7 shows the parameter-driven relationship between the hole feature and the cutter and tool path used to machine this feature. The cutter template and tool-path template are setup based on the best-practice knowledge in industry.

By using the process model, when a new feature type is added, the corresponding processes including cutters and machine motions can be generated. This is based on the shape, dimensions, and tolerances of the new feature type. When a new manufacturing resource is added, the manufacturing capability model will be updated to provide more solutions to meet the requirement of the process model.

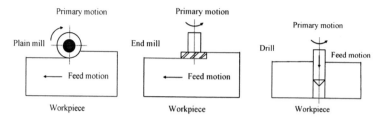

Figure 6.6. Cases of machine motions

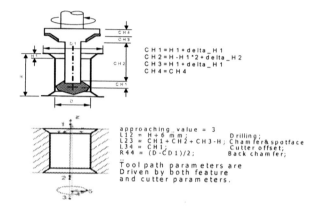

$$CH1 = H1 + delta\_H1$$
$$CH2 = H - H1*2 + delta\_H2$$
$$CH3 = H1 + delta\_H1$$
$$CH4 = CH4$$

approaching_value = 3
L12 = H + 6 mm;   Drilling;
L23 = CH1 + CH2 + CH3 - H; Chamfer & spotface
L34 = CH1;   Cutter offset;
R44 = (D - CD1)/2;   Back chamfer;
...
Tool path parameters are
Driven by both feature
and cutter parameters.

Figure 6.7. An example of cutter design and tool path design

6.2.3 Machine Tool Capability Modelling

As shown in Figure 6.5, mapping of manufacturing resources to features yields shape, dimension & precision, and position & orientation capabilities. The shape capability is fulfilled by the combination of the cutter and machine tool motions. Therefore, the motions in the process model need to be transformed into machine tool kinematic motions with accuracy.

In mass customisation, vendors may provide a variety of machine tools with similar functions. Therefore, an information model is needed to describe similarities and differences of the machine tool capabilities so that manufacturing engineers can make comparisons. The machine tool information structure is shown in Figure 6.8.

6.2.4 Setup Planning

In the research, the setup plan of a new part may be generated either by running an automated setup planning system or retrieving from existing setup plans in which the optimum process sequence and parameters have been established based on the best practice setup plans of similar parts. It also takes into consideration the manufacturing knowledge and available manufacturing resource capability. A general setup planning procedure is shown in Figure 6.9.

The main tasks of automatic setup planning include two steps: feature grouping and setup generation [6.17].

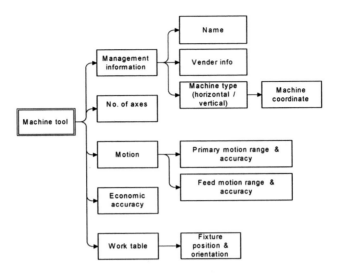

Figure 6.8. Machine tool information structure

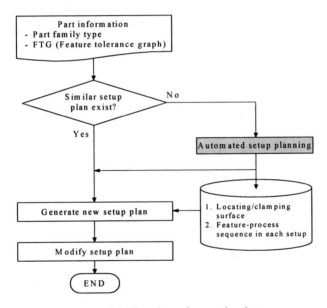

Figure 6.9. Flowchart of setup planning

Feature grouping. In order to minimise the inter-setup tolerance stackup, it is suggested to group those features that have close position, orientation or profile tolerance requirements in one datum frame. Based on the analysis on the FTG, the design datum frames, consisting of a geometric datum in three perpendicular directions, are identified so that initial feature groups can be constructed. The features in the same group have tight relationships and are recommended to be machined in one setup [6.3].

Although it is suggested that features with tight tolerance relationships are machined in one setup, the machine resource may not have the desired capabilities to machine the features in one setup. After all the selected manufacturing processes are attached to the features in the FTG, the processes need to be regrouped with the consideration of tool access directions. The FTG is expanded with the process information. Those groups that have the same datum frame and tool access direction can be reunited together and a datum machining surface graph, DMG1, is generated.

Setup generation. Manufacturing resource capability constrains the setup generation. Each group of features may be machined in one setup using standard manufacturing resources such as general machine tools and fixtures. While in mass customisation, in order to achieve a high production rate, multi-part fixtures and multi-axis CNC machines are widely used. Then, the number of setups can be greatly reduced and consequently the DMG2 is generated based on the available manufacturing resource capabilities.

The following problems need to be solved in setup generation:

- Reunite the feature groups based on available manufacturing resources. Those feature groups with the same datum frame are considered to unite into one setup.
- Determine the locating datum in each setup and pass this information to the fixture planning module. The locating datum is identified in each setup and passed to a fixture-planning module. The locating points and clamping surfaces are the outputs of the fixture-planning module.
- Determine the process sequence in each setup. The problem of process sequencing is transformed mathematically into a search for an optimal path to traverse each vertex in the DMG2 under specified constraints. The times of passing each vertex are determined by the number of processes linked to each feature.

The constraints are divided into strong and weak constraints. The former is the first priority to achieve and cannot be violated, while the latter comes from manufacturing experience and may provide optimal solutions.

The strong constraints may include: 1) Maintaining the manufacturing process sequence of each feature; 2) Maintaining the operation-dependent relationship in the graph. For example, planes prior to holes and holes prior to grooves; 3) Doing rough cuts first, semi and finish cuts in a prescribed order; and 4) Minimising the tool-change time and machine tool adjustment time (*e.g.* table-index time).

One example of a weak constraint might be that the cutter to mill the outboard flange could be combined with the cutter to drill, chamfer and back chamfer the mounting holes so that the tool-change time can be reduced. Figure 6.10 shows one solution for the process sequence of the example part.

6.2.5 Fixture Design in Computer-aided Manufacturing Planning

Fixture design issues are divided into conceptual fixture design, detailed fixture design and validation of fixturing performance. Conceptual fixture design includes fixture base selection and part layout on the fixture base. Detailed fixture configuration design is the generation of fixture components and configurations based on the best practice of the fixtures used for the parts in the same family. Validation of fixture performance is critical to generating permissible variation fixture designs. The following items may need to be verified: free from interference, chip shedding to avoid chip accumulation, locating accuracy, stability problems, clamping sequence, error proofing, pre-locating/pre-clamping, and ergonomic issues (*e.g.* load/unload accessibility). Figure 6.11 shows an example of variation fixture design generated by the computer-aided manufacturing planning system. The fixture configuration is generated based on the common fixtures used by its part family [6.18]. Part dimensions drive the dimensions of fixture components.

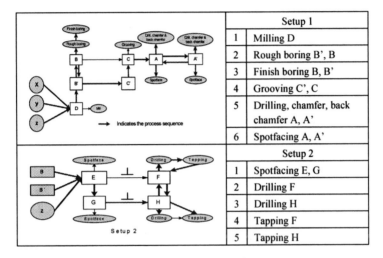

	Setup 1	
1	Milling D	
2	Rough boring B', B	
3	Finish boring B, B'	
4	Grooving C', C	
5	Drilling, chamfer, back chamfer A, A'	
6	Spotfacing A, A'	
	Setup 2	
1	Spotfacing E, G	
2	Drilling F	
3	Drilling H	
4	Tapping F	
5	Tapping H	

Figure 6.10. Process sequence in Setup 1

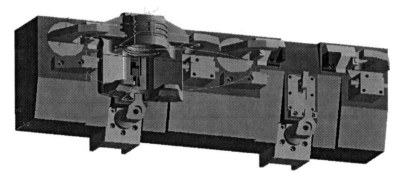

Figure 6.11. An example of variation fixture design

6.2.6 Manufacturing Plan Generation

Manufacturing plan generation is required for mass production and mass customisation. With the comparison of several alternative solutions of conceptual fixture design, a feasible manufacturing plan can be generated to determine:

- Process sequence to machine manufacturing features of the parts on the fixtures.
- Tool path to machine the features based on the process sequence.
- Minimal cycle time and manufacturing processes. Cycle time is the basic criterion to help choose the manufacturing plan. Cycle time is composed of cutting time, rapid travel time, tool-change time and machine tool table-index time. The cutting time is determined by the cutting parameters and the other time is related to the machine tool performance. Therefore, through the adjustment of cutting parameters, machine tool parameters and even the change of machine tools can help to reduce the cycle time and get a better manufacturing solution.
- Generate manufacturing documents. Documents are outputs generated to a company-specific format. Manufacturing planning information including setups, process sequence, machines, tooling and process parameters are stored in the documents. This helps users understand what the system does and what kind of information is used in the decision-making process.

Three solutions of conceptual fixture design for the Setup 1 are listed in Table 6.1 where the simplified part and fixture base information is presented. With the use of the same machine tool and process parameters, a different cycle time is achieved. It can be determined that the bridge is the best solution for Setup 1 that has the least cycle time. Figure 6.12 shows the process sequence and process parameters in the documentation.

Table 6.1. Alternative fixture solutions for Setup 1

Conceptual fixture design	Bridge	Rectangle plate	Round plate
Part layout			
Machine tool	DMV- 500 provided by Daewoo		
Cutting time/part (second s)	66.48		
Non-cutting time/part (s)	31.48	40.06	34.08
Cycle time/ part (s)	97.96	106.54	100.56

6.3 Automated Setup Planning

Setup planning plays a crucial role in CAPP to ensure product quality, while maintaining acceptable manufacturing cost. The tasks of setup planning consist of (1) identifying manufacturing features and corresponding manufacturing processes, (2) determining the number of setups, part orientation, locating datum and process sequence in each setup, and (3) determining the machine tools and fixtures.

The purpose of setup is to locate and fix a part on a machine tool so that machining can take place. There are three setup methods used to maintain tolerances

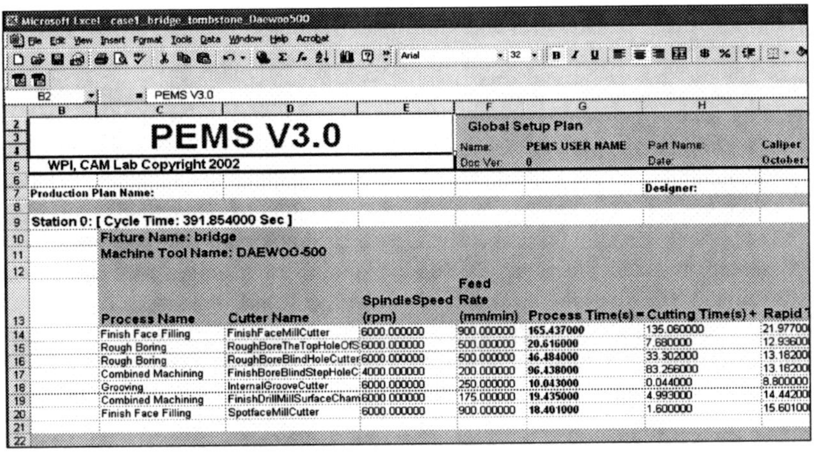

Figure 6.12. Standard document output

between two features: (1) Machining the two features in the same setup; (2) Using one feature as the locating datum and machine the other; and (3) Using an intermediate locating datum to machine the two features in different setups. These three methods are denoted as setup method I, II and III, respectively. It is concluded that setup method I may produce the least manufacturing error because no locating errors are involved [6.8]. Setup method II consists of one more locating error when two features cannot be machined in the same setup. Setup method III is used where a tolerance stackup is formed by every setup including the two features. Hence, to reduce the locating error effect, the setup planning priority may be given to minimise the number of setups and to process the maximum number of features that can be synchronously machined in one setup.

In the research of setup planning, the analysis of part information is always the starting point. Currently, graph-based representation has been recognised as an effective tool to describe the many-to-many relationships in part information and setup planning. An extended directed graph, including FTG and DMG, is used to represent part design tolerance specifications and operational tolerance relationships based on true positioning datum reference frames [6.3]. By the use of graphs, it is easy to track the tolerance generation routines among manufacturing processes.

Other than tolerance analysis, feature orientation, precedence constraints, kinematic analysis and force analysis have been considered in setup planning [6.6]. Several methodologies and algorithms have been proposed for setup planning, including a graph-matrix approach for rotational parts based on tolerance analysis [6.12][6.19], a hybrid-graph theory, accompanied by matrix theory to aid setup plan generation that was carried out on a 3-axis vertical milling centre [6.5], an approach for setup planning of prismatic parts with a Hopfield neural network where the algorithm converting feature sequencing problem to a constraint-based travelling salesman problem (TSP) [6.11], and a graph-based analysis and seven setup planning principles defined to minimise machining error stackup under a true positioning GD&T scheme [6.3]. Among all these strategies, limited manufacturing resource capabilities have been considered. It is hard to generate feasible setup plans when multi-axis CNC machines and multi-part fixtures are used. Furthermore, setup planning and fixture design are two closely related tasks. Setup planning is constrained by the fixture to be applied. But most researchers circumvent this problem by focusing on either setup planning or fixture design [6.5].

This section presents a systematic strategy on automated setup planning for non-rotational parts with the utilisation of flexible manufacturing resources. Three technical points are included: 1) Graph theory is applied to describe FTG and DMG, where the part design tolerance specification and operational tolerance relationships are presented in setup planning; 2) Setup planning of a single part is defined as transforming FTG to DMG based on tolerance and manufacturing resource capability analyses as well as the best practice in industry; and 3) In order to utilise the manufacturing resource capability effectively, the setup planning is extended to the station level by the use of multi-part fixtures and multi-axis machine tools.

6.3.1 Graph Theory and Application in Setup Planning

A graph is an ordered triple G = {V, E, I}, where V is a nonempty set of elements, E is a set disjointing from the elements in V, and I is an incidence map associated with each element of E [6.20]. Elements of V are called vertices of G, and elements of E are called edges (e_i) of G. If all the elements in E connect ordered pair of vertices, then G is called a directed graph. Let G be a graph and $v \in V$. The number of edges incident at v in G is called the degree of the vertex v in G and is denoted by $d(v)$. The in-degree $d^-(v)$ of v is the number of edges incident into v and the out-degree $d^+(v)$ is the number of edges incident out of v and the neutral-degree $d^0(v)$ is the number of undirected edges incident on v. A loop at v is to be counted twice in computing the neutral-degree of v. Hence,

$$d(v) = d^-(v) + d^+(v) + d^0(v) \tag{6.1}$$

A walk in a graph G is an alternating sequence W: $v_0 e_1 v_1 e_2 v_2 e_3 ... v_n e_n$ of vertices and edges beginning and ending with vertices in which v_{i-1} and v_i are the ends of e_i. The walk is closed if $v_0 = v_n$ and is open otherwise. A walk is called a trail if all the edges appearing in the walk are distinct. It is called a path if all the vertices are distinct. Thus a path in G is automatically a trail in G. The concepts of path and circuit are useful in setup sequencing and process sequencing.

New graphs can be generated by the use of the operations on graphs, which include add a vertex, remove a vertex, join two vertices, unite two graphs, and subtract graph1 from graph2.

6.3.2 Feature Tolerance Relationship Graph (FTG)

Part information is composed of the information of features and feature relationships. It can be represented by FTG. The relationships are the dimensions and tolerance specifications between features. In FTG, the vertices represent features, the edges represent dimensions and tolerances between features, and the incident maps represent the relationship types and values. Among them, a pair of unordered vertices represents the dimension tolerances, and a pair of ordered vertices represents the positional and orientation tolerances. In many cases, there may exist more than one tolerance between two features. Hence, the FTG of a part is a graph with undirected edge, directed edge, and multiple edges. It is not only a simple graph. Following is the mathematical representation of FTG:

$$G_{FTG} = \{F, T\} \tag{6.2}$$

where, $F = \{f_1, f_2, ..., f_n\}$ is a nonempty set of vertices of FTG. Each vertex represents one feature; $T = \{t_1, t_2, ..., t_m\}$ is a set of edges of FTG with relationship defined by $t_j = \{f_i, f_k, t_{type}, t_{value}\}$. Each edge associated with the features is the relationship type and the relationship value. If the relationship type is a dimension with or without tolerance, the edge is an undirected edge. If it is a positional or orientation tolerance, the edge is a directed edge and the first feature is the datum feature of the tolerance.

Figure 6.13(a) shows the terminology of a simplified calliper part in the automotive industry. Figure 6.13(b) shows its FTG that expresses the relationships between features. X, Y, Z are predefined datum surfaces that are mutually perpendicular to each other. Features A and A' are two holes used to mount callipers on the brake system. There exist required dimension and parallelism tolerances. The same situation exists between B and B'. The dimension tolerance is represented by an undirected edge and the parallelism is drawn by a directed edge, as shown in Figure 6.13(b). Hence, there are multiple edges between A and A', and B and B'.

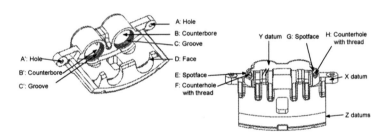

(a) Terminology of a calliper

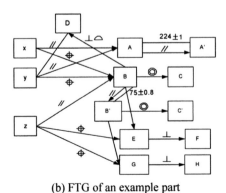

(b) FTG of an example part

Figure 6.13. FTG of a simplified calliper

Since features are associated with particular manufacturing methods, each of them may consist of several processes and each process may have its own tool approach directions. Therefore, the FTG is extended to link feature manufacturing processes onto the features. For a particular part, its FTG is unique, but it may have several extended FTGs since one feature may have alternative manufacturing methods. As a result, the task of setup planning is to design setups that perform all the manufacturing processes linked onto the features in the FTG. Figure 6.14 shows one of the extended FTGs of the example calliper.

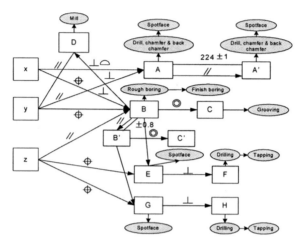

Figure 6.14. FTG with consideration of feature's processes

The extended FTG is mathematically represented as follows:

$$G_{FTG}^{E} = \{F, T\} \tag{6.3}$$

where, $F = \{f_1, f_2, ..., f_n\}$ is the feature set of a part. Each feature has its own manufacturing processes $f = \{p_{i1}, p_{i2}, ..., p_{io}\}$.

6.3.3 Datum and Machining Feature Relationship Graph (DMG)

Setup planning is to determine how many setups are needed to machine a part, what the datum features are in each setup, and how many processes can be finished in the setup. The information of setups should include datum features, manufacturing features and their processes. In order to fulfil the tolerance requirement between features, the errors caused by the manufacturing processes should also be recorded.

The information of setup planning can be represented by the relationship between datum features and manufacturing features, which is called the DMG. A DMG includes one or many sub-graphs and each sub-graph represents one setup. In a DMG, vertices are classified into two sets, the datum features (grey solid vertices) and manufacturing features (transparent vertices). An edge, which is associated with machining errors, marks the relationship between the datum feature and the target feature. A dashed line is used to connect the same feature in different setups. Figure

6.15 shows the DMG of the example calliper. With DMG, it is easy to track back the machining error stackup [6.6].

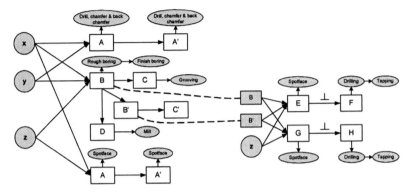

Figure 6.15. DMG of the calliper

The mathematical representation of DMG is as follows:

$$G_{DMG} = \{G^s_{DMG\ 1}, G^s_{DMG\ 2}, ..., G^s_{DMGn}\} \qquad (6.4)$$

$$G^s_{DMG} = \{D, F, Er\} \qquad (6.5)$$

where $D = \{D_1, D_2, D_3\}$ is datum features; $F = \{f_1, ..., f_m\}$ is feature sets associated with processes, $f_i = \{p_{i1}, ..., p_{io}\}$; and $Er_j = \{f_i, f_k, er_{type}, er_{value}\}$ is error specifications. It is true that $f_i \subset D \cup F, f_k \subset F$. A DMG is composed of sub-graphs that represent individual setups. Each setup consists of datum features, manufacturing features and machining errors generated in the setup. er_{type} is the same as t_{type} defined in Equation (6.2).

6.3.4 Automated Setup Planning

In this section, automated setup planning is divided into two sub-tasks: setup planning on the part level and on the station level in which fixtures and machine tools are selected to machine several parts sequentially on machine tools. Moreover, setup planning and fixture design are two close modules. Fixture planning may provide locating datum information to help the generation of the DMG, and station-level setup planning can give part orientation, part layout on fixture bases to facilitate fixture structural design. When fixture planning information is not available, the best practice of similar parts can be used to help setup planning. Figure 6.16 shows the architecture of automated setup planning.

Since the input and output information in setup planning for a part can be represented by FTG and DMG, respectively, the problem of setup planning is to transform the extended FTG into DMG based on tolerance and manufacturing capability analyses. Setup planning on the part level is carried out in three steps: feature grouping, setup generation, setup and process sequencing. In feature grouping, tolerance analysis is carried out to identify those features in FTG with

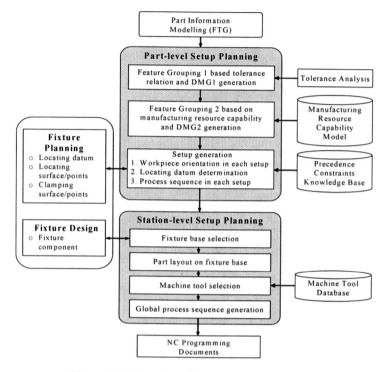

Figure 6.16. Overview of automated setup planning

tolerance relationships and suggest machining them in one setup. The locating datum of each feature group is also identified in feature grouping. The information generated in this step is represented by DMG1, which is a rough description of the setup plans. In the second step, it is the manufacturing resource capabilities that finally determine the number of setups needed, the setup sequence, workpiece orientation, features and the sequence of the features' machining processes in each setup. The information is represented by DMG2, which is the final result of the setup plans. The tolerance relationships in each setup are clearly shown in DMG2. Different manufacturing resource capabilities may lead to different setup plans because of the different manufacturing resource capability utilisation. In the last step, precedence constraints are applied to guide a walk through all vertices on DMG2 to determine the process sequence in each setup.

6.3.4.1 Part-level Setup Planning

Setup planning on the part level includes grouping features into setup, considering the manufacturing resource, and determining the process sequence in each setup.

Feature grouping based on tolerance analysis. In order to minimise inter-setup tolerance stackup, it is suggested to group those features with close position, orientation or profile tolerance relationship together and to be machined in one datum frame. An algorithm is developed to extract feature groups from FTG. The basic idea is to find datum features and machining features through calculation of the degree of vertices in FTG. The in-degree of the initial datum feature is 0. Thus, FTG is transferred into DMG1 and initial setups are generated.

Setup formation based on manufacturing resource capability. The next step of setup planning is to consider the manufacturing resource capabilities. First, features in DMG1 are attached with the manufacturing processes. Each process has its own specified tool approach direction (TAD). Those feature processes with the same datum frame and TAD can be reunited into one group. The TADs are given based on part coordinate system. For the group with more than one TAD, each TAD should be considered without violation of the feature-process sequence. Next, the machine tool capability is considered. For example, 3½-axis machine tools can provide more TADs than 3-axis machine tools so that the number of setups can be reduced.

Besides TAD, other machine tool capability measures are also considered, such as machine accuracy, table size and motion range, and combinations of fixture and cutting tool geometry. With these considerations, DMG1 is modified into DMG2. The setup sequencing is also determined in DMG2. The basic principle of setup sequencing is to ensure that a feature is machined before it is used as the locating datum or tolerance datum of other features. In this research, this is reflected in two principles:

Principle 1: The setup sequence must be arranged according to the sequence of datum features.

Principle 2: The setup sequence must be arranged according to the feature's pre-defined process sequences.

Process sequencing in each setup. The problem of process sequencing in each setup is transformed into a search for an optimal walk to traverse each vertex in DMG2 under specified constraints. The times of passing each vertex are determined by the number of processes linked to each feature. The constraints are divided into strong and weak constraints. The former is the first priority to achieve and cannot be violated, while the latter comes from manufacturing experience and may provide optimal solutions. The strong constraints may include, maintaining the manufacturing process sequence of each feature, maintaining the operation-dependent relationship in the graph. For example, planes prior to holes and holes prior to grooves, doing rough cuts first, semi and finish cuts in a prescribed order, and minimising the tool-change time and machine tool adjustment time. One example of a weak constraint might be that the cutter to mill a surface could be combined with the cutter to drill, chamfer and back chamfer a hole so that the tool-change time can be reduced.

6.3.4.2 Station-level Setup Planning

It is known that in the overall manufacturing time, cutting time only takes up a small portion. Non-cutting time, including tool-changing time, cutter rapid traverse time and machine tool table-index time, takes most of the cycle time. Hence, in mass production, the utilisation of multi-part fixtures may improve the productivity and reduce the cycle time. The station-level setup planning includes the following steps:

Machine tool selection. Candidate machine tools satisfy the entire requirement to machine tool capabilities from setup planning, including the number of axes of machine tools. The detailed machine tool planning is presented in the next section.

Conceptual fixture design and part layout design. Fixture design issues are divided into two steps: conceptual fixture design and detailed fixture design that includes the design of fixture structure and fixture components. Conceptual fixture provides ideas about which fixture base is used and how many parts are held on the fixture base as well as how the parts are oriented on the fixture. In mass production, fixtures are usually designed and fabricated by vendors. Hence, in setup planning, only conceptual fixture design is emphasised. The initial solution of the conceptual fixture design can be derived from the best practice design in industry. The part position and orientation on fixture bases need to be determined, which implies how much space would be left to accommodate fixture components. The conceptual fixture design is an extension of machine tool capabilities. There may be a mix of part-level setups in one station level setup. Therefore, after the generation of the initial solutions of conceptual fixture design, the machine tool capability needs to be re-evaluated: Whether it has enough space to accommodate the fixtures and parts, and whether it can access all the features and finish all the required processes. If the conditions are not satisfied, the fixture base and part layout may be reselected and adjusted, such as adjusting part position and orientation, or putting fewer parts on the fixtures, or even the machine tool is reselected.

Global process sequence and tool-path generation. In order to reduce the non-cutting time on each part, the processes that use the same cutters are suggested to be carried out together in sequence. A sequence is needed for all the manufacturing processes on the multi-part fixture. The corresponding tool paths are generated without interference with fixture components, machine tools, *etc.*

Cycle time calculation. Cycle time is a critical factor to help in choosing better setup plans in mass production. Hence, the estimation of cycle time is indispensable in station-level setup planning.

6.3.5 A Case Study

A sample part has been shown in Figure 6.13 with the FTG in Figure 6.14. Following is the result from the automated setup planning algorithm.

DMG1 generation. By applying the algorithm developed for setup formation, features x, y, z are found as initial datum features in step 2. In step 3, features A and B have three edges linking with x, y, z, respectively, and feature group (A, A'), (B, D), (B, C), (B, B', C') are identified as initial groups. In step 4, features E, F, G and H are identified beyond the above feature groups. Hence, an intermediate datum frame is needed. Based on the DMG rules (3 mutually perpendicular datums) and the best-practice knowledge of fixture planning, features B, B' and Z are identified as the intermediate datum frame and feature group (E, F) and (G, H) are constructed. The feature groups and corresponding datum frames are listed in Table 6.2. The DMG1 is shown in Figure 6.15.

DMG2 generation. The next step of setup planning is to consider the manufacturing resources capabilities. First, each process in DMG1 has its own specified TAD. Those feature processes with the same datum frame and TAD can be reunited into one group. Table 6.3 shows the results. The TADs are given based on a part coordinate system that is predefined on the example part.

The manufacturing resource is selected to carry out all processes in each group. Based on TADs, four setups are needed, in which group 3 can be carried out with

Table 6.2. Feature grouping based on tolerance analysis

	Manufacturing features	Datum
Group 1	A, A'	X, Y, Z
Group 2	B, D	X, Y, Z
Group 3	B, C	X, Y, Z
Group 4	B, B', C'	X, Y, Z
Group 5	E, F	B, B', Z
Group 6	G, H	B, B', Z

Table 6.3. Feature-process grouping based on tolerance analysis

	Manufacturing features	Machine tool	Datum	TAD
Group 1	A (Drilling, chamfer & back chamfer) A' (Drilling, chamfer & back chamfer) B, B' (Rough boring, finish boring) C, C' (Grooving)	3-axis machine	X, Y, Z	+X
Group 2	A (Spotface) A' (Spotface) D (Milling)	3-axis machine	X, Y, Z	−X
Group 3	A (Tapping) A' (Tapping)	3-axis machine	X, Y, Z	+X or −X
Group 4	E (Spotface) F (Drilling, tapping)	3-axis machine	B, B', Z	−0.6Y +0.8Z
Group 5	G (Spotface) H (Drilling, tapping)	3-axis machine	B, B', Z	0.6Y +0.8Z

group 2 through the use of the precedence constraint to maintain the feature manufacturing sequence. Figure 6.17 shows the corresponding DMG2. Compared with part FTG, it can be seen that there is a perpendicular tolerance requirement between features B and D. In this solution, B and D are machined in a different setup and a tolerance stackup between B and D is generated.

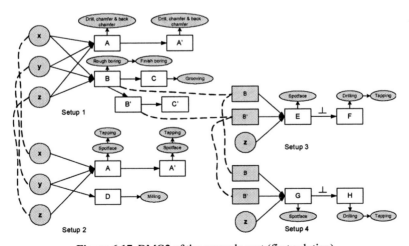

Figure 6.17. DMG2 of the example part (first solution)

However, if 3½-axis machining centres with a table-index function are available, the setup planning may generate another solution. It is assumed that the machine tool coordinate and part coordinate overlap; then groups 1, 2 and 3 can be finished in one setup by indexing machine table 180° and groups 3 and 4 in another setup by indexing machine table of 106°. The corresponding DMG2 is shown in Figure 6.18. In this solution, the number of setups has been reduced to 2. Features B and D are machined in the same setup so that there is no tolerance stackup between setups.

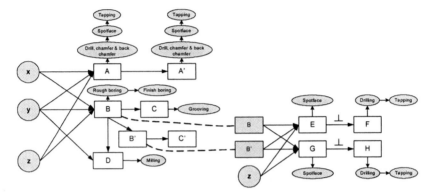

Figure 6.18. DMG2 of the example part (second solution)

Hence, in the setup planning of the example part, there are two datum feature sets (X, Y, Z) and (B, B', Z). Through the calculation of their degrees, the in-degree of X, Y, Z is $d(v) = 0$, while $d(B)>0$, $d(B')>0$, $d(z)>0$. Therefore, the setup sequence is from (X, Y, Z) to (B, B', Z).

Process sequence. One solution of process sequencing is shown in Figure 6.10, where 3½-axis machining centres are used in the production of the example part.

Station-level setup planning. Predefined fixture bases are used in the station-level setup planning. Corresponding requirements to the machine tool are generated and evaluated. Table 6.4 shows different fixture bases used in the two setups.

Table 6.4. Station-level setup planning

			Setup 1	Setup 2
Fixture base type			Bridge	Tombstone
Part layout on fixture base				
Machine tool requirements	No. of axis		3½	2½
	Moving range	X	800 mm	500 mm
		Y	363 mm	500 mm
		Z	765 mm	700 mm

6.4 Information Modelling

In today's advanced manufacturing, investments in automated production machinery and systems have increased steadily. These machines and systems place high demands on manufacturing planning, which serves as the bridge between product design and fabrication in order to convert design specifications into manufacturing instructions. Currently, CAD and CAM systems have become standard engineering tools in industry, but the effectiveness of CAMP systems is not fully satisfactory. The reasons lie in two aspects: One is the lack of correct information models of parts, planning methodologies, manufacturing processes, and resources [6.21]. There are two aspects to these information models, the conceptual models and the implemented or the computer models. It should be pointed out that finding conceptual models for manufacturing planning is very difficult because of the complex interaction between manufacturing planning and other activities in a manufacturing enterprise. Another reason is the fact that the scope of manufacturing planning is constantly changing, due to the new demands in product development practice. In recent years, a new production mode, mass customisation, has been introduced and widely applied in industries [6.1]. It allows customised products to be made to suit special customer needs, while maintaining near-mass-production efficiency. Compared to conventional mass production, mass customisation allows for more product variety in which products are grouped into families in industry. Some research was carried out in the design stage, in which product structures are decomposed into modules by the use of modularity principles [6.22]. The reuse of certain modules can simplify a new product design. In the production stage of mass customisation, the low cost is achieved primarily through the full utilisation of manufacturing process capability, in which multi-axis machining centres and multi-part fixtures are widely used. Hence, the difficulty of manufacturing planning for mass customisation is greatly increased due to the complexity of manufacturing resource capability analysis and utilisation. In order to pursue smaller turnaround time and increase the response speed to customer's needs, the modularity analysis in the design stage is expected to expand to the manufacturing planning stage. The tasks are executed by implementing interrelated modules. Some of the modules, including the planning methodologies and information modelling, are to be realised by the research of the CAMP. The others are designed for specific companies that have accumulated a variety of best-of-practice (BOP) knowledge. The reuse of planning methodologies and BOP will greatly reduce engineers' workload and increase their planning efficiency. As a result, the study on CAMP for mass customisation requires a clear structure of planning tasks, a redefinition of planning methodologies, and the establishment of correct information models, as well as the description and utilisation of BOP.

6.4.1 A Systematic Information Modelling Methodology

In the CAMP, the main challenge is to analyse the information involved in the manufacturing planning activities and to construct conceptual information models, which can facilitate the rapid generation of manufacturing plans, including the utilisation of manufacturing resources, according to changes of part design.

Information models are data structures that represent information content in part design and manufacturing. The main task of information modelling is to capture, describe, and maintain the information structure and information relationships in the CAMP. In this section, an object-oriented systems analysis (OSA) approach [6.23] is utilised to analyse and represent information models, and a systematic information modelling hierarchy is proposed to model both static and dynamic characteristics of information from a system perspective.

6.4.1.1 The Object-oriented Systems Analysis (OSA) Approach

Object-oriented (O-O) modelling is recognised as a powerful tool to model real-world systems. An object is an encapsulation of data and procedures (or methods) that operates on the data. A relationship establishes a logical connection among objects. An object could be an existing entity in the real world such as a part and a machine tool. The definition of objects implies the correlative relationships between the data and the procedures related to the data. The real world can be considered as a group of interacting objects. The interactions, including static relationships and dynamic relationships, are described according to the way that human beings think. Therefore, the main task of O-O modelling for a system is to identify objects and analyse their interactions within the system.

In this section, an OSA approach is used to analyse the information in CAMP: an object-relationship model (ORM) is used to represent the static relationships between objects. An object-behaviour model (OBM) describes the behaviour of individual objects and how objects respond to dynamically occurring events and conditions. An object-interaction model (OIM) expresses the interactions between objects.

Object-relationship model (ORM). An ORM is created to represent the static relationships between objects, which is described by an ORM diagram. A rectangle represents an object, and the variables of the object are shown in the lower rectangle. There are two basic relationships: Generalisation–specification relationship means the is-kind-of relationship, which is represented by a transparent triangle in an ORM diagram; and whole–part relationship indicates the is-part-of relationships, which is described by a solid-filled triangle. Users can define their own relationships with the specific relationship name attached to ORM diagrams.

Object-behaviour model (OBM). The objective of a behaviour model is to describe the way that each object in a system interacts, functions, responds, or performs. A behaviour model for an object is similar to a job description for an object. In this chapter, state nets are used to represent OBMs, which is composed of states, triggers and actions. A state, represented by a rounded rectangle, describes an object's status, phase, situation, or activity. The events and conditions that activate state transitions are called triggers. The activity that an object performs is called an action. A rectangle is divided into two sections, the top section contains a trigger description and the bottom section contains the actions.

Object-interaction model (OIM). The ORMs describe the static relationships among objects. The OBMs describe the behaviour of an object, but in isolation from other objects. An OIM model is used to describe the interaction among objects.

One object interacts with another in many different ways. For example, an object may send information to another, request information from another, alter another object, or cause another to do some actions. To understand object interaction, we must understand: 1) What objects are involved in the interaction; 2) How the objects act or react in the interaction; and 3) The nature of the interaction. Since objects are identified in ORMs, and the behaviour of each object is described in OIMs, a combination of ORMs and the state nets is used to create OIMs, in which a zigzag arrow is used to describe the interactions between objects. In general, the interactions indicated by zigzag arrows imply the user-defined correlative relationships associated to specific systems.

6.4.1.2 Systematic Information Modelling Hierarchy

When using the OSA approach to model a complex system, high-level abstraction of objects is applied to reduce complexity and make the information models easy to create, maintain, and display. A high-level object package groups relative objects and the relationships among the objects into a single object. The top-down approach is used to expand a high-level object into low-level objects and relationships. Figure 6.19 shows the hierarchical structure of system information models. The building of information models is split into three levels.

The definition of a system model contains domains that are subdivided into subsystems. The system model may be deduced from analysis of the system's high-level object interaction models. The definition of an information model contains objects that are subdivided into states. The definition of the state model describes the behaviour of objects.

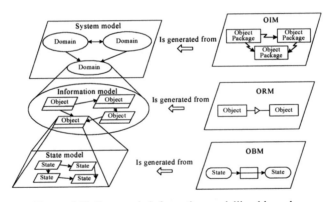

Figure 6.19. Systematic information modelling hierarchy

6.4.2 Information Model of CAMP for Mass Customisation

The tasks of CAMP have been carried out by four functional modules, which are shown by the grey round rectangles in Figure 6.20 [6.17]. Part information is composed of features and the tolerance relationships between the features. Part information modelling extracts features from part CAD models, with feature manufacturing strategies associated with the features. Setup planning is carried out based on tolerance and manufacturing resource capability analyses. Multi-part fixture design may be involved. Conceptual fixture design is used to determine part layout on the fixtures. Manufacturing plan generation is to determine the optimal process sequence and tool path. The information involved in CAMP is organised into three categories:

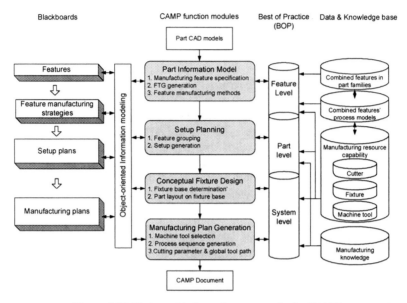

Figure 6.20. Tasks and information content in the CAMP

Manufacturing databases and knowledge bases. In CAMP, the information is considered and stored in the manufacturing data and knowledge bases. Combined features are defined based on particular part families. The parts in the same family may have the same type of combined features and feature relationships so that the part-family BOP can be used as the reference to generate new plans [6.17]. Combined features are associated with predefined manufacturing strategies, in which customised combination cutters, tool paths, and machine tool motion requirements are specified for particular part families. The designs of cutters and tool paths are based on prior experience and are stored in templates. Therefore, when the same combined feature is encountered, the existing experience can be reused.

Manufacturing resources include cutters, machine tools, and fixtures. Some of them are standard tools and can be brought from the market. The others are designed specifically for particular processes used in manufacturing plans. The capabilities of available manufacturing resources should be described and stored in a format that the CAMP can interpret and manipulate. Manufacturing rules and knowledge are extracted from BOP and applied in the automated reasoning mechanism such as automated determination of feature manufacturing strategy, setup planning, and manufacturing plan generation. Three levels of manufacturing knowledge are identified, general knowledge without regard to a particular shop, shop-level process details, and part-level information based on particular part-family production in a specific machine shop.

Best-of-practice (BOP). BOP for part families is the most important reference enabling engineers to design a new manufacturing plan. The specific decision-making strategies of part families are embedded in the BOP, which include strategies about how to deal with the correlative relationships between part design, part manufacturing, and the utilisation of manufacturing resource capabilities. Therefore, the decision-making strategies in the BOP must be identified first, and then the BOP should be described in a format that is accurate, complete, and unambiguous, so that it can be used by the CAMP system. In this chapter, information in BOP is divided into three levels: feature level, part setup planning level, and machine level.

Blackboards. Blackboards are used to store the shared information generated by the modules of the CAMP. It is in the blackboards that computers deal with the manufacturing information that is represented by information models. There are four blackboards in CAMP, which store features, features' manufacturing strategies, part setup planning and manufacturing plan information. The design of information models considers the following issues, information relationship, information integration, and information extendibility. The design of information models should pay attention to correlative relationships and try to avoid information redundancy in models. The design of information models should consider the overall information requirements of the CAMP system. Different functional modules may have different requirements for the same information model. With the consideration of the new demands in product development practice, the scope of the CAMP may change accordingly. Therefore, the information models should be extendable to accommodate more information content without damaging origin information content and information relationships.

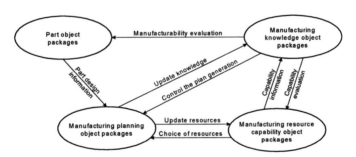

Figure 6.21. System models for the CAMP

The systematic information modelling methodology is used in this chapter to analyse and represent the information in the blackboards of the CAMP. As shown in Figure 6.21, four object packages are established to describe the primary information involved, as well as the interactions between these packages. The part information is the input, which is composed of features and the tolerance relationships between features, and the features' manufacturing strategies are linked with features. The manufacturing planning package includes feature-level, part-level and machine-level decision-making strategies. The manufacturing knowledge package provides the knowledge constraint to control the manufacturing planning behaviours. The manufacturing resource capability package provides the description of available manufacturing resources. Hence, breaking down these packages will result in a detailed study on the correlative relationships within the CAMP, which facilitate the use of the part-family BOP to help engineers rapidly design new manufacturing plans.

In part information representation, the procedure of establishing a process object for a feature is called feature-level decision making, whose OIM is shown in Figure 6.22. Similarly, the OIMs can be established for setup planning, manufacturing resource planning, and manufacturing planning generation.

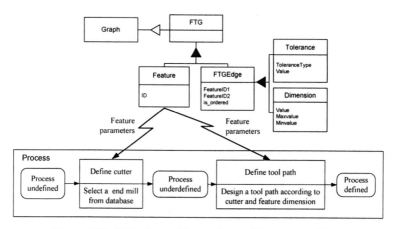

Figure 6.22. OIM of part object and one of the process objects

6.5 Summary and Discussions

The overall framework of a CAMP system for mass customisation has been developed and introduced in this chapter. The objective of the research is to provide a computerised tool for rapid design and simulation of manufacturing systems with emphasis on the utilisation of best-practice knowledge together with analysis in production planning. In the system, a feature-based part information model is used to represent part information. Combined features are parametrically represented and subsequently used in determining manufacturing methods and processes based on available manufacturing resources and capabilities. Graph-based automated setup planning has been extended to consider flexible manufacturing resources, multi-parts fixture configuration and process sequence optimisation. Finally, the standard manufacturing documentation is automatically generated to a company-specific format by the system.

To facilitate the CAMP system, a systematic information modelling technology is proposed to represent the correlative relationships from the system perspective. The OSA approach is used as the primary tool to describe the static and dynamic characteristics of information. Therefore, the correlative relationships within the CAMP between part design and manufacturing planning can be properly described, as can the information in BOP of part families. A tri-level decision-making mechanism is proposed by using a systematic information modelling technology. At the feature level, combined features and their manufacturing strategies are defined based on part families. At the part level, part information is represented by FTG, and setup planning information is described by DMG. Rules and constraints that are extracted from BOP control the transformation from FTG to DMG. At the system level, multi-part fixtures are utilised to reduce cycle time and to increase productivity. Part layout on multi-part fixture bases is also retrieved from BOP.

The research focuses on setup planning and station-level planning, while the process-level optimisation, such as tool path or process parameter, is not emphasised, but the results can be integrated. Fixture design and tolerance analysis are two important tasks of manufacturing planning and are discussed separately.

References

[6.1] Jiao, J., Ma, Q. and Tseng, M.M., 2001, "Towards high value-added products and services: mass customisation and beyond," *Technovation*, **21**, pp. 149–162.

[6.2] Shah, J.J., 1995, *Parametric and Feature-based CAD/CAM – Concepts, Techniques, and Application*, John Wiley and Sons, New York.

[6.3] Zhang, Y., Hu, W., Rong, Y. and Yen, D.W., 2001, "Graph-based setup planning and tolerance decomposition for computer-aided fixture design," *International Journal of Production Research*, **39**(14), pp. 3109–3126.

[6.4] Naish, J.C., 1996, "Process capability modelling in an integrated concurrent engineering system – the feature-oriented capability module," *Journal of Materials Processing Technology*, (6), pp. 124–129.

[6.5] Zhang, Y. and Feng, S.C., 1999, "Object-oriented manufacturing resource modelling for adaptive process planning," *International Journal of Production Research*, **37**(18), pp. 4179–4195.

[6.6] Huang, S.H., Rong, Y. and Yen, D., 2002, "Integrated setup planning and fixture design: issues and solutions," *NAMRC XXX*, West Lafayette, IN, May 21–24, pp. 589–596.

[6.7] Huang, S.H., 1998, "Automated setup planning for lathe machining," *Journal of Manufacturing Systems*, **17**(3), pp. 196–208.

[6.8] Rong, Y. and Zhu, Y., 1999, *Computer-aided Fixture Design*, Marcel Dekker, New York.

[6.9] Kiritsis, D., 1995, "A review of knowledge-based expert systems for process planning: methods and problems," *International Journal of Advanced Manufacturing Technology*, **10**(4), pp. 240–262.

[6.10] Zhang, H.-C. and Lin, E., 1999, "A hybrid-graph approach for automated setup planning in CAPP," *Robotics and Computer-Integrated Manufacturing*, **15**, pp. 89–100.

[6.11] Chen, J, Zhang, Y.F. and Nee, A.Y.C., 1998, "Setup planning using Hopfield net and simulated annealing," *International Journal of Production Research*, **36**(4), pp. 981–1000.

[6.12] Huang, S.H., Zhang, H.-C. and Oldham, W.J.B., 1997, "Tolerance analysis for setup planning: a graph theoretical approach," *International Journal of Production Research*, **35**(4), pp. 1107–1124.

[6.13] Britton, G., Thimm, G., Beng, T.S. and Jiang, F., 2002, "A graph representation scheme for process planning of machined parts," *International Journal of Manufacturing Technology*, (20), pp. 429–438.

[6.14] An, Z., Huang, S., Li, J., Rong, Y. and Jayaram, S., 1999, "Development of automated fixture design systems with predefined fixture component types: part 1, basic design," *International Journal of Flexible Automation and Integrated Manufacturing*, 7(3/4), pp. 321–341.

[6.15] Rong, Y. and Han, X., 2002, "Variation fixture design for part families," *Research Report*, Worcester Polytechnic Institute.

[6.16] Kang, Y., Rong, Y., Yang, J.-C., 2003, "Computer-aided fixture design verification, part 1: the framework and modelling; part 2: tolerance analysis; and part 3: stability analysis," *International Journal of Advanced Manufacturing Technology*, 21, pp. 827–849.

[6.17] Yao, S., Han, X., Rong, Y., Huang, S.H., Yen, D.W. and Zhang, G., 2003, "Feature-based computer aided manufacturing planning for mass customisation of non-rotational parts," In *Proceedings of the 23rd Computers and Information in Engineering (CIE) Conference*, Sept. 2–6, Chicago, IL, DETC2003/CIE–48195.

[6.18] Rong, Y. and Han, X., 2003, "Computer-aided reconfigurable fixture design," In *Proceedings of CIRP 2nd International Conference on Reconfigurable Manufacturing*, Aug. 20–22, Ann Arbor, MI.

[6.19] Bai, Y. and Rong, Y., 1993, "Machining accuracy analysis for computer-aided fixture design," *Modelling, Monitoring & Control Issues in Machining Processes*, ASME WAM, New Orleans, LA, Nov. 28–Dec. 3, PED–64, pp. 507–512.

[6.20] Balakrishnan, R. and Ranganathan, K., 2000, *A Textbook of Graph Theory*, Springer, New York.

[6.21] ElMaraghy, H.A., 1993, "Evolution and future perspectives of CAPP," *Annals of the CIRP*, 48, pp. 739–752.

[6.22] Magrab, E., 1997, *Integrated Product and Process Design and Development*, CRC Press LLC, Boca Raton, FL.

[6.23] Embley, W.D., Kurtz, D.B. and Woodfield, N.S., 1992, *Object-oriented Systems Analysis – A Model-driven Approach*, Yourdon Press, New Jersey.

7

Scheduling in Holonic Manufacturing Systems

Paulo Sousa[1], Carlos Ramos[1] and José Neves[2]

[1] GECAD R&D Group, School of Engineering, Polytechnic Institute of Porto, Portugal
Emails: psousa@dei.isep.ipp.pt, csr@dei.isep.ipp.pt

[2] Escola de Engenharia, Universidade do Minho, Braga, Portugal
Email: jneves@di.uminho.pt

Abstract
This chapter presents some of the issues concerning holonic manufacturing systems. It starts by presenting the current manufacturing scenario and trends and then provides some background information on the holonic concept and its application to manufacturing. The current limitations and future trends of manufacturing suggest more autonomous and distributed organisations for manufacturing systems; holonic manufacturing systems are proposed as a way to achieve such autonomy and decentralisation. After a brief literature survey, a specific research study is presented to handle scheduling in holonic manufacturing systems. This work is based on task and resource holons that cooperate with each other based on a variant of the contract net protocol that allow the propagation of constraints between operations in the execution plan. The chapter ends by presenting some challenges and future opportunities for research.

7.1 Introduction

In recent years, changes occurring in society and economy gave rise to a new economic model characterised by adjectives such as digital, global, competitive and customer focused [7.1]–[7.4]. As opposed to the post-war economy, stable markets, nationwide competitors, heavy hierarchical organisations and isolation in business, give place to dynamic markets, worldwide competition, network organisations, and cooperation and alliances in business [7.5].

Manufacturing has changed (and will continue to change) [7.6]–[7.9]; there is a shift from mechanisation and mass production to flexible manufacturing and product customisation as well as customised "digital" services. Innovation is no longer neglectable and knowledge has become the primary growth factor as opposed to capital and labour.

The manufacturing company of the future [7.3][7.6]: will use intelligent processes and flexible tools to achieve new dimensions of flexibility and reactivity; will support its decisions with knowledge-based systems; and will operate on worldwide networks of plants, suppliers, delivery and service centres, paying attention to change and discontinuity in order to achieve competitive advantage.

As observed in [7.10], rigid, static and hierarchical manufacturing systems are expected to be replaced by adaptable and reconfigurable ones that are more resilient to disturbances and changes. Furthermore, the limitations of current systems along with market trends, motivated the birth of distributed manufacturing systems where autonomous and flexible manufacturing entities cooperate in a coherent and coordinated manner [7.9][7.11][7.12].

Some researchers have then proposed *intelligent manufacturing systems* as a solution to these problems. Monostori [7.13] notes that the term was coined by Hatvani and Nemes in 1978 [7.14] sketching the new generation of manufacturing systems. Such systems, using the results of the (at that time) young field of artificial intelligence, would be able to solve problems without precedents, handle disturbances and unforeseen events even in the presence of incomplete information [7.15]. Intelligent manufacturing systems give a futuristic view of manufacturing with innate abilities to react correctly and promptly to external events, differing from traditional ones in its ability to adapt to a world of constant change without the need for external intervention.

Several methodologies have been proposed to realise the intelligent manufacturing system concept; some are based on the evolution of species, others in mathematics, or even philosophy. Such systems (bionic, fractal and holonic, respectively) have some common grounds, especially the fact that all of them are distributed in nature, giving special emphasis to the autonomy and cooperation of its constituent parts.

Holonic manufacturing systems try to take advantage in the manufacturing activities of the same processes and organisations that holonic structures provide to organism and societies; *i.e.* stability in the face of disturbances, flexibility to change, and efficient usage of resources [7.10][7.16].

7.2 Background

7.2.1 Holonic Systems

Holonic systems are based on the notion of a holon [7.17], a term coined by the Hungarian philosopher Arthur Koestler. *Holon* is a neologism derived from the Greek *holos* (meaning whole) and the English suffix *on* (meaning part). A holon is a dual entity with recurring structure acting both as a part and a whole.

Koestler proposes the term after analysing the parable of the two watchmakers by Herbert Simon [7.18][7.19] as a way to explain the structures observed in nature and society. Simon uses the parable of the two watchmakers to explain that complex systems evolve much more rapidly if there are stable intermediate forms (resilient to disturbances and easily replaceable by other elements) and that the ability to build large complex systems depends on the environment (frequency and gravity of disturbances, *e.g.* technology changes, rush orders).

Koestler establishes the link between the parable and the holonic concept, since a holon is an *autonomous*, *stable* and *cooperative* unit of a bigger whole [7.17]. *Autonomous* because it has the ability to build and execute its own plans; *stable* because it reacts in an adequate way to disturbances; and *cooperative* since it works with other holons on mutually developed and agreed plans.

A holon is a paradox entity since it is autonomous but it also partially controls its constituent holons as well as is partially controlled by other holons, where it is one of the constituents [7.1][7.16].

To counterpose with hierarchy, Koestler proposes the term holarchy as the fundamental structural organisation of holons. A holarchy is an ordered system (as opposed to chaotic) of interrelated parts where each part is itself a holarchy. A holarchy is a (temporary) cooperative group of holons to achieve a common goal. This self-organisation allows the creation of dynamic associations according to objectives or the state of the environment [7.20]–[7.22]. The holarchy defines the rules of cooperation, thus limiting the holons' autonomy. An interesting characteristic is that a holon can belong to several holarchies at the same time (in the same way a person can belong to several groups of interest) [7.20][7.23].

7.2.2 Holonic Manufacturing Systems

The "Intelligent Manufacturing Systems" international research project started in October 1989 with the goal to produce a science of manufacturing able to cope with the requirements of the 21st century and provide country- and organisation-independent solutions [7.24]–[7.26]. The project set up an agenda to reach for excellence in the manufacturing sector [7.27]. The project was divided into six feasibility test cases, of which "Holonic Manufacturing Systems" (HMS) was a part. From the outset of this project a new long-term research project dedicated solely to HMS was started.

A holonic manufacturing system applies the holonic concept to the manufacturing arena, creating a dynamic and decentralised manufacturing process, where humans are effectively integrated, to allow dynamic and continuous change [7.10]; It is an holarchy that grasps the complete set of manufacturing activities from design to sales to achieve an agile manufacturing company [7.16][7.28].

A holon in an HMS is defined as an autonomous and cooperative entity with an information processing part and eventually a physical part (in the case of holonic machines) [7.16]. In an HMS, every element of the system, such as machines, tools, humans, are treated as holons. Since a holon is structurally recurrent, the factory itself is also a holon.

The characteristics that distinguish HMS from other similar approaches are:

- autonomous, cooperative, reusable self-configurable elements;
- recursive structure;
- no centralised control;
- integration of human tasks in manufacturing cells.

HMS try to answer requirements posed by the demand of a small series of a great variety of products [7.21], which is believed to be a dominant pattern in the future [7.29]. It is expected that HMS delivers cost reduction due to changes and disturbances; enhanced ability to automatically recover from unplanned stops; more efficient use of the production capacity; and greater satisfaction of human operators [7.20].

The holonic organisation of the system allows HMS to combine the high efficiency and predictability of hierarchical systems with the robustness and agility of heterarchical systems due to the self-configuration and self-organisation ability of the holons [7.28].

Typical problems in manufacturing (*e.g.* scheduling) are extremely complex, with a large dimension and are extremely dynamic. This dynamism influences the solution to the problem, leading to the need to achieve a new solution in real time. This type of scenario is a good candidate for distributed computation and artificial intelligence, namely *distributed artificial intelligence* (DAI) and *multi-agent systems* (MAS). These type of systems are "suited for modular, decentralised, dynamic, complex and ill-applications" [7.30], showing "a large number of interactions among components" [7.31]. Complex, flexible and configurable systems are also the focus of attention of holonic systems. Since these approaches are mainly distributed, it is necessary to regulate the interactions among its constituents, empowering the units of organisation (*e.g.* holons) with negotiation abilities (*e.g.* Contract Net [7.32][7.33]).

In the manufacturing arena, *holonic manufacturing systems* apply the holonic concept to the manufacturing enterprise, allowing the existence of a dynamic and decentralised manufacturing process where changes are applied dynamically and continuously. The following table (from [7.11]) relates the list of characteristics a distributed manufacturing system (DMS) should have with the holonic concept.

Table 7.1. Relation between HMS and distributed manufacturing systems

Characteristic of DMS	Holonic concept support
Distribution	A holon is by definition distributed
Decentralisation	Autonomy
Autonomy	A holon is by definition autonomous
Dynamic	Dynamic holarchies
Reactivity	Must be implemented in each holon
Flexibility	Autonomy and dynamic holarchies
Adaptability	Autonomy and dynamic holarchies
Agility	Dynamic holarchies and cooperation

As one can see, the autonomous and cooperative nature of a holon, combined with dynamic organisation of holarchies can guarantee most of the identified characteristics. Thus, the holonic paradigm offers the necessary abstractions and mechanics to build the next generation of manufacturing systems.

7.3 Applications of Holonic Manufacturing Systems

The Katholieke Universiteit Leuven, Belgium, was responsible for the HMS sub-project within the IMS project. They built a seminal holonic assembly station as a feasibility study for the HMS concept [7.10][7.34].

In their approach, each physical resource (machining, assembly and transport) is represented by a holon with both a software and hardware component. Besides these holons, the system also comprises (software-only) holons for scheduling, planning and control. The objective was to analyse the cooperation between the scheduling and the control holons. The scheduling holon uses a traditional centralised scheduling algorithm to propose tasks to the control holon that tries to obey that schedule; on unforeseen conditions the control holon would behave autonomously and change the schedule as necessary, asking for a new schedule later on.

From their research, Bongaerts *et al.* [7.34] later identified the minimum set of holon types for a minimalist implementation of a manufacturing system, namely: products, resources and orders. With further study, they proposed the PROSA taxonomy [7.22][7.23] which is regarded as an archetype for any holonic manufacturing system.

PROSA defines four types of holons (Figure 7.1):

- *Product* – encapsulate the product model, *i.e.* project and process plans;
- *Resources* – encapsulate the physical resources (eventually, with the corresponding hardware control unit);
- *Orders* – represent manufacturing tasks to be executed by the system;
- *Auxiliary* – additional functionalities; usually work as advisors for the other holons (*e.g.* optimal centralised scheduling algorithm to be used in regular operation).

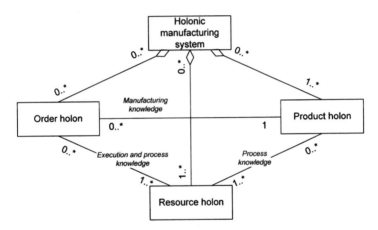

Figure 7.1. PROSA reference architecture

The PROSA reference architecture was initially used by [7.28] and [7.21] for the implementation of specific systems, but several other studies apply this reference architecture when modelling their systems. Different implementations will differ in the algorithms used in each holon as well as specific organisation (holarchies) for the holons in the system.

Gou *et al.* [7.35] propose a highly modular holonic architecture composed of *Product* holons (composed of *part* holons) for representing manufacturing orders; *Cell* holons (composed of *machine* and *cell coordinator* holons) for representing the resources; and *Factory* holons as the high-level holarchy. They perform scheduling by decomposing the problem into sub-problems using Lagrangian relaxation.

Leitao and Barbosa [7.36] present a holonic control system with a predictive disturbance handling approach that transforms the traditional "fail and recover" practices into "predict and prevent" practices, allowing improvement of the control system performance.

Walker *et al.* [7.37] compare performance results of their holonic scheduling system (which evolves the rules used to generate the schedule instead of the schedule itself) with pure heuristic scheduling heuristics and randomly generated mixed heuristics.

Kotak *et al.* [7.38] present a framework for holonic control based on the JADE platform. It allows for the integration of human experts to override the decision of the holonic controller.

Jarvis *et al.* [7.39] present and evaluate the technical feasibility of a strategy for the incremental introduction of holonic manufacturing principles into existing production control systems.

Huang *et al.* [7.40] propose a holonic framework (and respective control mechanisms) for virtual enterprises with the ability to guarantee stability, predictability and global optimisation of the system performance.

Holonic manufacturing systems have been applied to several aspects of the manufacturing business with a focus on scheduling and control. Given the holistic view promoted by the holonic concept, there is a considerable number of works that present architectures to model the "complete" enterprise, covering several or all of the manufacturing functions as well as enterprise integration and virtual enterprises and supply-chain management.

For a wider survey of applications of holonic manufacturing systems please refer to [7.41]. For a survey on distributed manufacturing systems (including agent-based manufacturing) please refer to [7.11][7.42][7.43].

7.4 An Approach: the *Fabricare* Holonic System

7.4.1 General Description

This work tries to present an integrated view of a manufacturing system, based on the distributed manufacturing systems paradigm, which is able to accomplish the requirements for post-twentieth-century society [7.44].

In this solution the main entities in the manufacturing process are modelled as holons, each one contributing with a small parcel of the overall system's functionality. To demonstrate the previous concepts, a prototype system has been developed, named *Fabricare*, which is the Latin word for "to manufacture". The *Fabricare* project uses a holonic architecture (Figure 7.2) where several key functions of the manufacturing process are identified and modelled as holarchies [7.45][7.46]. These holarchies are composed of "basic" holons such as *Product*, *Task* and *Resource*, each one representing a core entity in the manufacturing system.

The core holons are grouped together in holarchies representing the major function of the manufacturing system. The *design holon* (DH) aggregates *product holons* (PH), *customer holons* (CH) and *supplier holons* (SuH) for the design activity. The *process planning holon* (PPH) aggregates *resource holons* (RH) and *product holons* (PH). The *production planning holon* (ProdPH) aggregates *resource holons* and *sales holons* (SaH) to execute material requirements planning and capacity planning. The *scheduling holon* (SH) aggregates *resource holons* and *task holons* (TH) to dynamically schedule manufacturing orders. Additionally, there is a *directory service* (DS) that acts as a repository for the system providing information about running holons.

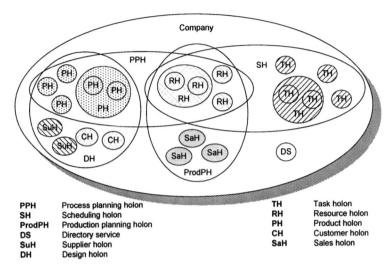

PPH	Process planning holon	TH	Task holon
SH	Scheduling holon	RH	Resource holon
ProdPH	Production planning holon	PH	Product holon
DS	Directory service	CH	Customer holon
SuH	Supplier holon	SaH	Sales holon
DH	Design holon		

Figure 7.2. The architecture of the *Fabricare* project

7.4.2 Description of Major Holons

The key entities participating in the scheduling process (activity chosen as the test case) are the physical resources and the manufacturing orders. These two entities are represented in the system by resource holons and task holons, respectively. This section details the internal knowledge base and behaviour of each of these holons.

Every holon in the system has information about itself such as *name*, *type*, *relationships* and *holarchies* it belongs to. Each holon may have one or more *types* that define a class of behaviours through a mechanism of inheritance [7.44][7.47]. Furthermore, each holon is also endowed with the ability to handle and reason about incomplete information [7.44] using a notation based on [7.47] and [7.48]. This notation allows for: explicit negative information; unknown information; mutually exclusive information; and forbidden information. Additionally, a meta interpreter was developed to infer the truth value of a question posed to the holon's knowledge base.

The scheduling activity is performed by task and resource holons by means of cooperative negotiation (see Section 7.4.3 Negotiation Protocol). Figure 7.3 shows the conceptual model of the *Fabricare* system from the task's and resource's perspective.

(1) Task Holon
A task holon represents a manufacturing order to execute a certain quantity of a specific product on the shop floor. The objectives of this kind of holon are to schedule the order and afterwards to monitor its execution [7.49][7.50].

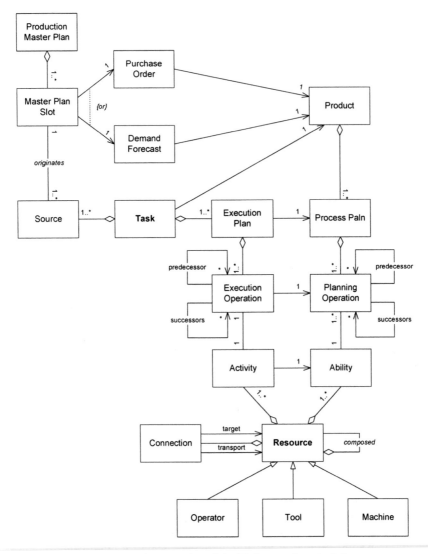

Figure 7.3. Conceptual model of *Fabricare* for scheduling

Each *task* holon has a set of internal attributes that represents the manufacturing order data such as order number, requested product, demanded quantity, due date and the selected production plan.

Its lifecycle (Figure 7.4) begins when the manufacturing order is created (either to fulfil a customer order or to balance stocks). During its existence, the *task* holon will negotiate with resource holons the execution of the operations needed to perform the ordered product. The holon will cease to exist when the order is fulfilled or cancelled.

After obtaining information about the order, the *task* holon negotiates with resource holons using CNCPP (contract net with constraint propagation protocol) –

see Section 7.4.3. The holon will then wait for the bids and evaluate them, in order to select one (if possible). If it is not possible to schedule the order, the *task* holon will recombine the resources and perform a new negotiation. A renegotiation may also be necessary if the order's condition change, *e.g.* anticipated due date, delayed, *etc.* The evaluation of bids is performed by taking into account a prioritised list of criteria. The following criteria have been implemented in the prototype: (i) first valid solution; (ii) least-costly solution; and (iii) greatest slack until due date. The cost of a solution is determined by the cost of performing the specified operations in a specific resource.

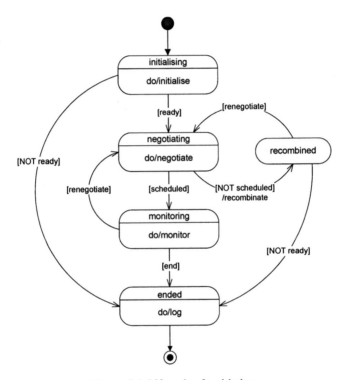

Figure 7.4. Lifecycle of *task* holons

(2) *Resource Holon*

A *resource* holon represents the current state of a physical resource on the shop floor. The resource's list of activities is called the agenda, starting with what to do and when. The resource is able to perform operations necessary to execute products (*e.g.* drill). A *resource* holon can represent a single resource or a work cell composed of several resources [7.49][7.50].

The *resource* holon has a set of internal attributes that represent general information about the physical resource in a manufacturing plant. It also has knowledge about its own abilities (machining functions the resource is able to perform (*e.g.* drill)) and the committed activities with specific tasks (*i.e.* the resource's agenda).

The objective of a *resource* holon is to control the physical equipment, provide information about its abilities and status to the system and manage the scheduled activities. Its lifecycle is very long, since it is expected that a resource is fully operational for long periods of time. During its existence, the *resource* holon executes the commands sent by the resource controller and negotiates with task holons the scheduling of manufacturing orders.

During initialisation, the holon builds its initial agenda, registers in the directory service and joins the several holarchies it belongs to (*e.g.* scheduling, process planning). The negotiation process of the holon is guided by the execution of the CNCPP state machine (Figure 7.5) for each conversation currently taking place with task holons. Upon receipt of a service request (for the possible execution of one or more operations), the *resource* holon will check its availability and engage in negotiation with other *resource* holons for propagating constraints between the operations. After calculating its feasible intervals for each request, it will send a bid and wait for the task holon's reply (accept or decline). There is a cost mechanism associated with operations and resources such that a *resource* holon replies to a task holon with a bid specifying the price (in abstract units) of performing that operation.

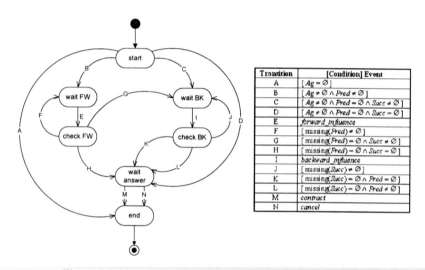

Transition	[Condition] Event
A	$[Ag = \varnothing]$
B	$[Ag \neq \varnothing \wedge Pred \neq \varnothing]$
C	$[Ag \neq \varnothing \wedge Pred = \varnothing \wedge Succ \neq \varnothing]$
D	$[Ag \neq \varnothing \wedge Pred = \varnothing \wedge Succ = \varnothing]$
E	*forward_influence*
F	$[missing(Pred) \neq \varnothing]$
G	$[missing(Pred) = \varnothing \wedge Succ \neq \varnothing]$
H	$[missing(Pred) = \varnothing \wedge Succ = \varnothing]$
I	*backward_influence*
J	$[missing(Succ) \neq \varnothing]$
K	$[missing(Succ) = \varnothing \wedge Pred = \varnothing]$
L	$[missing(Succ) = \varnothing \wedge Pred \neq \varnothing]$
M	*contract*
N	*cancel*

Figure 7.5. State diagram of a *resource* holon for one negotiation

7.4.3 Negotiation Protocol

In order to regulate the interaction between *Fabricare*'s holons, a protocol is used, aiming to achieve cooperation. The key entities participating in the scheduling process (activity chosen as the test case) are the physical resources and the manufacturing orders. These two entities are represented in the system by resource holons and task holons, respectively.

For the scheduling of a task's sub operations, the *task* holon will negotiate with *resource* holons, using an extension of the Contract Net protocol [7.32][7.33] with a cooperation phase between service providers (*i.e.* resource holons). The *resource* holons will use constraint propagation in order to guarantee the relationships among different operations that aim at the same task. This new protocol is called *Contract Net with Constraint Propagation Protocol* (CNCPP) [7.45][7.46].

First, when a new task arrives at the system (via *task launcher*), it will obtain information from the process planning holon about the product's alternative plans and will choose one based on a set of criteria given by the scheduling holon based on the plant current status. The *task* holon will then announce the job by broadcasting each individual operation to every holon able to perform that operation. This information is obtained by querying the *directory service*, thus the set of resources participating in a specific negotiation is built at runtime considering the available resources at that specific moment. This gives the system a certain degree of dynamism in what concerns the system's geometry (number and type of running holons).

After receiving a request, each resource holon will be "forward influenced" by its predecessors and will forward influence its successors. Likewise, each resource holon will be backward influenced before making its bid, and will backward influence its predecessors. The forward- and backward-influence operations make adjustments to the beginning and end of each resource's agenda of free-time intervals in accordance with the agenda of predecessor and successor resources; *i.e.* remove from the list of free time intervals that overlap with busy intervals (for the same order) of other resources [7.51]. The set of resources participating in a specific negotiation is built at the beginning of that negotiation by the task holon. If new resource holons are added to the system they will not be considered for running negotiations but will enter future ones.

Figure 7.6 shows the messages exchanged, during forward influence, for a process plan with 3 operations done in sequence using different resources for each operation and having an alternative resource for the first operation.

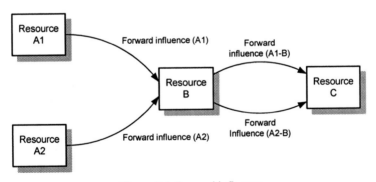

Figure 7.6. Forward influence

The sequence of steps performed by a resource holon is: (i) analyse its agenda and determine free-time windows to execute that operation; (ii) wait for a "forward influence" from predecessor resources and use that information to eliminate time windows and adjust the lower limit of other time windows in order to cope with precedence relationships; (iii) send that new list of time windows to the successor resources; (iv) wait for a "backward influence" from successor resources in order to adjust the upper limit of its time windows; (v) send this final list (the resource's bid) to predecessor resources and present a bid to the task holon (containing the list of free-time intervals and the cost of performing that operation).

Each resource is able to make a final bid (feasible in its own agenda and respecting the constraints) to the task holon. If there are alternative resources, a bid is made for every combination. Figure 7.7 shows the backward-influence and bidding phases of the protocol.

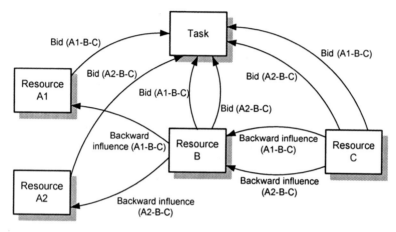

Figure 7.7. Backward influence and bidding

After receiving all the bids from resource holons, the task holon will evaluate them. The evaluation of bids is performed by taking into account a prioritised list of criteria. The following criteria have been implemented: first valid solution; least-costly solution; and greatest slack until due date. The cost of a solution is determined by the cost of performing the specified operations in a specific resource.

Since multiple tasks can be negotiated at the same time, conflicts may arise if some resources are used in the same time interval for different tasks. In these scenarios, resource holons have an indecision problem [7.45] since they cannot guarantee the delivery of both tasks. In order to overcome this problem, a solution is proposed that involves a pre-negotiation step in the protocol (Figure 7.8). Before starting negotiation, each task holon will ask for authorisation from the scheduling holon, which maintains a list of negotiating resources and respective time windows. Only in the case of non-overlapping, will a "green light" be given to the negotiation.

The system is prepared for overlapping functionality on the resource holons, *i.e.* different resource holons can perform the same operation (*e.g.* drill). A task holon will receive the production plan for a product with an indication of the necessary operations, and will request them to every resource holon able to perform each operation. This causes a combinatorial explosion in the number of exchanged messages between holons (with a $O(r^n)$ complexity, where r is the number of resources and n the number of operations). To decrease the number of messages, a modification was made in the protocol, in which combinations from predecessor resources are clustered before sending them to successor resources [7.49], thus changing the complexity of the problem to class $O(n)$, which respects the number of exchanged messages. However, the problem of a combinatorial explosion in the search space still exists (possible solutions for this problem are being addressed).

The same protocol can be used for renegotiating the task in the case of machine breakdowns or other unexpected events. In that case, the task holon will be informed by the resource holon where the event originated, and will begin negotiations for the operation(s) previously contracted to that resource with other resource holons following the same steps previously described. At this time, if rescheduling is not possible, the task will be completely abandoned, and a manual scheduling of that task must be engaged.

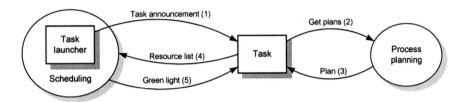

Figure 7.8. Avoiding conflicts in CNCPP

7.4.4 A Prototype

The *Fabricare* scheduling suite (Figure 7.9) is composed of several applications and was developed using SICStus Prolog for the holon's part of the system, and Visual Basic for the user interface [7.49][7.50].

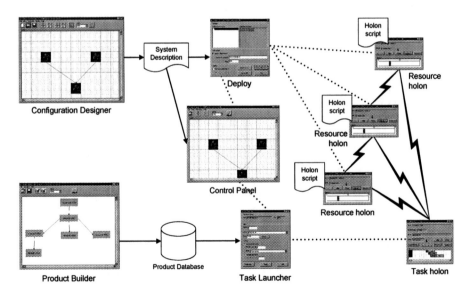

Figure 7.9. Scheduling application suite of *Fabricare*

The *Configuration Designer* allows specification of the resource holons in the factory plant, and to some extent, represents graphically the physical layout of the resources. The system description is read by the *Deployment* tool, which launches *resource holons* on the desired machines. Each resource holon is composed of a kernel and individual characteristics and behaviours specified in the holon script written in Prolog representing the "mental" state of the holon (*e.g.* resource's agenda), as well as specific clauses for the *Fabricare* holon kernel (*e.g.* name).

The *Control Panel* is the interface to the system's operation, monitoring and controlling running holons. This tool also allows the user to launch tasks (manufacturing orders) in the system by invoking the *Task Launcher* tool, which prompts the user for data about the order and dynamically creates a *task holon* for that order. One last tool in the suite is the *Product Builder*, which allows graphical generation of a product's process plan. Each operation in a process plan corresponds to a parameterised invocation of a resources' ability.

Additionally, there are the following support holons:

- *Blackboard Server* that allows the communication between holons via a shared tuple-space (*i.e.* shared memory);
- *Directory Service Holon* maintains an in-memory database of registered holons (their location and capabilities), acting as a yellow pages directory for the system, where each holon can advertise its functions and query for service providers;
- *Process Planning Holon* is responsible for providing information on a product's process plans based on a set of selection criteria passed by the task holon and the current executing environment; the production plans for each product are read from the product database generated with the "Product Designer" tool; and
- *Scheduling Holon* maintains a list of negotiating resources and respective time windows to avoid negotiation conflicts.

SICStus PROLOG's Linda coordination technology [7.52][7.53] is used for the *Blackboard Server*. Exchanged messages between *task holons* and *resource holons* are Prolog clauses written to and read from the Linda tuple-space. Each clause has identifiers indicating the origin and the destination holon, as well as a tag that makes it possible for each party to identify related messages.

The system is very dynamic in what concerns its holons, *i.e.* resource holons depend on the system description file; task holons depend on the existing tasks (dynamic events). Each negotiation uses the set of holons that are present and available at that time, thus giving the system a high degree of adaptability to the dynamic nature of the manufacturing arena (*e.g.* resource in maintenance or overloaded).

The holons are extended logic programs written in Prolog with the ability to handle negative and incomplete knowledge [7.44]. The decision procedure is not yet totally driven by this kind of knowledge; however, real-life scenarios where only partial information is available have been identified [7.44] and modelled (*e.g.* resource holons will use this information to generate low-commitment schedules into their agendas).

The New Prototype System

We are currently migrating the *Fabricare* system to a new platform, in this case the Microsoft® .Net 2.0 Framework and using the Microsoft Message Queuing (MSMQ) component of the windows operating system for asynchronous communications [7.54]. The main rationale for this migration is the use of a mainstream development environment instead of Prolog and Linda. Furthermore, the current version of the system uses synchronous communications that involved some tricks and timer-based pooling of messages, as well as a single centralised message board (tuple-space) for all the messages exchanged in the system.

The new implementation offers a more natural way to develop each holon, since each holon owns its own queue (which may be distributed) and the message handling is now done in an event-driven way, allowing the holon's main thread to execute other operations and be interrupted only when a new message arrives.

The solution (Figure 7.10) is comprised of several projects that have been grouped according to their area of concern: kernel, user interface and management suite.

The kernel of the system has been divided into seven major projects:

- *Agents* – contains the agent/holon's object model and interfaces for all the agents/holons in the system (*i.e.* task, resource, directory service, process planning and scheduling). It also defines the classes used by the notification mechanism for a holon's lifecycle events.
- *AgentsImplementation* – contains the implementation of each service defined in an agent interface. These implementations are independent of the communication mechanism.
- *Messages* – contains the message's data structures.
- *MessageHandling* – contains auxiliary classes for building and translating message contracts to agent contracts and *vice versa*.
- *Communications* – contains generic interfaces for abstracting the communication mechanism and allowing for the independent evolution/ substitution of the agent's implementation and communication mechanism.
- *Communications.Messaging* – an actual implementation of a communication mechanism. In this case using the MSMQ component of the windows operation system.

The current user interface of the system is a simple console application (project `Fabricare.Launcher`) that is capable of executing the service corresponding to each holon role in the system: directory service; process planning; scheduling holon; task holon; and resource holon.

In order to show some output and progress information to the user, a notification mechanism is used in which the application registers an event handler to output to the console.

The development of a new graphical user interface or the adaptation of the existing GUI applications to call the new kernel is still being considered. We think the latter approach will probably be taken since the GUI is not the main focus of our work. The adaptation of the existing code will need some wrapper classes to hide the differences between the Prolog representation and the new .Net representation (for example, the resource's agenda was returned as a string containing a Prolog list of tuples, while the new version returns a collection of `AgendaEntry` objects).

As a goal, it was decided at the outset that the system should be flexible to accommodate different communication mechanisms (or at least allow for an easy substation of the underlying communication mechanism). For the moment, we decided to use the MSMQ component of the windows operating system as it provides an asynchronous and reliable way for transmitting messages. The choice of using messaging instead of remoting was due (1) to the ease of use for asynchronous operation and (2) to the fact that the future Windows Communication Foundation subsystem of the .Net Framework will be based on the messaging paradigm (which makes messaging a more secure bet for future evolutions).

We factored out the functionality for low-level MSMQ interaction and the actual message processing in the `MessageHandling` project. A set of classes called message processors is defined as responsible for the translation of the message contract and agent contract and the routing of the message request to the actual object implementing the agent algorithm. These message processor objects are created by the MSMQ Service objects (*e.g.* `ResourceHolonService`). This allows for a reuse of this message-processing functionality independent of the actual communication mechanism in use.

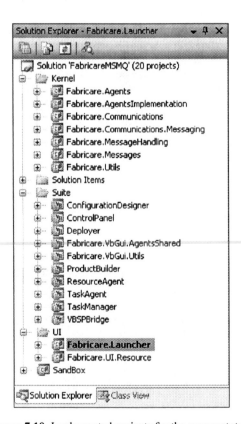

Figure 7.10. Implemented projects for the new prototype

7.4.5 Experiments

This section presents some of the experiments made with *Fabricare* essentially that concerns the scheduling procedure (comparing *Fabricare*'s results with the original results from [7.55]). A more comprehensive set of experiments can be found in [7.56].

Almeida presents a scenario with three tasks (following the plans in Figure 7.11) for the production of two items of each product. The initial conditions of the problem are described in [7.55] and given in Table 7.2.

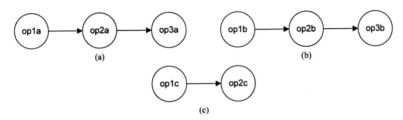

Figure 7.11. Production plans for test scenario #1

Table 7.2. Test conditions for test scenario #1

Resource	Operation	Duration	Initial agenda
M1	op1a	2	
	op3b	1	[(5, 7), (11, 13), (18, 20)]
	op1c	2	
M2	op2a	3	[(8, 10), (15, 18)]
	op1b	2	
M3	op3a	1	
	op2b	1	[(1, 5), (16, 19)]
	op2c	1	

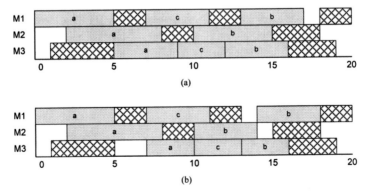

Figure 7.12. Results of test scenario #1

Figure 7.12(a) shows the results obtained by Almeida and Figure 7.12(b) shows the results obtained by *Fabricare* using the "first valid solution" selection criteria. The new scheduling procedure allows for the use of new time intervals since it does not use a static mapping (*behaviours* according to [7.55] and [7.51]), and generates the feasible intervals by considering buffers between the resources.

Another experiment was made considering a task with a sequential production plan ($op1 \rightarrow op2 \rightarrow op3$) for the production of two items. The initial agenda for each resource as well as the cost associated with each operation for the several resources are presented in Table 7.3.

Table 7.3. Test conditions for scenario #2

Resource	Operation	Duration	Cost	Initial agenda
R1	op1	1	100	[(1, 2)]
R2	op2	1	100	[(6, 7)]
R3	op3	3	300	[(7, 10)]
R4	op1	1	90	[(1, 2), (3, 6), (10, 12), (17, 18)]
R5	op2	1	100	[ ]

This experiment considers the existence of alternative resources for the realisation of *op1* (R1 and R4) as well as for the realisation of *op2* (R2 and R5), hence generating four alternative solutions shown in Table 7.4. Furthermore, the cost of executing an operation varies according to the resource it is performed in.

Table 7.4. Results of test scenario #2

Solution	Cost	Resource	Proposed intervals	Partial cost
R1-R5-R3	1000	R1	[(2, 20)]	200
		R5	[(3, 21)]	200
		R3	[(10, 24)]	600
R1-R2-R3	1000	R1	[(2, 20)]	200
		R2	[(7, 21)]	200
		R3	[(10, 24)]	600
R4-R5-R3	980	R4	[(6, 10), (12, 17), (18, 20)]	180
		R5	[(7, 21)]	200
		R3	[(10, 24)]	600
R4-R2-R3	980	R4	[(6, 10), (12, 17), (18, 20)]	180
		R2	[(7, 21)]	200
		R3	[(10, 24)]	600

The selection criterion used was "least-costly solution", which eliminated the first two proposals. The other two proposals have identical costs (and identical schedules since R5 had no initial commitments), thus the task holon used the "first valid solution" criterion as a secondary selection criterion, resulting in the following assignment (including slack): R4/(6, 10), R2/(7, 21) and R3/(10, 24).

For relatively simple production plans (maximum 5 operations) with some alternatives (maximum 8 alternative plans), the performance obtained was quite good since solutions were given (on average) in less than 1.10 seconds. The test bench consisted of a network of Microsoft Windows workstations and the results were measured from the control panel component and include network latency and communications overhead – thus, this is the system response time from the user point of view. These results allow a convergence of the solution considering soft real-time constraints to be expected. Obviously, more complex production plans and several alternative resources may produce an exponential explosion of the solution space and result in the degradation of the system performance.

7.5 Conclusions

In this chapter we addressed the ability to build and maintain computer-supported manufacturing systems able to cope with recent (and expected future) requirements, giving the social-economic context of the new digitised, customised, global society.

As proposed by [7.6] and [7.3], the next generation of manufacturing systems should have characteristics such as distribution, decentralisation, autonomy, dynamism, reactivity, flexibility, adaptability and agility. In a way, the answer to this question can be given by organising the manufacturing system into "small" units structured according to the holonic theory. As presented in Section 7.2, the holonic paradigm does an almost perfect match with the set of desired characteristics of next-generation manufacturing systems.

The chapter also presented the *Fabricare* system that resembles the distributed nature of manufacturing, thus allowing for a natural modelling of the real system. The *Fabricare* system combines resource-based holons with task-based holons that cooperate explicitly via a derivation of the Contract Net Protocol in order to provide scheduling functionality to the system. The system also proposes a company-wide architecture based on the holonic concept.

Challenges and Future Opportunities

Even though there is plenty of work being done on the subject of holonic manufacturing systems, there is still plenty more to do. Some of the challenges and/or future opportunities faced by researchers are (in no particular order):

- *Real-life applications* – unfortunately, there are only a few HMS implementations by companies in real-life scenarios (*e.g.* see Section 5.1 of [7.41]). This would be useful to demonstrate the real added value of the concept but also to allow for improvements in the concept itself. The question of "how to convince the industry" is a challenge that almost every research faces and is not particular to HMS. Another issue is related to how to tackle the implementation with the existing system; although some research has already been done in the integration of HMS with existing FMS (*e.g.* see Section 5.5 of [7.41]), there is still plenty of research needed here, especially considering the business strategy perspective and not the technological one. By tackling the problem from a business strategy perspective would provably provide arguments that would make it easier to convince the industry on adoption of the HMS concept.

- *Effective integration of human operators* – the holonic manufacturing concept promises to "effectively integrate human operators in the manufacturing process" [7.16] but that promise is still to be realised. In part this is due to the fact that no real implementations have been done and the demo implementations and proof-of-concept have mainly focused on automated tasks. This is obviously a difficult issue to handle and as such one can perfectly understand why it has not yet been treated. This will provably require holons with bidirectional person–machine interfaces in order to give information and commands to the operator as well as gather input and commands from the operator. The existence of human-performed operations poses new difficulties for the scheduling algorithms, and the holonic nature of adaptability and local decision (autonomy) can probably provide greater efficiency than centralised approaches to scheduling and control.

- *Effective use of (dynamic) holarchies* – the developed HMS should leverage the distinctive concept of recursive composition of holons not only at the conceptual design level, but also at the implementation level. Not only static holarchies are defined at the design time, but also dynamic holarchies (created, evolved and deleted) according to the current goals of the system.

- *Real plug-and-play capabilities* – to allow third-party interoperability between holons in a system. This is a long-term goal that would allow different vendors to provide holonic cells or machine that would easily plug-and-play to an existing shop floor with minimum configuration work. Obviously, some standards are necessary for such a goal to be realised. An area of interest to enforce such a vision is the area of ontologies and ontology mapping to allow the different holons to communicate with each other.

- *Holonic methodologies* – there is a lack of (semi-) formal methodologies for the design and modelling of holonic systems that can provide best practices and guidelines when building new systems. Currently, each author/research group uses its own process for designing and modelling the system but there is no accepted body of knowledge on how to do it. This is in part due to the fact that HMS is still in its infancy. However, as critical mass and knowledge is accumulated there is an opportunity (and need) to build a set of core principles and processes to help and guide the development of new systems.

- *Holonic frameworks* – partially related to the previous bullet, there is also a lack of holonic frameworks that would help the development of new systems (from a technical point of view). These frameworks could perhaps be built on top of some agent frameworks since there is a lot of common ground between holons and agents. The main characteristics that such a framework would provide would be for the dynamic composition of holons and related enforcement of holarchies as well as integration of software and hardware. Such frameworks would be even more powerful as standards came along (*e.g.* IEC 61499). Additionally, there is a need for core services and tools such as directories, monitors, blackboards, integrated development environments and debuggers.

- *Application in different industries* – apply the concept (at least in the form of proof-of-concept) not only to different policies (job shop, flow shop) but also to different industries (*e.g.* chemical, pharmaceutical, electronics) to see if there are specific characteristics that prove the concept not viable. Such a study would also provide great input to the holonic methodologies (namely particular methodologies according to industry type).

- *Full company-wide implementation* – due to limited resources, most research teams focus their work on a specific subject area (*e.g.* control), but the power of the holonic concept comes (at least in part) from the holistic view it provides. As such, the application of the concept to the whole company (covering not only production but also other functions) could be valuable. It would even provide an enhanced ability to support new forms of business organisation (*e.g.* virtual enterprises) by applying the same principles of cooperation and composition that are applied at lower levels (with the necessary adaptations).

- *Incomplete information* – due to the distributed nature of an HMS and the nature of the manufacturing business itself, it is only natural that sometimes, some of the information that is necessary is not available. Better mechanisms and methodologies need to be developed to handle incomplete information scenarios (especially in the mid/long-term planning phase of manufacturing).

Some of the challenges presented above are not particular to HMS, but general to many research areas; as one can see there is still much to be done.

Acknowledgment

The author would like to acknowledge FCT, FEDER, POCTI, POSI, POCI and POSC for their support to R&D Projects and GECAD R&D Group.

References

[7.1]   Kelly, K., 1999, *New Rules for the New Economy: 10 Radical Strategies for a Connected World*, Penguin, USA.
[7.2]   Kidd, P., 1994, *Agile Manufacturing, Forging New Frontiers*, Addison-Wesley.
[7.3]   NGM, 1997, *Next Generation Manufacturing – A Framework for Action*, Next Generation Project Report, Agility Forum.
[7.4]   Schonfeld, E., 1998, "The customised, digitised, have-it-your-way economy," *Fortune*, **138**(6), pp. 114–124.
[7.5]   Atkinson, R. and Court, R., 1998, *The New Economy Index: Understanding America's Economic Transformation*, Technology, Innovation, and New Economy Project; Progressive Policy Institute, November 1998.
[7.6]   National Research Council, 1998, *Visionary Manufacturing Challenges for 2020*, National Academic Press, Washington, D.C.
[7.7]   Hunt, V., 1989, *Computer-Integrated Manufacturing Handbook*, Chapman & Hall.
[7.8]   Kusiak, A., 1990, *Intelligent Manufacturing Systems*, Prentice-Hall, Inc.
[7.9]   Solberg, J. and Kashyap, R., 1993, "ERC research in intelligent manufacturing systems," In *Proceedings of the IEEE*, **81**(1), pp. 25–41.

[7.10] Valckenaers, P., Bonneville, F., Van Brussel, H., Bongaerts, L. and Wyns, J., 1994, "Results of the Holonic control system benchmark at K.U. Leuven," In *Proceedings of International Conference on Computer Integrated Manufacturing and Automation Technology*, pp. 128–133.

[7.11] Sousa, P., Silva, N., Heikkila, T., Kallingbaum, M. and Valcknaers, P., 2000, "Aspects of co-operation in distributed manufacturing systems," *Studies in Informatics and Control Journal*, 9(2), pp. 89–110.

[7.12] Tharumarajah, A., Wells, A. and Nemes, L., 1996, "Comparison of the Bionic, Fractal and Holonic manufacturing concepts," *International Journal of Computer Integrated Manufacturing*, 9(3), pp. 217–226.

[7.13] Monostori, L., 1997, "Editorial," In *Proceedings of the Second World Congress on Intelligent Manufacturing Processes & Systems* (IMP&S'97), Monostori, L. (ed.), pp. vi–vii.

[7.14] Hatvani, J. and Nemes, L., 1978, "Intelligent manufacturing systems – A tentative forecast," In *Proceedings of the VIIth IFAC World Congress*, Niemi, A. (ed.), **2**, pp. 895–899.

[7.15] Hatvani, J., 1983, "The efficient use of deficient knowledge," *Annals of the CIRP*, **32**(1), pp. 423–425.

[7.16] Valckenaers, P., Van Brussel, H., Bongaerts, L. and Wyns, J., 1997, "Holonic manufacturing systems," *Journal of Integrated Computer Aided Engineer*, **4**(3), pp. 191–201.

[7.17] Koestler, A., 1967, *The Ghost in the Machine*. Hutchinson & Co, London, UK.

[7.18] Simon, H., 1962, "The architecture of complexity," In *Proceedings of the American Philosophy Society*, **106**(6), pp. 467–482.

[7.19] Simon, H., 1969, *The Sciences of the 'Artificial*, MIT Press, Cambridge, MA.

[7.20] Höpf, M., 1994, "Holonic manufacturing systems – The basic concept and a report of IMS Test Case 5," In *Sharing CIM Solutions*, Knudesen, J. *et al.* (eds.), IOS Press.

[7.21] Langer, G., 1999, *HoMuCS: A Methodology and Architecture for Holonic Multi-cell Control Systems*, PhD Thesis, Technical University of Denmark, Department of Manufacturing Engineering.

[7.22] Van Brussel, H., Wyns, J., Valckenaers, P., Bongaerts, L. and Peeters, P., 1998, "Reference architecture for Holonic manufacturing systems: PROSA," *Computers in Industry – Special Issue on Intelligent Manufacturing Systems*, **31**(3), pp. 255–276.

[7.23] Wyns, J., 1999, *Reference Architecture for Holonic Manufacturing Systems – the Key to Evolution and Reconfiguration*, PhD Thesis, Production and Automation Division, Katholieke Universiteit Leuven, Leuven, Belgium.

[7.24] Hayashi, H., 1993, "The IMS international collaborative program," In *Proceedings of the 24th International Symposium on Industrial Robots*, Japan Industrial Robot Association.

[7.25] Yoshikawa, H., 1993, "Intelligent manufacturing systems: technical cooperation that transcends cultural differences," *Information Infrastructure Systems for Manufacturing, IFIP Transactions*, **B-14**, pp. 19–40.

[7.26] ISC, 1994, *IMS: Intelligent Manufacturing Systems – A Program for International Cooperation in Advanced Manufacturing*, International Steering Committee Final Report (adopted at ISC6, Hawaii, USA, 24–26 January 1994; available at http://ksi.cpsc.ucalgary.ca/IMS/ISC/ISC94_Contents.html).

[7.27] Barram, D., 1994, "Keynote address," *North American Symposium – Preparing for the Full-Scale IMS Program*, June 23, 1994, Dallas, Texas, USA. (available at http://ksi.cpsc.ucalgary.ca/IMS/Dallas/IMS_Barram.html)

[7.28] Bongaerts, L., 1998, *Integration of Scheduling and Control in Holonic Manufacturing Systems*, PhD Thesis, Production and Automation Division, Katholieke Universiteit Leuven, Leuven, Belgium.

[7.29] van Leeuwen, E. and Norrie, D., 1997, "Holons and holarchies," *IEE Manufacturing Engineer*, pp. 86–88.

[7.30] Parunak, H., 1998, "What can agents do in industry and why?" In *Proceedings of the Second International Conference on Co-operative Information Agents* (CIA'98), Paris, France, 3–8 July 1998, pp. 1–18.

[7.31] Kouiss, K., Pierreval, H. and Mebarki, N., 1997, "Using multi-agent architecture in FMS for dynamic scheduling," *Journal of Intelligent Manufacturing*, **8**, pp. 41–47.

[7.32] Davis, R. and Smith, R., 1983, "Negotiation as a metaphor for distributed problem solving," *Artificial Intelligence*, **20**(1), pp. 63–109.

[7.33] Smith, R., 1980, "The contract net protocol," *IEEE Transactions on Computers*, C-**29**(12), pp. 1104–1113.

[7.34] Bongaerts, L., Wyns, J., Detand, J., van Brussel, H. and Valckenaers, P., 1996, "Identification of manufacturing holons," In *Proceedings of the 1st European Workshop on Agent Oriented Systems in Manufacturing*, Albayrak, S. and Bussmann, S. (eds.), pp. 57–73.

[7.35] Gou, L., Luh, P. and Kyoya, Y. 1998, "Holonic manufacturing scheduling: architecture, cooperation mechanism, and implementation," *Computers in Industry*, **37**, pp. 213–231.

[7.36] Leitao, P. and Barbosa, J., 2006, "Disturbance detection, recover and prediction in holonic manufacturing control," In *Proceedings of IEEE Workshop on Distributed Intelligent Systems: Collective Intelligence and Its Applications* (DIS'06), pp. 133–138.

[7.37] Walker, S., Brennan, R. and Norrie, D., 2005, "Holonic job shop scheduling using a multi-agent system", *IEEE Intelligent Systems*, **20**(1), pp. 50–57.

[7.38] Kotak, D., Wu, S., Fleetwood, M. and Tamoto, H., 2003, "Agent-based holonic design and operations environment for distributed manufacturing," *Computers in Industry*, **52**(2), pp. 95–108.

[7.39] Jarvis, J., Jarvis, D. and McFarlane, D., 2003, "Achieving holonic control: an incremental approach," *Computers in Industry*, **51**(2), pp. 211–223.

[7.40] Huang, B., Gou, H., Liu, W., Li, Y. and Xie, M., 2002, "A framework for virtual enterprise control with the holonic manufacturing paradigm," *Computers in Industry*, **49**(3), pp. 299–310.

[7.41] Babiceanu, R. and Chen, F., 2006, "Development and applications of holonic manufacturing systems: a survey," *Journal of Intelligent Manufacturing*, **17**, pp. 111–131.

[7.42] Shen, W. and Norrie, D., 1999, "Agent-based systems for intelligent manufacturing: a state-of-the-art survey," *Knowledge and Information Systems*, **1**(2), pp. 129–156.

[7.43] Shen, W., Wang, L. and Hao, Q., 2006, "Agent-based distributed manufacturing process planning and scheduling: a state-of-the-art survey," *IEEE Transactions on Systems, Man and Cybernetics*, **36**(4), pp. 563–577.

[7.44] Sousa, P., Ramos, C. and Neves, J., 2000, "Manufacturing entities with incomplete information," *Studies in Informatics and Control Journal*, **9**(2), pp. 79–88.

[7.45] Sousa, P. and Ramos, C., 1998, "A dynamic scheduling holon for manufacturing orders," *Journal of Intelligent Manufacturing – Special Issue on Agent based Manufacturing*, **9**(2), pp. 107–112.

[7.46] Sousa, P. and Ramos, C., 1999, "A distributed architecture and negotiation protocol for scheduling in manufacturing systems," *Computers in Industry – Special Issue on Life Cycle Approaches to Production Systems: Management, Control and Supervision*, **38**(2), pp. 103–113.

[7.47] Neves, J., Machado, J., Analide, C., Novais, P. and Abelha, A., 1997, "Extended logic programming applied to the specification of multi-agent systems and their computing environment," In *Proceedings of the 1997 IEEE International Conference on*

Intelligent Processing Systems, Beijing, China, pp. 159–164.

[7.48] Traylor, B. and Gelfond, M., 1993, "Representing null values in logic programming," In *Proceedings of the International Logic Symposium* (ILPS'93), Vancouver, Canada, pp. 341–352.

[7.49] Sousa, P., Ramos, C. and Neves, J., 2003, "The *Fabricare* scheduling prototype suite: agent interaction and knowledge base," *Journal of Intelligent Manufacturing*, **14**(5), pp. 441–455.

[7.50] Sousa, P., Ramos, C. and Neves, J., 2004, "The *Fabricare* system," *Production Planning & Control*, **15**(2), pp. 156–165.

[7.51] Ramos, C., Almeida, A. and Vale, Z., 1995, "Scheduling manufacturing tasks considering due dates: a new method based on behaviours and agendas," In *Proceedings of the International Conference on Industrial and Engineering Applications of Artificial Intelligence and Expert Systems*, pp.745–751.

[7.52] Carriero, N. and Gelernter, D., 1989, "Linda in context," *Communications of the ACM*, **32**(4), pp. 444–458.

[7.53] Carriero, N. and Gelernter, D., 1989, "How to write parallel programs: a guide to the perplexed," *ACM Computing Surveys*, **21**(3), pp. 323–357.

[7.54] Sousa, P., 2006, "Using asynchronous messaging for agent-based development," *WSEAS Transactions on Computers*, **5**(6), pp. 1285–1292.

[7.55] Almeida, A., 1995, *Escalonamento Dinâmico de Tarefas Industriais Sujeitas a Prazos de Entrega*, Dissertação de Mestrado, Faculdade de Engenharia da Universidade do Porto, Setembro de 1995. (MSc Thesis in Portuguese)

[7.56] Sousa, P., 2000, *Agentes Inteligentes em Sistemas Holónicos de Produção*, Tese de Doutoramento, Universidade do Minho, Escola de Engenharia, Dezembro, 2000. (PhD Thesis in Portuguese)

8

Agent-based Dynamic Scheduling for Distributed Manufacturing

Weiming Shen and Qi Hao

Integrated Manufacturing Technologies Institute
National Research Council of Canada, London, Ontario, Canada
Emails: weiming.shen@nrc.gc.ca, qi.hao@nrc.gc.ca

Abstract

Manufacturing enterprises are facing great challenges to improve their production efficiency and profitability in order to survive in a globally competitive market. Agent-based manufacturing scheduling technology provides a promising way to address these challenges. This technology can well address both dynamic changes and unpredictable disturbances in the manufacturing shop floor locally without disruptions to regular production – a problem that cannot be solved by traditional centralised planning and scheduling systems. It can significantly improve equipment-utilisation rates, thereby improving the efficiency and productivity of manufacturing enterprises. This chapter provides a brief research literature review on agent-based distributed manufacturing scheduling, presents some of our recent R&D results in this area, and discusses key issues in deploying this technology in industry.

8.1 Introduction

Manufacturing enterprises are facing great challenges as their ways of doing business evolve rapidly in response to emerging technologies and growing competitive pressures. In order to survive, manufacturing enterprises must find innovative solutions to improve their flexibility and agility in response to unpredictable changes or disturbances, while maintaining their performance measured by cost, quality, and productivity.

In the MASCADA report, production disturbances are distinguished from production changes [8.1]. A production change is defined as an alteration to the production conditions, which is intentionally performed by the plant; while a production disturbance is an unanticipated change to production conditions with a negative effect on the (manufacturing) process performance. The important characteristic that distinguishes changes from disturbances is pinpointed on the question whether the production facility intends the change. However, a disturbance in one area of production can cause a change in another area, and *vice versa*. Both changes and disturbances may occur due to internal or external sources. Therefore, in this context, we do not distinguish between these two concepts, and consider they have the same metaphoric meaning.

The dynamism of shop-floor activities is mainly featured by their unpredictable occurrences and short time spans. According to our survey of Canadian manufacturing industry, the manufacturing equipment-utilisation rate is very low in many manufacturing enterprises, which is primarily due to the lack of software tools to address dynamic changes and disturbances from both internal and external manufacturing environments. In fact, even for some manufacturing systems operating "at near optimal productivity under normal conditions, they fail to sustain that performance under process disturbances of any kind" [8.2]. Therefore, the companies that are likely to survive would be those that respond better to the unpredictable and volatile global environments. It is imperative that manufacturers not only cope with frequent product changes and fluctuating demands, but more importantly, reduce the impact of disturbances on their regular manufacturing practices. Agent-based dynamic manufacturing scheduling technology provides a promising way to address these challenges. An agent is a software system that communicates and cooperates with other software systems to solve a complex problem that is beyond the capability of each individual software system [8.3]. Unlike traditional manufacturing scheduling systems in which a centralised scheduler is used, an agent-based scheduling system supports distributed dynamic scheduling in a way that intelligent agents are used to represent manufacturing resources (such as machines, operators, robots). As a result, each agent is able to make local decisions and collaborate with others on behalf of its representative physical resource.

In the following sections of the chapter, we will discuss the complexity of manufacturing scheduling problems, provide a brief literature review, introduce the *iShopFloor* framework, present the agent-based dynamic manufacturing technology, describe two implemented prototypes, and discuss key issues in deploying the technology in industry as well as future R&D directions.

8.2 Complexity of Manufacturing Scheduling Problem

Scheduling problems are not exclusive to manufacturing environments. Similar situations happen routinely in publishing houses, universities, hospitals, airports, transportation companies, *etc.* Scheduling problems are typically NP-hard, *i.e.* it is impossible to find an optimal solution without the use of an essentially enumerative algorithm and the computation time increases exponentially with problem size. Manufacturing scheduling is one of the most difficult scheduling problems [8.4][8.5].

A well-known manufacturing scheduling problem is the classical job-shop scheduling where a set of jobs and a set of machines are given. Each machine can handle at most one job at a time. Each job consists of a chain of operations, each of which needs to be processed during an uninterrupted time period of given length on a given machine. The purpose is to find a best schedule, *i.e.* an allocation of the operations to time intervals on the machines, which has the minimum duration required to complete all jobs. The total number of possible solutions for a classical job-shop scheduling problem with n jobs and m machines is $(n!)^m$ [8.5].

The problem becomes even more complex in the following situations:

- When other manufacturing resources such as operators and tools are also taken into account during the scheduling process. In a real manufacturing enterprise, production managers or human schedulers need to consider all kinds of manufacturing resources in the enterprise/shop floors. For a classical job-shop scheduling problem with n jobs, m machines and k operators, the total number of possible solutions could be $\left(\left(n!\right)^m\right)^k$ [8.6].

- When both manufacturing process planning and scheduling are to be done at the same time. Traditional approaches separating process planning and scheduling can obtain sub optimal solutions at two separate phases. The global optimisation of a manufacturing system is only possible when process planning and scheduling are integrated.

- When unforeseen dynamic situations are considered. In a job-shop manufacturing environment, rarely do things go as expected. Scheduled jobs may be cancelled and new jobs may be inserted. Certain resources can become unavailable and additional resources can be introduced. A scheduled task can take more or less time than anticipated, and tasks can arrive early or late. Other uncertainties include power-system failures, machine failures, operator absence, unavailability of tools and materials. An optimal schedule, generated after considerable effort, may become unacceptable because of unforeseen dynamic situations in the shop floor; therefore, a new schedule has to be generated to restore performance. This kind of rescheduling problem is also called dynamic scheduling or real-time scheduling.

8.3 Literature Review

Manufacturing scheduling is an optimisation process that allocates limited manufacturing resources over time among parallel and sequential manufacturing activities. This allocation must obey a set of rules or constraints that reflect the temporal relationships between manufacturing activities and capacity limitations of a set of shared resources. The allocation also affects a schedule's optimality with respect to criteria such as cost, lateness, or throughput [8.6].

Centralised/hierarchical job-shop schedulers tend to be complex and hardly applicable in real time, dynamically changing circumstances. On the contrary, agents were welcome in manufacturing because they helped realise important properties such as autonomy, responsiveness, adaptiveness, openness, distributed control, distributed decision making, self-organisation, and cooperation. Hence, many tasks related to manufacturing, from engineering design, process planning and scheduling, shop-floor control, to enterprise integration and collaboration, can be implemented as multi-agent systems by utilising agent-based technologies. Within the past two decades, researchers have applied agent-based approaches in attempts to resolve the manufacturing planning and scheduling problems [8.7]. In fact, this area represents one of the most active research topics in agent-based manufacturing.

We have done an extensive literature review on applications of software agents to manufacturing in general [8.8] and to manufacturing process planning and scheduling in particular [8.7]. Here, we will only provide a brief overview of the state-of-the-art of agent-based manufacturing scheduling, particularly of those developments and deployments in industrial settings.

Shaw may be the first person who proposed using agents in manufacturing scheduling and factory control. He suggested that a manufacturing cell can subcontract work to other cells through a bidding mechanism [8.9][8.10]. YAMS (yet another manufacturing system) [8.11] is another example of an early agent-based manufacturing system, wherein each factory and factory component is represented as an agent. Each individual agent has a collection of plans as well as knowledge about its own capabilities. The Contact Net protocol [8.12] is used for inter-agent negotiation. Below are some representatives among significant R&D efforts reported in the literature.

In AARIA [8.13], different kinds of manufacturing capabilities (*e.g.* people, machines, and parts) are encapsulated as autonomous agents. Each agent seamlessly inter-operates with other agents in and outside the factory boundary. AARIA used a mixture of heuristic scheduling techniques: forward/backward scheduling, simulation scheduling, and intelligent scheduling. Scheduling is performed by job, by resource, and by operation.

Lu and Yih [8.14] proposed a framework that utilises autonomous agents and weighted functions for distributed decision making in elevator manufacturing and assembly. This system dynamically adjusts the priorities of sub-assemblies in the queue buffer of a cell by considering the real-time status of all sub-assemblies in the same order.

Yen and Wu [8.15] proposed an agent-based market-driven scheduling architecture to integrate all kinds of existing manufacturing scheduling systems over the Internet. Each standalone scheduling system is endowed with agent features by migrating a legacy system into an Internet-based scheduling agent. A number of heterogeneous scheduling agents collaborate or compete with each other for scheduling tasks using a market-driven protocol – Vickrey Auction.

Other good examples of agent-based manufacturing shop-floor scheduling can be found in [8.16]–[8.18]. Recent interesting research work includes market-based negotiation protocols [8.19]–[8.21], agent-based integration of manufacturing process planning and scheduling [8.7], combination of agent-based approaches with traditional scheduling techniques such as heuristic search methods, performance matrix, Petri nets, neural networks, genetic algorithms, and simulated annealing [8.22][8.23].

Similar research efforts are also reported in the HMS (holonic manufacturing systems) research community [8.24][8.25]. For example, Sousa and Ramos [8.26] proposed a dynamic scheduling system architecture that consists of task holons and resource holons. The Contract Net protocol is also adopted to handle temporal constraints and deal with conflicts. Sousa *et al.* [8.27] further proposed an extended Contract Net protocol with constraints propagation for explicit representation of the precedence relationships between operations of a task (with a cooperation phase between service providers). This extension has some novelty compared with other variants of the Contract Net protocol.

Despite the fact that significant research efforts have been made in this area during the past two decades, only a few testbeds and industrial applications have been developed and reported in the literature. LMS [8.28] has been used in commercial production at IBM. The manufacturing scheduling and control testbed

(MASCOT) [8.29] was a simulated testbed for manufacturing scheduling and control. The shop floor agents (SFA) project at NCMS [8.30] was focused on application of agent-based systems for shop-floor scheduling and machine control. The project members include three manufacturers (AMP, GM, and Rockwell Automation/Allen-Bradley) and a software company (Gensym). The objective was to develop agent-based systems to support the three industrial scenarios, one sponsored by each of the manufacturing members of the project [8.31]. The exploitation of agent-based technology in production planning was addressed in the ExPlanTech project [8.32]. Five levels of production planning were distinguished in a reconfigurable enterprise (RE), in which decentralised decisions where made by planning agents at different planning horizons, while global planning decisions are achieved by means of coordination and negotiation among them. The negotiation model was compared with a centralised solution, showing the benefits of the agent-based approach [8.33]. It provides a communication infrastructure, a shared ontology, a shared interface based on that ontology, and a base set of realistic modules for the design, integration and operation of agile enterprises. Other industrial applications include: a multi-agent software system RIDER (real-time decision making in manufacturing) for a cable producing company and for a carpet manufacturer [8.34]; an agent-based dynamic scheduling in steel production [8.22]; and a collaborative control system for mass-customisation manufacturing [8.35]. As pointed out by Monostori *et al.* [8.36], the elements of agent-based system architectures are distributed, usually have no access to global information. Therefore, optima cannot be guaranteed by such systems. We believe that agent technology is particularly suitable for dynamic scheduling (or schedule repair) rather than static global optimisation problems. This technology can improve responsive behaviours of a manufacturing system under unpredictable changes. Moreover, in an agent-based system, since schedules are determined through the negotiation process participated in only by a number of involved agents, the impact of a change, if any, on the whole production process is minimal.

8.4 *iShopFloor* Framework

Our agent-based dynamic manufacturing scheduling technology is part of the intelligent shop floor (*iShopFloor*) framework we proposed and developed during the past six years [8.37] and has recently been further developed according to the FIPA specifications [8.38].

Figure 8.1 shows the *iShopFloor* concept. *iShopFloor* stands for *intelligent shop floor* and focuses on the application of distributed artificial intelligence (particularly intelligent agents) to the shop floor for distributed intelligent manufacturing process planning, scheduling, sensing and control. A detailed description of the *iShopFloor* concept is presented in [8.37].

iShopFloor focuses on the development of a real-time distributed decision-making system in the shop floor through implementing distributed intelligence of software agents. An "intelligent" shop floor is defined as being:

- able to accept high-level input (*e.g.* STEP files) for production planning and scheduling;

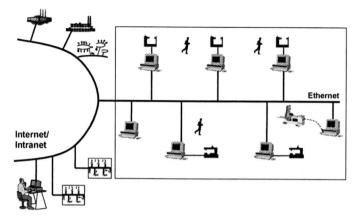

Figure 8.1. *iShopFloor* concept

- able to outsource in case of overloaded or system failures;
- ready for Web-based remote monitoring, control, diagnosis and maintenance;
- composed of "intelligent" manufacturing devices/resources (*e.g.* intelligent machines, smart sensors, robots, and AGVs).

Whereas these "intelligent" devices/machines are able to:

- "plug and play";
- make decisions based on local knowledge and in collaboration with other devices;
- accept high-level commands (*e.g.* STEP files) for bidding and control execution;
- negotiate with other devices in case of system failures;
- proactively send important data to related servers or other devices.

The *iShopFloor* framework is intended for implementing Internet-based distributed manufacturing control systems in different kinds of situations, particularly in lower-volume production of complex products/parts. Such systems work in a dynamic world and are able to keep pace with changes in real time. *iShopFloor* also provides an open architecture that is extendable both to enterprise integration and supply-chain management as well as to lower-level intelligent machine control. On the one hand, an intelligent shop floor is a component of a networked enterprise, which is in turn a component of a virtual enterprise or a supply chain. An individual shop floor is connected with other shop floors within the same enterprise as well as other enterprise business systems through a local network, an Intranet or the secured Internet. On the other hand, a shop floor is further composed of networked manufacturing resources/devices including intelligent machines (through embedded open-architecture controllers or connected computers), transportation vehicles (through wired or wireless communication devices), and operators (through graphical user interfaces).

In terms of agents, *iShopFloor* framework is composed of three types of agents: resource agents, product/part agents, and service agents [8.39]. Resource agents are used to represent (model or wrap) manufacturing devices on the shop floors and are the primary class of agents in the *iShopFloor* environment. Typically, resource agents are intelligent machine agents, but they can also be used to represent cells, workstations, robots, human operators (interfaces for human operators), AGVs (automatically guided vehicles), conveyers, and sensors, *etc.* The resource agents are usually connected through an Ethernet. In some cases, wireless connections may also be needed. While the detailed internal architecture of a resource agent is beyond the scope of this chapter, we provide a brief description here. A typical intelligent machine agent includes the following modules:

- a network interface for connecting the physical machine to the network;
- a communication interface for processing incoming and outgoing messages;
- an initiation module for agent initiation so that a machine agent can "plug and play";
- an interface to lower-level path planning and motion control for machining time estimation, and to download commands for execution in the real-time mode;
- a local planning module with related knowledge bases, databases, and reasoning mechanisms for preparing "intelligent" bids and negotiating with other agents;
- a learning module for updating the agent's knowledge about its environment, other agents, ongoing projects and products.

Product/part agents are used to represent products/parts (or assemblies). Here, we introduce only part agents. Part agents are stationary (static) agents during the manufacturing process planning and scheduling stage, while becoming mobile when the production starts. A "mobile" part agent will move to the computer connected to the machine working on the corresponding part. This approach is proposed for two reasons. First, it aims to reduce the communication overhead on the shop-floor network. Normally, there are a lot of data exchanges between the part agent and the machine agent during the manufacturing process. Moving the part agent to the computer with the machine agent reduces the communication overhead significantly. Second, by keeping the part agent on the same computer as the machine agent, communications may be kept private. When a manufacturing task finishes at one machine, the part agent will move on to the next computer connected to the machine working on the next manufacturing task. When the part is completed and the part agent moves out from the shop floor, all important data/information related to the production of this part, including machine IDs/names, operator IDs/names if applicable, processing time, finish tolerance, and failures that occur, will be recorded in the final product database for future use.

Service agents are adopted to facilitate the communication, collaboration and coordination among resource agents and product/part agents. Different kinds of service agents have been proposed and developed in the literature. In *iShopFloor*, service agents include *directory facilitators* (adapted from FIPA [8.38]), *scheduling mediators/coordinators*, and *ontology servers*.

iShopFloor has a highly distributed generic system architecture [8.39]. From this point of view, *iShopFloor* architecture is very similar to the heterarchical control architecture proposed by Duffie and Prabhu [8.40]. However, Duffie *et al.*'s approach focuses more on the arrival time control with a primary objective to produce parts as per the prescribed schedule. The primary goal of *iShopFloor* is to implement Internet-enabled intelligent shop floors to rapidly respond to changing manufacturing environments and customer demands so as to increase the enterprise's productivity and profitability.

All *iShopFloor* agents (including resource agents, product/part agents, and service agents) are connected to the network independently. All the service agents, for example, *directory facilitators* and *scheduling mediators/coordinators*, though providing certain kinds of services/functionalities, do not have any control over resource agents.

Since an agent-based dynamic scheduling mechanism is generally realised through agent negotiations, *iShopFloor* provides multiple negotiation protocols in order to adapt to different shop-floor situations. Details of agent-based dynamic manufacturing scheduling technology will be described in Section 8.5.

XML has been chosen as an important enabling technology for the design and deployment of the *iShopFloor* concept [8.41]. XML is used in *iShopFloor* in three ways: (1) communication messages, (2) persistent object storage, (3) initialisation and configuration files. Resource agents communicate with service agents by serialising messages into XML form and transporting them across TCP/IP sockets. Messages are then parsed into XML documents, verified to be well formed, and then mapped into appropriate Java objects. All agents are multithreaded to support multiple requests at a time.

8.5 Agent-based Dynamic Manufacturing Scheduling

The *iShopFloor* framework provides an open architecture for easy extension to enterprise integration and supply chain as well as to lower-level intelligent machine control. As described in [8.37], the proposed agent-based manufacturing scheduling approach is intended for implementing distributed manufacturing planning, scheduling and control systems in different kinds of shop-floor situations, with particular attention paid to enhancing the system's responsiveness and adaptiveness to the shop-floor dynamics.

Using the physical decomposition approach [8.8], a number of agents are used to represent entities in the physical world, such as operators, machines, tools, fixtures, products, parts, features, and operations. There exists an explicit relationship between an agent and a physical entity. The intelligence of shop floors is hence presented by the intelligence of responsive agents on the Internet. They can participate in different levels of decision-making processes on behalf of their physical devices, whether it is at the shop-floor control level, the enterprise management level, or enterprise collaboration level. An agent-based manufacturing scheduling system supports a distributed dynamic scheduling mechanism in a way that each agent is able to locally handle the schedule of its corresponding physical party. Global decisions of scheduling are coordinated through autonomous and spontaneous decisions made by individual agents. Since an enterprise has other business processes apart from shop-floor level scheduling, service agents at different decision levels are responsible for agent coordination. In Figure 8.1, at the shop-floor level, there is a *scheduling coordinator* (a kind of service agent) for coordinating the shop-floor manufacturing scheduling processes. In a broader view, a shop-floor coordinator connects with other shop floors through a common network. Such an architecture may also be extended across an enterprise's boundaries to reach its suppliers and customers.

As long as the three kinds of agents are modelled (namely, resource agents, product/part agents, and services agents), negotiation-based scheduling mechanisms can be implemented in different ways to meet different shop-floor situations/scenarios. Such negotiation protocols being implemented include Contract Net [8.12] and modified Contract Net protocols, auction-based protocols, game theory-based protocols, and optimised market-based protocols [8.6]. We proposed an adaptive negotiation approach for grid computing that is applicable to other distributed applications [8.42].

A simple scenario is shown in Figure 8.2, where there are only a small number of agents including part agents and resource agents (machine agents and operator agents). All part agents and resource agents are registered with the directory facilitator. No scheduling mediator/coordinator is required on this occasion. Once a production order comes in, a part agent is generated to take care of the order. The part agent first needs to decompose the production order into manufacturing operations (tasks). After finding a list of capable machine agents through the directory facilitator, it will then negotiate with these machine agents and allocate operations/tasks to the winning machines directly. When an operator is required for performing a manufacturing operation/task, the machine agent will find the operators with the required skill sets and negotiate with the operator agents to finalise their commitments similarly through direct communication.

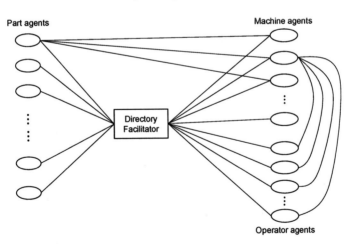

Figure 8.2. A simple scenario of dynamic scheduling using agent-based negotiation

If there are a large number of products/parts (jobs) and a large number of manufacturing resources (represented by resource agents), scheduling mediators/coordinators may be required (Figure 8.3) and other global scheduling approaches may be integrated as well. As shown in Figure 8.3, we have integrated genetic algorithms (GA) with agent-based negotiation for complex scheduling scenarios, particularly advance scheduling, *i.e.* using genetic algorithms to generate an optimal sequence of operations (or production orders), while using agent-based negotiation to allocate the generated operations in the sequence. The rescheduling processes are also occurring with considerations of one operation at a time. From this aspect, agent-based negotiation is only used for real-time manufacturing scheduling in order to reduce the response time. That is why we call our agent-based scheduling technology the "dynamic scheduling" approach, which is a field that agent technology mostly fits into – finding locally feasible solutions in a short time rather than doing global optimisation.

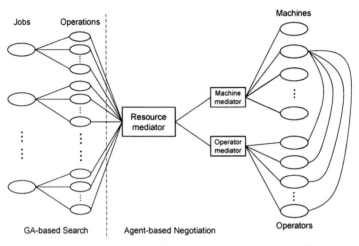

Figure 8.3. A complex scenario of scheduling using both a genetic algorithm and agent-based negotiation

The approach in Figure 8.3 is different from the traditional GA-based search methods for scheduling optimisation in that agent-based negotiation approaches are used to schedule each individual task among manufacturing resources on shop floors so the search space is significantly reduced. For example, in an extreme situation of a classical job-shop scheduling problem with n jobs, m machines and k workers, the total number of possible solutions could be $((n!)^m)^k$ as mentioned above. By using this approach, the search space for genetic algorithms is reduced to $(n \times m)!/(m!)^n$.

It should be noted that the GA-based operation sequencing and agent-based negotiation are not two separate processes, they are totally integrated. During the GA-based operation sequencing, each possible solution of an operation sequence (or each feasible sequence of operations) is evaluated by the agent-based negotiation method, *i.e.* an optimal schedule in this possible operation sequence situation is generated using an agent-based approach. A lowest cost is obtained using a cost function by taking into account the selected machines, workers, penalties for order delays, *etc.*, which serves as the fitness for this possible operation sequence solution. Similarly, in the situation of real-time dynamic scheduling or rescheduling, an agent-based negotiation approach can also be used to compare and evaluate different possible solutions, particularly when all manufacturing resource on the shop floor are almost at full capacity and most manufacturing tasks need to be rescheduled. The GA-based search engine can also be "turned off" in real-time scheduling situations.

An assumption in the above scenario is that the process planning is done before the manufacturing scheduling process, and that the number and the sequence of each job's operations are defined during the process planning phase. However, in real shop-floor situations, it is possible to change the number and sequence of each job's operations according to the capability and availability of the shop-floor resources. This is related to the integration of manufacturing process planning and scheduling [8.8].

8.6 Agent Framework – AADE

AADE is an engineering-oriented agent framework that provides agent-based engineering application developers with reusable agent-oriented classes (templates) that share useful relationships. To build the framework, the agent-level issues and domain-dependent ones must be identified and programming abstractions must be encapsulated. Based on our investigation on the FIPA specifications [8.38] and the existing agent frameworks that implement FIPA specifications, we expect that our agent framework can benefit from the characteristics of JADE [8.43], FIPA-OS [8.44] and ZEUS [8.45].

AADE is originated from our previous work on the development of an agent-based framework for the rapid prototyping of collaborative product design and engineering environments [8.46]. This environment is proved to be flexible enough for multi-agent systems development. In the next section, we will give an example of an agent-based scheduling system that we developed as a Web service for the support of inter-enterprise manufacturing resource sharing. The current version of AADE implements a semantically closed agent community in which all agents communicate with each other within the agent framework using a dedicated content language expressed in XML.

Figure 8.4 shows the agent abstraction model in AADE, which is somewhat similar to other agent architectures in the literature. Processing logics are separated from data representation by using different illustrations, solid vs. dashed line boxes, respectively. There are four major packages developed in an AADE agent, namely, the *communication package*, the *messaging package*, the *general function package*, and the *utility package*. Using separate socket listening and sending threads, the communication package handles all incoming and outgoing messages for an agent. Multiple-message queues are designed for secure attainability and maintainability of messages. In the messaging package, the Message Handling module deals with the payload of messages that are represented in FIPA ACL (agent communication language), while the Conversation Control module maintains a conversation list, enforces messaging sequence of FIPA protocols, and controls the states of conversations. The corresponding ACL Parser is responsible for understanding the meaning of content in the message payload after the XML parser (in the utility package) distils the payload information. General functionalities defined in FIPA specifications that all agents commonly share are packed as the general function package. Typical functions include agent registration, agent management, and routine maintenance schedules that an agent needs to execute, *i.e.* logging, error handling, backup and restore, *etc*. In the utility package, we provide agents with atomic software engineering facilities, such as database operations, file operations, FTP client, email handling, XML parsing, and zipping classes (for message security and encoding). Application-specific agents can be constructed by customising the interfaces of the generic agent model. The parts that need to be specialised are called "application-specific modules", which are represented by the grey boxes in Figure 8.4. These application-specific modules include user interfaces that manipulate the agent, application interfaces that wrap legacy or external programs, and a problem-solving core that embeds an agent's reasoning and intelligent capacities.

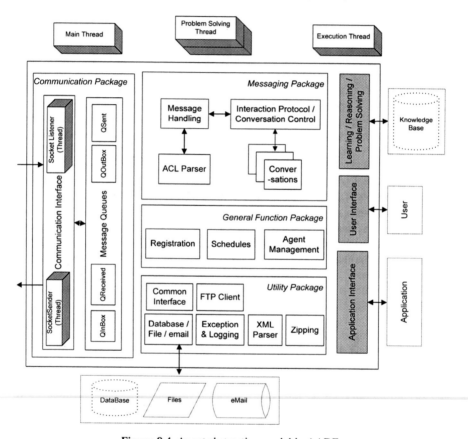

Figure 8.4. Agent abstraction model in AADE

A detailed description of the AADE framework and its architecture can be found in [8.46]. Key features of AADE include:

- *Multi threading*. Each agent deals with multiple threads at the same time to handle concurrent tasks. Figure 8.4 shows briefly the threads that an agent in AADE should have using boxes with dark shadow. Two threads (SocketSender and SocketListener) are committed to the socket communication, one thread (Execution Thread) takes care of the execution of the wrapped external program, while the Main Thread and Problem Solving Threads take the roles of periodic scheduling, event-driven scheduling, coordinating, and domain-dependent tasks' scheduling.

- *Multi-layered message frame and message handling.* The structure of message (messageFrame is used as a terminology in AADE) contains multi-layered information. An AADE message envelope is added on the standard FIPA envelope to enable advanced messaging functions. For example, msgLevel is used to differentiate the priority level of a message so that urgent messages in the incoming and outgoing mailboxes can be treated differently. Similarly, the message-handling mechanism involves a multi-layered architecture.
- *Peer-to-peer agent communication.* The lower-level agent communication is conducted through a peer-to-peer transport protocol, TCP/IP socket. This is a strategy that represents more precisely the agent's "loose coupling" and "autonomous" concepts compared with the tightly coupled client-server RMI invocation used in JADE and FIPA-OS. The communication package is designed to be within the functionalities of each individual agent rather than having a shared MTS (the Message Transport Service defined in FIPA) to handle the intra-agent platform messages. In other words, we expect the MTS to deal with the cross agent platform messages only.
- *Originated from and serving to the engineering fields.* The agent knowledge base, problem-solving abilities, configuration file, registration information, and more specially, the inclusion of an engineering data management agent (EDM agent) can greatly accelerate the development process of engineering multi-agent systems. The related supporting facilities in the utility package contribute to this feature as well, such as facilities dealing with engineering files and file transportation (FTP client/server).

8.7 Proof-of-concept Prototypes

8.7.1 Agent-based Dynamic Scheduling in *iShopFloor*

The prototype of *iShopFloor* and its agent-based dynamic scheduling system were implemented in a previous version of the AADE agent framework. All communications among agents are implemented using Java client/server socket communication. Each intelligent machine agent connects to the shop-floor network with a unique IP address and a port number. Each machine agent functions both as a client and a server. It functions as a client when it proactively communicates with other agents. It sends messages to other agents for registration to the shop-floor directory facilitator, for negotiation with other agents, for sending manufacturing progress information to product/part agents, and for sending machine state information to other servers. It also functions as a server when it receives requests from other agents (*e.g.* requests for machine information/data/state, and call-for-bid messages).

Because of the generic architecture and common software modules, all machine agents (for different types of machines) use the same software. There is no need for coding before installing the software for a machine agent. Only the data related to machine specifications, IP address and port numbers, *etc.*, need to be inputted and/or edited before "plugging" it to the network.

A graphical user interface (GUI) is developed for machine agents. Figure 8.5 shows an example of the actual prototype implementation. A picture of the machine is displayed on the GUI – a very useful feature when the computer is not located close to the corresponding machine, or when one computer is used to connect multiple machines.

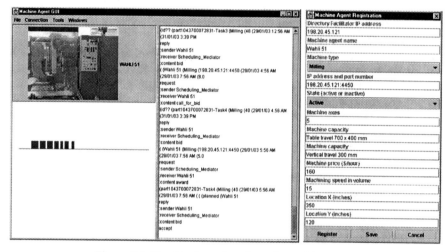

Figure 8.5. Graphical user interface of a machine agent

Service agents have been implemented in two versions: standalone application version and Web-based version. In the Web-based version, service agents are implemented as servlets [8.41].

Product/part agents are currently represented using Java objects. Each agent keeps track of the status and interdependency of tasks, and can move from one resource agent (or service agent) to another. They are serialised into XML form for communication between resource and service agents.

The current prototype implementation only considers intelligent machine agents as shop-floor resources. Other types of resources (*e.g.* robots, AGVs, conveyers, and human operators) are to be considered and implemented in later prototypes.

8.7.2 Real-time Scheduling Service for Enterprise Collaboration

This section presents an example of application of the AADE framework to a multi-agent shop-floor resource scheduling service that supports enterprise collaboration. The implemented software prototype demonstrates two levels of collaboration: inter-enterprise collaboration for manufacturing resources sharing and the collaboration among machines in the shop floor for dynamic scheduling of manufacturing resources within one enterprise. Details of this prototype can be found in [8.47].

Figure 8.6 shows the system architecture of the implemented prototype at the inter-enterprise level. Three enterprises are providing manufacturing services on the

Internet through a Web portal. Each enterprise is encapsulated as an individual Web service and it owns a number of expensive manufacturing resources offering services that can be shared with other enterprises. An ontology agent is designed as a complement to the public UDDI registry where advanced match-making services can reside. Any customer can place an order (a bundle of manufacturing tasks) through the Web portal. Each enterprise will submit a bid if its agent-based dynamic scheduling system is able to allocate the necessary manufacturing resources to meet the manufacturing requirements (the entire bundle of manufacturing tasks) of the order. The customer then makes a decision based on the bids received from service providers (enterprises) and signs an agreement with the selected enterprise.

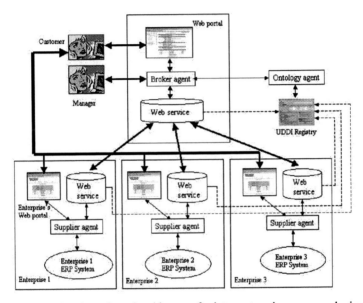

Figure 8.6. Service-oriented architecture for inter-enterprise resource sharing

An agent-based manufacturing scheduling system is implemented behind the Web service for dynamic scheduling of manufacturing resources (those expensive machines only) within the enterprise (Figure 8.7). The implemented agents include an interface agent, a mediator agent and five machine agents.

An *interface agent* provides connectivity of an enterprise's internal systems to external business processes. It performs functions such as forwarding messages, receiving tasks, returning bids, and controlling access security. In fact, the interface agent performs functions similar to an application gateway and is a joint point of external scheduling service of an enterprise. The interface agent decides which agent is responsible for a received message and forwards the message to it.

A *mediator agent* is a coordination agent in an enterprise that is responsible for the scheduling of manufacturing sources. The scheduling process is realised through the coordination of the mediator and the participating resource agents. As shown in Figure 8.7, when a task is received, a mediator agent takes three steps to schedule the task:

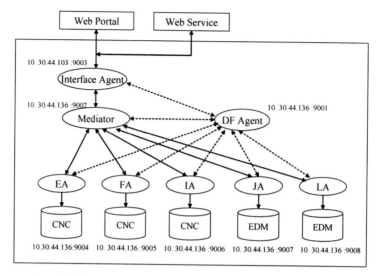

Figure 8.7. An agent-based manufacturing resource scheduling service

1. Task decomposition is a process of building a set of goals for the multi-agent system to accomplish. In this step, a received order is decomposed into a set of tasks and each task may be further decomposed into subtasks. After decomposing the task, the mediator agent needs to assign tasks to appropriate resource agents.
2. Scheduling can be considered as a negotiation process that mediator agent sends a call-for-bid request to all potential resource agents and makes its final decisions after receiving and evaluating the bids from resource agents.
3. After award/confirmation, resource agents will commit to the awarded tasks when the mediator agent receives contract messages from the interface agent.

Resource agents are used to represent manufacturing resources (expensive machines in this case) that an enterprise would like to provide as external manufacturing services. When a resource agent receives a call-for-bid request from the mediator agent, it will compute a bid according to its schedules and status, and return the bid after calculation. Once its proposed bid is awarded, the resource agent will lock the task into its schedule and prepare for task execution.

The *directory facilitator* (DF) provides various registration-related services to other agents, including keeping an up-to-date agent registry, informing all registered agents with updated information, and performing lookup and matchmaking services on a request basis.

The solid links in Figure 8.7 show the scheduling negotiation process between the mediator and participating machine agents, while the dashed links show the registration and lookup processes between the DF and all other active agents.

The AADE framework has been used for the implementation of the agent-based dynamic manufacturing scheduling Web service of an enterprise. The Contract Net protocol is used both at the inter-enterprise collaboration level for the selection of a service provider and at the enterprise level for the allocation of manufacturing resources.

8.8 Key Issues in Technology Deployment in Industry

After six years of research and prototype development, our agent-based dynamic manufacturing scheduling technology is ready for deployment in industrial settings. We are negotiating with industrial partners on the implementation of pilot systems within real manufacturing shop floors, particularly in job-shop environments within the tooling industry.

However, some practical issues need to be carefully addressed when deploying this technology in industry.

(1) Integration with legacy systems
As advanced planning and scheduling systems appear to be the current trend and practices in industries, we expect that integrating with these systems will be a major challenge to agent-based approaches. By advanced planning and scheduling systems, we refer to the commercial ERP, MRP or other planning software on the market. It is a common view from empirical studies that this kind of software is very costly in investment and difficult to implement [8.48]. Moreover, ERP software packages form the treasure of a company in that they embed so many interweaved functionalities and large amounts of running data. Except for the fact that the agent-based dynamic scheduling system is implemented as a standalone system (for those small companies that do not have existing ERP/MRP systems in place), this integration is essential and time consuming. In some cases, a certification may be required to integrate it with a previously implemented commercial software system.

(2) Integration with physical devices
For the same reason as the first integration issue, this is a requirement of hardware accessibilities. Tracking and recording real-time data from the physical world of shop floors is a missing link of ERP systems. Without automatic data collection from manufacturing processes, production planning, monitoring, and quality control at the shop-floor level become an unreachable goal. For this reason, some manufacturing plants try to pursue manufacturing execution systems on the market, be it a simple DNS connectivity, or a comprehensive MES solution. Our agent-based manufacturing scheduling system, using physical agent-composition architecture, can be mapped to physical devices and has the capacity to receive real-time signals directly from these devices or indirectly from existing real-time data-collection systems (*e.g.* SCADA – *supervisory control and data acquisition* systems and/or RFID – *radio frequency identification* systems). Therefore, real-time information can be handled by our system and then congregated and fed back to ERP systems as necessary.

(3) Human adaptation and cooperation
Implementation of agent-based dynamic manufacturing scheduling technology within a manufacturing enterprise may bring some significant changes to the business operations, particularly to the way of managing manufacturing resources

(operators, equipment, materials, *etc.*) and operating the equipment. Without the cooperation of human managers and operators, there is no way for the agent-based manufacturing scheduling system to work properly and effectively since the cooperation among manufacturing resources (including humans, hardware and software) is the key to this technology. However, it may take some time for managers and operators to fully accept this technology. Some education and training are essential during this human adaptation process in order for the company to successfully apply this technology in its business.

(4) *Support of a stable and reliable agent framework*
There are many agent-development tools or platforms on the market. Some open-source frameworks are also available. However, working with a specific agent platform has some drawbacks. First, the code is portable only if delivered with the host platform (*e.g.* JADE). Secondly, the limitation of documentation and licensing problem hinders its adaptability to engineering/industrial applications. Many industrial customers would like to protect their proprietary systems that cannot be guaranteed by codes under open-source agreement. Thirdly, the code efficiency is also a serious problem. Because the purpose of agent platforms such as JADE is developed from the viewpoint of a pure computer science discipline, this target may not be balanced and optimised to meet the special requirements of engineering applications. The agent-based manufacturing dynamic scheduling technology presented in this chapter builds on a proprietary agent-development platform AADE that can facilitate the development of reliable, stable, and scalable industrial applications.

(5) *Cost and return on investment*
Manufacturing enterprises, particularly those small-and-medium-sized enterprises (SMEs), are always having trouble justifying the return on investment for a new software system. While the pressure of increasing productivity and profitability is forcing SMEs to adopt new technologies and advanced software systems, expensive software solutions are not affordable to them. The proposed agent-based scheduling system has been designed in a way to allow a company to choose not only among functional modules, but also the size of applications. The cost will be relative to the number of resource agents (related to the number of manufacturing resources to be included in the dynamic scheduling system). So a smaller manufacturer, with fewer agents, will pay less for the software tool than a much larger manufacturer, and can add more agents later when the company grows. Therefore, this technology becomes more attractive to small-and-medium-sized manufacturing enterprises.

8.9 Conclusions and Future Work

The agent-based dynamic manufacturing scheduling technology presented in this chapter can well address dynamic changes and disturbances on the shop floor locally without disruptions to the regular production – a problem that cannot be solved by traditional planning and scheduling systems because these changes and disturbances cannot be predicted in advance.

This technology is expected to fill a major gap in the manufacturing software industry to address the real-time scheduling issues. It allows manufacturing enterprises to: (1) respond to market demands quickly so as to win the increasingly competitive market; (2) respond to changes and disturbance on the shop floor quickly so as to increase their productivity through optimising equipment-utilisation rates. Successful application of this technology in manufacturing enterprises will bring significant economic benefits to these enterprises.

This technology can be used in different manufacturing environments, but particularly in those environments with frequent changes (including design/engineering changes and production changes, *e.g.* machine failures, tool breakages, and missing/wrong parts). It can be applied in various manufacturing sectors including automobile (particularly heavy duty vehicles) and aerospace industries. Furthermore, since it is a generic platform technology, it can also be easily extended and applied to other industrial sectors with similar scheduling needs, *e.g.* transportation scheduling, airport scheduling, medical facility scheduling, and inventory management.

Compared with other agent-based manufacturing scheduling systems reported in the literature, this technology is unique in its engineering focus and robust fault tolerance. Close collaborations of the project team with industrial partners during the development of this technology make the technology closer to industrial needs than other similar frameworks/platforms. The use of the proprietary agent framework AADE can also avoid open-source concerns for industrial applications.

Our project team members have designed and implemented key models and software modules based on AADE. Generic interfaces have been designed and developed to integrate various software applications and hardware devices. The current version of the software framework is compliant with the FIPA Specifications – that are becoming international standards for agent-based software applications.

We have developed software prototype systems in simulated environments within our research facilities. The applications have been at both the shop-floor level and the enterprise level.

We are collaborating with industrial partners to validate the technology in industrial settings and are looking for a software company in the manufacturing sector to commercialise the technology. Further R&D work is underway to address some practical issues for industrial applications:

- integration with existing/commercial ERP systems;
- integration with existing real-time data-collection systems.

For those manufacturing enterprises without any ERP/MRP systems in place, it would be easy to deploy an agent-based dynamic manufacturing scheduling system as a standalone application. Only some specific interfaces are to be developed to access related databases and for users to interact with the system. Efforts to develop such interfaces are minimal (in terms of a few person-months).

In order to integrate with existing software applications currently used in manufacturing enterprises (which is highly desired), significant efforts are required (in terms of a few person-years), particularly when a certification is required for integrating/interfacing with some of these commercial software applications.

Acknowledgment

We would like to acknowledge the important contributions of our guest workers including Giuseppe Stecca, Christopher Brooks, Shuying Wang, and Akbar Siami Namin in the development of the software prototype systems.

References

[8.1]   Valckenaers, P., 1998, "WP1 dissemination report: analysis and evaluation of change and disturbances in industrial plants," MASCADA ESPRIT LTR 22728.

[8.2]   Chong, C.S., Sivakumar, A.I. and Gay, R., 2003, "Simulation-based scheduling for dynamic discrete manufacturing," In *Proceedings of the 2003 Winter Simulation Conference*, pp. 1465–1473.

[8.3]   Shen, W., Norrie, D.H. and Barthès, J.P., 2001, *Multi-Agent Systems for Concurrent Intelligent Design and Manufacturing*, Taylor and Francis, London, UK.

[8.4]   French, S., 1982, *Sequencing and Scheduling: An Introduction to the Mathematics of the Job-Shop*, John Wiley & Sons, New York, NY.

[8.5]   Bagchi, T.P., 1999, *Multiobjective Scheduling by Genetic Algorithms*, Kluwer Academic Publishers, Norwell, MA.

[8.6]   Shen, W., 2002, "Distributed manufacturing scheduling using intelligent agents," *IEEE Intelligent Systems*, 17(1), pp. 88–94.

[8.7]   Shen, W., Wang, L. and Hao, Q., 2006, "Agent-based distributed manufacturing process planning and scheduling: a state-of-the-art survey," *IEEE Transactions on Systems, Man, and Cybernetics, Part C*, 36(4), pp. 563–577.

[8.8]   Shen, W., Hao, Q., Yoon, H.J. and Norrie, D.H., 2006, "Applications of agent systems in intelligent manufacturing: an updated review," *International Journal of Advance Engineering Informatics*, 20(4), pp. 415–431.

[8.9]   Shaw, M.J. and Whinston, A.B., 1983, "Distributed planning in cellular flexible manufacturing systems," *Technical Report*, Management Information Research Centre, Purdue University.

[8.10] Shaw, M.J., 1998, "Dynamic scheduling in cellular manufacturing systems: a framework for networked decision making," *Journal of Manufacturing Systems*, 7(2), pp. 83–94.

[8.11] Parunak, V.D., 1987, "Manufacturing experience with the contract net," In *Distributed Artificial Intelligence*, Huhns, M.N. (ed.), Pitman Publishing Ltd., London, England, pp. 285–310.

[8.12] Smith, R.G., 1980, "The contract net protocol: high-level communication and control in a distributed problem solver," *IEEE Transactions on Computers*, C-29(12), pp. 1104–1113.

[8.13] Parunak, V.D., Baker, A.D. and Clark, S.J., 1997, "The AARIA agent architecture: an example of requirements-driven agent-based system design," In *Proceedings of the First International Conference on Autonomous Agents*, Marina del Rey, CA.

[8.14] Lu, T.P. and Yih, Y., 2001, "An agent-based production control framework for multiple-line collaborative manufacturing," *International Journal of Production Research*, 39(10), pp. 2155–2176.

[8.15] Yen, B.P.C. and Wu, O.Q., 2003, "Internet scheduling environment with market driven agents," *IEEE transactions on Systems, Man and Cybernetics – Part A: Systems and Humans*, 34(2), pp. 281–289.

[8.16] Daouas, T., Ghedira, K. and Muller, J.P., 1995, "How to schedule a flow shop plant by agents," In *Applications of Artificial Intelligence in Engineering*, Computational Mechanics Inc., Billerica, MA, pp. 73–80.

[8.17] Kouiss, K., Pierreval, H. and Mebarki, N., 1997, "Using multi-agent architecture in FMS for dynamic scheduling," *Journal of Intelligent Manufacturing*, **8**(1), pp. 41–47.

[8.18] Trentesaux, D., Tahon, C. and Ladet, P., 1998, "Hybrid production control approach for JIT scheduling," *Artificial Intelligence in Engineering*, **12**(1), pp. 49–67.

[8.19] Goldberg, D., Cicirello, V. and Dias, M.B., 2003, "Task allocation using a market-based planning mechanism," In *Proceedings of AAMAS'03*, Melbourne, Australia, **2**, pp. 996–997.

[8.20] Lee, Y.H., Kumara, S.R.T. and Chatterjee, K., 2003, "Multi-agent-based dynamic resource scheduling for distributed multiple projects using a market mechanism," *Journal of Intelligent Manufacturing*, **14**(5), pp. 471–484.

[8.21] McDonnell, P., Smith, G., Joshi, S. and Kumara, S.R.T., 1999, "A cascading auction protocol as a framework for integrating process planning and heterarchical shop floor control," *International Journal of Flexible Manufacturing Systems*, **11**(1), pp. 37–62.

[8.22] Cowling, P.I., Ouelhadj, D. and Petrovic, S., 2004, "Dynamic scheduling of steel casting and milling using multi agents," *Production Planning & Control*, **15**(2), pp. 178–188.

[8.23] Shen, W., 2002, "Implementation of genetic algorithms in agent-based manufacturing scheduling systems," *Integrated Computer-Aided Engineering*, **9**(3), pp. 207–218.

[8.24] Deen, S.M., 2003, *Agent-Based Manufacturing – Advances in the Holonic Approach*, Springer-Verlag, Germany.

[8.25] Mařik, V., McFarlane, D. and Valckenaers, P., 2003, *Holonic and Multi-Agent Systems for Manufacturing*, Springer-Verlag, Germany.

[8.26] Sousa, P. and Ramos, C., 1997, "A dynamic scheduling holon for manufacturing orders," *Journal of Intelligent Manufacturing*, **9**(2), pp. 107–112.

[8.27] Sousa, P., Ramos, C. and Neves, J., 2003, "The *Fabricare* scheduling prototype suite: agent interaction and knowledge base," *Journal of Intelligent Manufacturing*, **14**(5), pp. 441–456.

[8.28] Fordyce, K. and Sullivan, G.G., 1994, "Logistics management system (LMS): integrating decision technologies for dispatch scheduling in semiconductor manufacturing," In *Intelligent Scheduling*, Zweben, M. and Fox, M. S. (eds.), Morgan Kaufman Publishers, San Francisco, CA, pp. 473–516.

[8.29] Parunak, V.D., 1993, "MASCOT: a virtual factory for research and development in manufacturing scheduling and control," *Technical Memo 93-02*, Industrial Technology Institute.

[8.30] NCMS, 1998, "Shop floor agents," *Technical Report*, National Centre for Manufacturing Sciences.

[8.31] Parunak, V.D., 1996, "Workshop report: implementing manufacturing agents," *Technical Report*, National Centre for Manufacturing Sciences.

[8.32] Riha, A., Pechoucek, M., Vokrinek, J. and Marik, V., 2002, "ExPlanTech: exploitation of agent-based technology in production planning," In *LNAI 2322*, Springer, pp. 308–322.

[8.33] Bruccoleri, M., Lo Nigro, G., Perrone, G., Renna, P. and Noto La Diega, S., 2005, "Production planning in reconfigurable enterprises and reconfigurable production systems," *Annals of the CIRP*, **54**(1), pp. 433–436.

[8.34] Váncza, J. and Márkus, A., 2000, "An agent model for incentive-based production scheduling," *Computers in Industry*, **43**(2), pp. 173–187.

[8.35] Papakostas, N., Mourtzis, D., Bechrakis, K., Chryssolouris, G., Doukas, D. and Doyle, R., 1999, "A flexible agent based framework for manufacturing decision-making," In *Proceedings of the 9th Flexible Automation and Intelligent Manufacturing*, Tilburg, the Netherlands, pp. 789–800.

[8.36] Monostori, L., Váncza, J. and Kumara, S.R.T., 2006, "Agent-based systems for manufacturing," *Annals of the CIRP*, **55**(2), pp. 697–720.

[8.37] Shen, W., Lang, S. and Wang, L., 2005, "*iShopFloor*: An Internet-enabled agent-based intelligent shop floor," *IEEE Transactions on Systems, Man and Cybernetics-Part C*, **35**(3), pp. 371–381.

[8.38] FIPA, Foundation for Intelligent Physical Agents, http://www.fipa.org/.

[8.39] Shen, W., Lang, S., Korba, L., Wang L. and Wong, B., 2000, "Reference architecture for Internet-based intelligent shop floors," In *Network Intelligence: Internet-Based Manufacturing, Proceedings of SPIE*, Berry, N.M. (ed.), **4208**, pp. 63–71.

[8.40] Duffie, N.A. and Prabhu V.V., 1996, "Heterarchical control of highly distributed manufacturing systems," *International Journal of Computer Integrated Manufacturing*, **9**(4), pp. 270–281.

[8.41] Shen, W., Wang, L. and Lang, S., 2005, "XML based message services for Internet-enabled agent-based intelligent shop floors," *International Journal of Agile Manufacturing*, **8**(1), pp. 27–42.

[8.42] Shen, W., Li, Y., Ghenniwa H. and Wang, C., 2002, "Adaptive negotiation for agent-based grid computing," In *Proceedings of AAMAS2002 Workshop on Challenges in Open Agent Environments*, Bologna, Italy, pp. 32–36.

[8.43] Bellifemine, F., Poggi, A. and Rimassa, G., 2001, "Developing multi-agent systems with a FIPA-compliant agent framework," *Software Practice and Experience*, **31**(2), pp. 103–128.

[8.44] Poslad, S., Buckle, P. and Hadingham, R., 2000, "The FIPA-OS agent platforms: open source for open standards," In *Proceedings of the 5th International Conference on the Practical Application of Intelligent Agent and Multi-Agent Technology*, Manchester, UK, pp. 355–368.

[8.45] Nwana, H.S., Ndumu, D.T., Lee L.C. and Collis, J.C., 1999, "Zeus: a toolkit for building distributed multi-agent systems," *Artificial Intelligence Journal*, **13**(1), pp. 129–186.

[8.46] Hao, Q., Shen, W. and Zhang, Z., 2005, "An autonomous agent development environment for engineering applications," *International Journal of Advanced Engineering Informatics*, **19**(2), pp. 123–134.

[8.47] Wang, S., Shen, W. and Hao, Q., 2005, "Implementing inter-enterprise workflow management using agent-based web services," In *Proceedings of ASME IMECE'05*, Orlando, FL, IMECE2005-79242.

[8.48] Motwani, J., Subramanian, R. and Gopalakrishna, P., 2005, "Critical factors for successful ERP implementation: exploratory findings from four case studies," *Computers in Industry*, **56**(6), pp. 529–544.

9

A Multi-agent System Implementation of an Evolutionary Approach to Production Scheduling

Scott S. Walker, Douglas H. Norrie and Robert W. Brennan

Schulich School of Engineering, University of Calgary
2500 University Dr. N.W., Calgary, AB T2N 1N4, Canada
Email: rbrennan@ucalgary.ca

Abstract
In this chapter we describe a multi-agent systems (MAS) approach to manufacturing job-shop scheduling that evolves the rules by which schedules are created rather than the schedule itself. The system is tested using a benchmark agent-based scheduling problem and performance results are compared with pure heuristic scheduling heuristics and randomly generated mixed heuristics.

9.1 Introduction

Shop-floor scheduling is an inherently difficult manufacturing task, and as a result, has served as a good motivation for the application of distributed artificial intelligence in the manufacturing domain. The traditional approach to shop-floor scheduling is to use a single, centralised scheduler, which as [9.1] notes, is analogous to using a large sequential computer program to solve the problem. If there are no disturbances on the shop floor (*e.g.* rush orders, missing tools or material, machine processing time variability or failure), this approach is very effective and can generate optimal solutions to the scheduling problem. Real manufacturing systems are not deterministic, however. Also, they consist of physically distributed materials and resources that behave in a highly concurrent and dynamic fashion. As a result, centralised schedules quickly become obsolete and must be either ignored or frequently updated: this is not always possible if the schedule generation and dissemination time is great.

In this chapter we describe a multi-agent systems (MAS) approach to manufacturing job-shop scheduling that evolves the rules by which schedules are created rather than the schedule itself. Unlike traditional scheduling methods, which often assume infinite capacity for a resource for calculation simplicity, these resources cannot be overscheduled. Implementation of these scheduling rules is accomplished through Mediator Agents [9.2], which coordinate the needs of the various Order Agents and Resource Agents towards achieving an optimum schedule.

The goal here is to apply holonic systems [9.3] and multi-agent systems [9.4] principles towards achieving an intelligent scheduling system.

We begin this chapter with some background on holonic systems and intelligent job-shop scheduling. Next, we describe the implementation of our proposed evolutionary scheduling approach. We then report the results of using this approach to tackle a benchmark agent-based scheduling problem proposed by [9.5], and conclude with a discussion of the effectiveness of the proposed approach.

9.2 Background

The scheduling approach described in this chapter combines ideas from holonic manufacturing systems (HMS) [9.3] and intelligent systems. In this section, we begin with a brief overview of the research on HMS and scheduling, and then focus more specifically on intelligent scheduling.

9.2.1 HMS Architectures and Scheduling

Christensen [9.6] describes the architectural requirements of holonic systems, comparing the holonic architecture to other possible forms: centralised, proper hierarchical, modified hierarchical (some peer-to-peer interaction), and heterarchical (exclusively peer-to-peer interaction). In the generic activity model of a holon, Christensen shows the information processing and physical processing activity of a holon; the material flow between the humans, intelligent control system, and processing system; and, the interfaces between these human, control, and processing elements.

Bussman [9.7] shows the commonality of the visions of MAS and HMS for manufacturing, emphasising that the vision of MAS deals with a subset of activities focusing on the control system, whereas HMS is also strongly concerned with the human workers and physical equipment in the holonic system. He argues that "agent-oriented techniques are an appropriate technology for designing and implementing the information processing part of a holonic manufacturing system", and briefly presents an agent-oriented architecture for a holon, concerned with the information processing of a holon (decision making, organisational techniques, behavioural control, cooperation techniques, and communication techniques), and its interfaces for physical communication and physical control.

Van Brussel et al. [9.8] have described a *product-resource-order-staff* architecture (or PROSA), defining three basic holons for manufacturing control (including planning and scheduling): resource, product and order holons. The resource holon consists of a "production resource of the manufacturing system, and an information processing part that controls the resource," remembering, of course, that a holon is hierarchical in nature. Likewise, the "product holon holds the process and product knowledge to assure the correct making of the product with sufficient quality", and "comprises functionalities that are traditionally covered by product design, process planning, and quality assurance." Finally, the "order holon represents a task in the manufacturing system", i.e. any type of order. "Product holons and resource holons communicate process knowledge, product holons and order holons exchange production knowledge, and resource holons and order holons share process execution knowledge."

Van Brussel *et al.* also make the following two conclusions about their architecture. First, "PROSA can be regarded as a generalisation of the two former approaches [*i.e.* hierarchical and heterarchical control architectures]"; and second, "PROSA also introduces several important innovations: the system structure is decoupled from the control algorithm, logistical aspects are decoupled from technical ones, and PROSA allows [one] to incorporate more advanced hybrid control algorithms."

Related to scheduling, "the architecture foresees staff holons to assist the [three] basic holons in performing their work … a type of external expert that gives advice. …The centralised scheduler of a shop is an example of a staff holon."

Research in holonic scheduling has its roots in intelligent scheduling techniques ([9.9], [9.10] and the references therein) and distributed factory control algorithms ([9.11]–[9.13] and the references in [9.14]). The key areas where research has been initiated in holonic scheduling are as follows, taken in part from [9.15]: flexible manufacturing systems [9.16][9.17], assembly lines [9.18], job shop [9.19], assembly and machining cells [9.20][9.21], continuous process lines [9.22][9.23], plant-wide maintenance [9.24], and generic scheduling methods [9.25][9.26].

McFarlane and Bussman [9.15] describe some key themes in current HMS scheduling research, including (1) central vs. distributed problem formation, (2) local decision-making/computational techniques, (3) cooperative interaction strategy, (4) interchange mechanism, and (5) degree of central coordination. The scheduling approach presented in their thesis addresses item (1) above with central problem formation; item (2) with local decisions from a control point of view; item (3) with a reactive implicit cooperation scheme; item (4) with straightforward request–reply type protocols and the passing of relevant serialised data structures; and, item (5) with complete central coordination of the scheduling function (all holonic scheduling approaches being reported as having some degree of central coordination).

9.2.2 Intelligent Job-shop Scheduling

For the research reported in this chapter, a layered approach to the job-shop scheduling problem is used. More specifically, a low-level, reactive approach to scheduling is used initially for the general job-shop scheduling problem. Next, a higher-level, more deliberative layer is added to provide better optimisation, once a reactive kernel for job-shop scheduling has been found that has reasonable levels of quality of results, flexibility, speed, and scalability.

Such a reactive scheduling approach could fall under three broad categories: evolutionary algorithms for scheduling, such as genetic algorithms, heuristic rules for dispatching and/or scheduling or, specialised algorithms for scheduling that move over the schedule, adjusting it.

Genetic algorithms have been applied extensively to scheduling, for example [9.27], often with reasonable success. Heuristic rules have also been applied extensively to scheduling, with varied success [9.28]. Specialised algorithms exist (and schedule representations that facilitate these algorithms), such as reactive rescheduling [9.29], least-commitment scheduling [9.30], and others.

Any of the above would be applicable, however, the authors' desire is to produce a so-called "intelligent system", meaning one that can learn something from the environment and then apply it to future actions. A specialised algorithm might fit with this goal, but one would need a deep understanding of job-shop scheduling to develop such an algorithm; this is not the focus of this research, but rather to apply holonic [9.3] and agent systems [9.4] principles to the shop scheduling (and control) problem. This may be an excellent place to apply genetic programming to evolve an algorithm that can perform effective job-shop scheduling, although this potential approach was deemed beyond the scope of this research. A simple genetic algorithm would not fulfil the intelligent system goal, as such a solution is brittle and would only apply to one state of the system at one particular time; it definitely does not learn anything about the system in general for future use.

On the other hand, a system of generalised heuristic rules could potentially learn a small amount about the environment, namely which rule is most appropriate to which situation. Also, [9.31] mentions "probabilistic dispatching" or "Monte Carlo methods" for choosing between various heuristic rules, saying that "with a sensible choice of biasing probabilities it does seem to work well, substantially better than the deterministic application of a priority rule" for dispatching. Heuristic rules fired sequentially for dispatching are extremely well suited to the reactive side of the continuum of holonic and agent-based systems, and have been applied to dispatching or queue-selection decisions in holonic manufacturing systems (though not to advance scheduling). Perhaps heuristic rules could also be applied to scheduling, if fired in parallel to varying degrees to assign priorities to operations, jobs, resources, *etc.*, which can then be used for scheduling. The weight of each rule would be something that is learned from the environment, and that could also change over time to represent the changing state of the environment. First discussed by the authors in [9.32], this seems like an interesting avenue of approach to the problem of reactive job-shop scheduling. This is discussed, along with the rest of the system implementation, in the next section.

9.3 Implementing the Agent-based Scheduling System

9.3.1 The Benchmark

Within the domain of manufacturing scheduling and control, a benchmark needed to be chosen in the interests of being able to compare results with other researchers. The real-world problem selected is the scheduling and control benchmark presented in [9.5] although, in the limit of the project scope, the benchmark had to be modified somewhat, and expansion of the system capabilities to fulfil the full benchmark is suggested as future work.

The test environment contains four types of resources, each having non-overlapping capabilities, with the third resource type as the bottleneck. The benchmark demands two of each type of resource, or eight in total. Transportation between resources is accomplished through automated guided vehicles (AGVs) that, for simplicity, are assumed to be always available and require zero transportation time.

It should be noted that this transportation assumption, though included in the benchmark, is not strictly necessary. Any transportation activities can themselves be considered resources, and included accordingly in the approach presented in this chapter. Note, however, that there are additional complexities in this approach; for example, operation times for transportation operations have variable setup times, based upon the transporter's physical location at the end of the previous job. Another example is the ability of a transportation resource to perform more than one operation simultaneously.

There is a fixed set of job types, each with fixed precedence constraints and fixed operation processing times as given in Table 9.1.

Table 9.1. Job types

Job type	Operation (process time (min) / operation type)			
	1	2	3	4
J1	6 / 1	8 / 2	13 / 3	5 / 4
J2	4 / 1	3 / 2	8 / 3	3 / 4
J3	3 / 4	6 / 2	15 / 1	4 / 3
J4	5 / 2	6 / 1	13 / 3	4 / 4
J5	5 / 1	3 / 2	8 / 4	4 / 3

The benchmark defines the following experiments, called Plant Scenarios: (1) deterministic processing times, with no unexpected events, (2) stochastic variability of processing times, and (3) non-bottleneck random breakdowns included. Other variations are also defined by the benchmark, called *Operational Scenarios*: (1) Just-In-Time (JIT) scheduling, with jobs dispatched regularly with random (uniform distribution) due dates, and (2) JIT scheduling, with some randomly distributed rush jobs (*i.e.* As-Soon-As-Possible or ASAP scheduling).

Another scenario that would complete this operational scenario set would be one involving exclusively ASAP scheduling. The development perspective taken was that the system should be able to handle exclusively JIT scheduling and exclusively ASAP scheduling, at which point the problem of JIT scheduling with rush orders should intrinsically be solved.

This system has been designed so as to support conformance to the test environment and to implement all operational and plant scenarios as defined in the benchmark; however, at its current level of implementation, the system deviates from the benchmark in two important ways.

The first deviation from the benchmark is in the test environment requirement for two of each type of resource. The implementation currently allows only one of each resource type at this time, or four in total, though it has been designed to allow more than one of each resource type in a following development stage.

The second deviation is in the operational scenarios requirement for JIT scheduling. The implementation currently allows only ASAP scheduling, though it has been designed to allow JIT scheduling in a following development stage. Also, the required object data, method stubs, and method call parameters exist for almost all aspects of JIT scheduling, but final implementation has been postponed at this time. The reasoning behind implementing ASAP scheduling before JIT scheduling (rather than *vice versa*) revolved around the desire to produce as reactive a scheduler as possible for as reactive a manufacturing system as possible. To conform to the principles of holonic systems, a reactive kernel of manufacturing functionality is most appropriate. Then, additional layers of increasingly less reactive, more deliberative system capabilities can be added in the spirit of Ref. [9.33].

9.3.2 The System Architecture

A description of the system architecture of this implementation includes three main aspects: (1) the holonic architecture, created primarily for system analysis purposes, (2) the multi-agent architecture, created primarily for system design purposes, and (3) FIPA-OS – the multi-agent platform used to simplify development.

The holonic architecture and the FIPA-OS implementation are discussed in detail in [9.34]. In this section, we focus on the multi-agent architecture that was developed to implement our scheduling system. Our general architecture follows [9.35][9.36], with some modifications.

The scheduling is accomplished through interactions within a clustering of several agents: an order agent (OA), resource agent (RA) for each resource represented in the shop, and one resource scheduling dynamic mediator agent (RSDMA) or more. The OA is implemented as the software component of the corresponding order holon (OH), with possible variations of this holon-to-agent mapping. An RA corresponds to a lowest-level holon in the RSH hierarchy (*i.e.* machine level), with possible implementations of higher-level holons (*i.e.* work cell, shop floor, *etc.*) not considered at this time.

This cluster of agents is not a physical cluster, but rather is a logical cluster formed in order to manufacture the products needed to fill orders. The resource scheduling agent is a mediator; it directs orders to the desired resource(s), and it facilitates the processing of these orders on the resources and the interactions and constraints between the resources (*i.e.* scheduling). Because this mediator is linked with its dynamic virtual cluster, it is correctly termed a dynamic mediator agent as described in [9.2]. This dynamic virtual cluster thus requires a dynamic mediator agent, in this case a resource scheduling dynamic mediator agent (RSDMA). This RSDMA is the means by which the system's scheduling algorithms are implemented. It coordinates the local optimisation required by each RA for its resource and by the OA for its jobs, along with the global optimisation required by the work cell and shop floor. The scheduling approach is intended to use an evolving system of rules to accomplish locally/globally "towards optimal" scheduling that can be described as having emergent intelligence. The exact mechanisms and algorithms by which this is accomplished are discussed in Section 9.3.3.

A customer, when ordering, will be interacting with some standard interface, either manually via a Webpage for example, or automatically with his agent via an interface agent. If the scope is limited to exclude the issues of supply-chain management (and its holons), we can call the holon with which the customer interacts the order holarchy. This OH is populated in part by one or more sales agent (SA) and, in the event of more than one sales agent, a sales mediator agent (SMA) that is a static mediator of the type described in [9.2]. As illustrated in Figure 9.1, the customer first interacts with the SMA that directs him to the proper SA (which may have to be created first). The SA may need to confirm cost and manufacturability with the product knowledge holon; it would query the product knowledge mediator agent (PKMA), another static mediator, to find the correct product knowledge agent (PKA) for this task.

Upon receiving a confirmed order, the SA will spawn an order agent (OA), using the knowledge about the product contained in the PKA, or may interact with an already existing OA, passing it the required product information. This order agent is hereafter responsible to the SA for the order, as well as any progress tracking and final delivery. It needs to find the resources that may fulfil the manufacturing requirements of its products, and then interact with this subset of the resource scheduling holarchy. Again, a static mediator is used: the OA, through the resource scheduling mediator agent (RSMA), is put in contact with the RSDMA of the relevant area or shop. The RSDMA sends operations to each relevant resource agent (RA), and manages the scheduling task. The interactions with the scheduling and control holarchies are as previously described.

This chapter focuses on the Scheduling Dynamic Virtual Cluster shown above, and considers only one of these clusters: *i.e.* essentially one work cell or area of the shop. Only the agents within that cluster were implemented. The sales holons and product-model knowledge holons are simulated by a scripted release of preset job types, as defined previously in the benchmark problem. Since there is only one cell/group in this scenario, a resource scheduling mediator agent (RSMA) is not required at this development stage. Finally, the execution control agents (ECA) and control execution agents (CEA) are simulated with an update thread within the resource agent, which internally updates the state and knowledge of the agent, while taking into account the current experimental scenario or scripted randomness in both operation processing times and machine breakdowns. The agent types actually implemented are the order agent (OA), the resource scheduling dynamic mediator agent (RSDMA), and the resource agent (RA), with capabilities and activities as already described.

9.3.3 The Scheduling Algorithm

The resource scheduling dynamic mediator agent (RSDMA) is responsible for, among other things, the scheduling/rescheduling task of a particular area of the shop. It implements a reactive scheduling algorithm that must conform to the design principles of holonic systems [9.3]. The particular algorithmic method designed for this purpose is described here.

This focus on the reactive quality of the scheduler comes from the philosophical roots of holonic systems and the philosophical roots of nouvelle AI (and reactive agents) systems. Basically, a reactive core is desirable for such systems (equivalent to human subconscious), in order to obtain the desired qualities already outlined. More deliberative capabilities are useful at higher levels and should wrap such a reactive core (equivalent to human consciousness). The core of the control system as developed by the authors is a reactive one, including both reactive control capabilities and the ability to reactively schedule and reschedule jobs/operations within the job shop. As a result, a new form of scheduling is proposed as described in the remainder of this section.

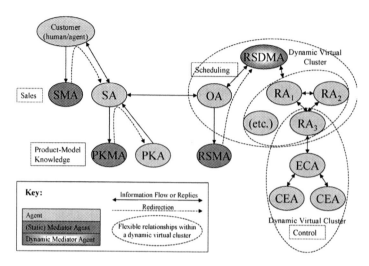

Figure 9.1. Summary of sales/ordering, scheduling, and control

9.3.3.1 A Mixed-heuristic Priorities Approach

The implementation approach to the scheduling problem is to view the system as a number of flexible slots (*i.e.* available resource times) and a number of jostling items (the orders' job times that need to be scheduled), with the scheduling mechanism acting as the Brownian motion excitation that causes the items to fall into low-energy job slots (jobs to be scheduled for a certain time on a certain resource). Unlike traditional scheduling methods that often assume infinite capacity for a resource [9.37] for calculation simplicity, these resources cannot be overscheduled. Also, the jobs that fit best within a resource time slot – as determined by a variety of priority-calculation (*i.e.* scheduling) rules – will be least likely to be jostled out of their time slots, whereas those with the worst fit will be the most likely to be bumped from their slots.

The maximal framework of rules that could be used is created using groups of "tuples" – as outlined in Table 9.2 for completeness – with six order and resource time scheduling rule groups. For example, a rule that can be generated from group one is: "The order with the least number of operations remaining (absolute set point) has increased precedence." These rules are used to rebalance the jobs, either from the orders' perspectives or from the resources' perspectives. Each rule has a real-valued weighting, with some rules weighted more heavily than others. Also, the encoding of each rule with set point(s) requires one or more real-valued variables. These real-valued weights and encoding variables define an individual scheduler.

Not all of the possible rules were used in this initial implementation. Balancing is done only from the point of view of the Jobs, not the Resources, removing the bottom half of Table 9.2. Only the fine granularity rules (*i.e.* job) are used, and not the coarse granularity rules (*i.e.* order or company), removing two-thirds of the top of the table. And, only ASAP scheduling is implemented (JIT scheduling is not currently allowed), removing rule group two from the table. Also, since the increase and decrease in precedence due to these rules is implemented as a simple multiplier (*i.e.* equivalent to a function that is a line through the origin) and not some more complicated function, the Least/Most tuple is entirely redundant for this implementation. Therefore, six core rules were ultimately used in this particular implementation at the current time.

Table 9.2. Tuples for generating order scheduling rules

Jobs		
{Job/order/company} with {least/most} {number/time} of operations		
1.{Remaining/completed}	2.{Early/tardy}	3.Slack
has {increased/decreased} precedence.		

Resources		
{Machine/cell/factory} with {least/most} {number/time} of operations		
4.Scheduled	5.Early/tardy	6.{Slack/retooling}
has {increased/decreased} precedence.		

The default scheduling mode is First Come First Served (FCFS). When jobs cannot be sorted strictly by precedence (*i.e.* two or more jobs have equal precedence), the fallback sorting criteria is the order of job creation, or FCFS.

The first two rule-generating tuples in Table 9.2, *i.e.* those related to Jobs, are essentially a reformulation of some standard scheduling heuristics, plus some related to slack time and slack "operations" within a Job (slack operations are empty spaces in between the actual operations within the Job's schedule). The heuristic rules so generated include the most common heuristics used in scheduling, as shown in the next section. A nice bonus from formulating the common job-related scheduling heuristics in this way is that a group of resource-related scheduling heuristics also become obvious, as shown, although these have not yet been applied to the implemented scheduler at this time.

9.3.3.2 The Scheduling Algorithm

The scheduling and rescheduling is intended to occur as continuously as possible. For the scenarios as tested, rescheduling occurs at a period that is 2/3 the length of the shortest operation. This means that rescheduling occurs every 2 minutes, as the shortest operation is 3 minutes long. (In actuality, since the experiments are run at 20 times real time, the shortest operation is 9 seconds and rescheduling occurs every 6 seconds.) This period roughly matches the maximum time required to complete rescheduling for the scenarios as run, as seen in early experiments, given the available computing power of the workstations used and the multiple of normal time (*i.e.* 20 times) at which experiments are run. It is expected that rescheduling at a period smaller than the length of the shortest operation should logically be expected to give reasonably representative results; this was also indicated by early experiments. In actuality, to be certain, one would need to rerun all experiments in normal time, with the shortest possible scheduling period (*i.e.* 6 seconds in this case) in order to be certain how normal-time experiments would perform, although the time required for this was prohibitive under the circumstances, and so this was not done at this time.

The scheduler is part of the *resource scheduling dynamic mediator agent* (RSDMA) that requests an updated schedule (containing up-to-date information about job completions, times, *etc.*) from each *resource agent* (RA) for each scheduling period. These schedules are passed in an agent-to-agent text message as serialised objects (an automatic Java capability, whereby binary objects are serialised to and from a text string representation in a regular way). Note that since the RAs are obviously still active while rescheduling is occurring, the scheduler can be considered to have "old" information. Therefore, to avoid rescheduling an operation that has already begun to be processed, operations within a RA schedule are "locked" to scheduling when they are within two full scheduling periods of being executed.

The scheduling algorithm itself consists of two parts: firing the mixed heuristic rules, and populating a new schedule. The firing of the mixed heuristic rules involves first calculating all of the required rule inputs for each job (through its operations) at the current time, and then multiplying the weight of each rule (varies between zero and one) by the input value for that rule (for example, processing time remaining) and adding the contributions from each rule to determine a total precedence for that job. These total precedence values for all of the jobs are then used to reorder the jobs (in descending order of total precedence) for the second step of populating the new schedule.

Populating the new schedule first involves taking all operations that are "locked" to scheduling and copying them from the old schedule into the new schedule. Then, the jobs that have non-locked operations are all added to the new schedule one at a time in descending order of precedence. All operations from a particular job are added together, beginning with the earliest operation attempting to start as near as possible to the present in the case of ASAP scheduling, or beginning with the latest operation attempting to finish as close as possible to the due date in the case of JIT scheduling. (Recall, though, that JIT scheduling is not fully implemented and has been left for future work at this time.)

It is worth noting that this scheduling algorithm scales well: most of the algorithm scales in proportional time (in terms of operations or jobs), with the worst-behaving portion of the algorithm (sorting jobs in order) scaling in $n\log(n)$ time (in terms of jobs). On the whole, therefore, this algorithm scales in $n\log(n)$ time, "n" being the number of jobs.

The schedule is generated as an active schedule, meaning that an operation can leap directly to the best spot for itself where there is still room in the schedule. The generated schedule is not a non-delay schedule, since resources that are able to process some operation at a particular time may be forced to wait idly, depending on where the operations have placed themselves in the schedule.

Valckenaers *et al.* [9.31] note that non-delay schedules are, on average, better schedules. This is somewhat intuitive, since it seems reasonable to never keep machines waiting if they can process an operation. It may have been a better choice for this scheduling algorithm to attempt to produce non-delay schedules for this reason, especially since an overriding philosophy behind the development of this scheduler is the desire only to find schedules that are towards optimal, and not necessarily truly optimal. However, the choice to create active schedules was based on the desire to populate jobs into the schedule from both directions, depending on whether the job was ASAP or JIT, in which case a non-delay schedule does not make much sense, whereas an active schedule does.

Once the schedule is fully populated, it is returned to each of the RAs, which then attempt to follow them as closely as possible.

9.3.3.3 Evolutionary Tuning of the Scheduler

The original and ultimate intent was to employ an evolutionary approach for the tuning of a scheduler (*i.e.* the tuning of its rules' real-valued weights and encoding variables that define the individual scheduler). The founding father category of the rest of the domain of evolutionary algorithms (including genetic algorithms) is known as evolution strategies [9.38], specifically involving the evolutionary tuning of real-valued weights that are both problem-related parameters and evolution constants such as crossover and mutation rates, and having its own unique techniques and approaches; these would apply well to this problem, since the evolutionary strategies algorithm tunes itself over time. It should be emphasised that the goal is to evolve the rules by which schedules are created, and not the schedule itself. In this way, the cumulative investment of the evolutionary calculations is retained in the tuning of the scheduling rules. This is unlike a traditional genetic algorithm scheduler that evolves the schedule itself, and in which there is no accumulation of evolutionary "equity" in the system. This is an important distinction, because the latter is merely an algorithm, whereas the former can be expected to implicitly recognise and retain patterns (*i.e.* a learned set of behaviours) and therefore can be described as a learning intelligence, however limited, according to our previous definition. The evolutionary scheduler is illustrated in Figure 9.2.

9.3.3.4 Scheduling Mediator Agents

Dynamic mediator sgents [9.2] implement these scheduling rules – which coordinate the needs of the various *order agents* and *resource agents* towards achieving an optimum schedule – in a reactive manner. One benefit to using both agents and the outlined scheduling approach is that there is no need to implement a rescheduler as well, because with this approach scheduling is identical to rescheduling. Another benefit is the ease with which the scheduling approach can be modified, since the scheduling is completely contained within the *scheduling mediator agents*, which can be dynamically recommissioned into the system. This important benefit means that the value of the method of employing scheduling mediator agents is not tied to the success of the scheduler itself; they are decoupled. The specifics of the scheduling approach taken are felt in hindsight to be somewhat naïvely chosen, in that they do not necessarily capture the complexities and inter-relationships present in the job-shop scheduling problem; however, this will be discussed in more detail later.

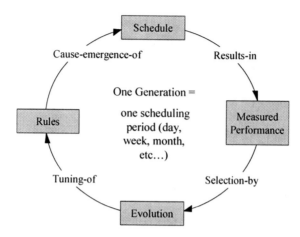

Figure 9.2. Intelligent scheduling using evolved rules

Two types of activities are envisioned in the *resource scheduling dynamic mediator agents* (RSDMA) in this reactive scheduler: one for balancing the jobs and another for load balancing between the resources. These activities occur concurrently, reactively applying the rules to modify elements of the schedule(s) for those orders and machines with the greatest need, as measured by the calculated precedence for each. In this way, the scheduler uses the evolved/evolving rules to dynamically and (somewhat) chaotically (re)generate a towards-optimal schedule. As only one of each type of resource is implemented currently, there is no need for any load-balancing activities, though the system has been designed to allow this in the next stage of development.

9.3.3.5 Scheduling Comparisons

This implementation's system of heuristics and priority rules for scheduling will be compared against some common scheduling heuristics, such as: FCFS (first come first served), LWKR (least work remaining), MWKR (most work remaining), SPT (shortest processing time), LOPNR (least operations remaining), and MOPNR (most operations remaining).

The testing of the scheduler against these common heuristics is straightforward, as each of these common heuristics can be implemented by activating only one (or possibly two) of the scheduling rules described previously, rather than in combination as outlined.

For example, implementing the LWKR heuristic is a matter of turning off all rules except the "{job} with the {least} {time} of operations {remaining} has {increased} precedence".

As another example, recall that the default scheduling mode of the precedence rule scheduler is FCFS; this decision rule is used to decide between jobs when two or more each have a precedence that is equal to the other. In order to implement a purely FCFS system, one merely turns off all other precedence rules, and allows the default FCFS scheduling to take over fully.

9.4 Experiments

The end purpose of the collective set of experiments contained within this section is to show, for real-world scenarios and in real time, that: (1) a mixed-heuristic scheduler can be used effectively for job-shop (re)scheduling, with reasonable results, in both deterministic and stochastic scenarios, and (2) an evolutionary algorithm can be used effectively to continually tune this mixed-heuristic scheduler towards an optimal configuration as the conditions in the job-shop change over time.

9.4.1 Summary of the Experimental System

The experiments reported in this chapter are performed on a 20/4/G/* scheduling problem: 20 jobs, 4 resources, general job shop, with the "*" signifying optimising in more than one performance measure at once. (Since there is no explicit limit on the number of jobs, and the number of resources could be changed with only the barest of modifications, one could possibly more accurately describe the system as solving the $n/m/G/*$ scheduling problem.)

9.4.1.1 Experimental Test Scenarios

The experiments conducted are based upon the benchmark presented in [9.5] and described in Section 9.3. In each experiment, 20 jobs are dispatched into the system. There are five job types, as described in Table 9.1; these are dispatched in an ascending and repeating order, as in J1, J2, J3, J4, J5, J1, J2, *etc.*

There are three stages of experiments, as follows: Stage 1 – Deterministic scenario, Stage 2 – Stochastic scenarios, and Stage 3 – Evolutionary scenario. Stages 1 & 2 use both pure heuristics that are essentially formed from the core scheduling rules of the scheduling engine, and mixed heuristics with their six core rule weights generated randomly (once). Experimental Stage 3 attempts to evolve combinations of the core scheduling rules that result in useful mixed-heuristic schedulers. (These basic and mixed heuristics are described in the next section.) Stages 1 & 2 also serve to verify the system in preparation for Stage 3, as well as to allow comparison between system response for deterministic and stochastic scenarios. Stage 3 serves to test the ability of this learning holonic system to learn and migrate towards optimal scheduling methods for a continually varying realistic scheduling problem or scenario. The results of the Stage 1 experiments can be found in [9.34]: in the remainder of this section we focus on the Stage 2 & 3 experiments.

This is a production system (*i.e.* not a simulation) that is intended to run in real time, so the experiments within each stage would each require hours to complete. However, the MAS (including the scheduler) require far less than the entire processor of a current 2 to 3 GHz workstation running Linux in text-only command-line mode (*i.e.* operating system is performing no windowing functions). As implemented, without any optimisation of the code to speed up execution, it was found in pre-testing that this system could be speeded up by a factor of 20 over normal time execution, without ill effects (*i.e.* inter-agent messages timing out, rescheduling frequency dropping too low, *etc.*). At this rate, rescheduling is run at a periodic interval that is 2/3 the length of the shortest operation in the scenario, which was felt to be enough to approximate "continuous" rescheduling in most circumstances.

Unless stated otherwise, each stage has 30 schedulers (15 pure heuristics and 15 random mixed heuristics) each running 30 repetitions for each experimental stage, for a total of 900 different experiments run for each stage. Even speeded up by a factor of 20, and also run in parallel on 20 separate workstations, each partial experimental stage still requires being run overnight (or slightly longer) to complete. Implementing this system as a truly multi-agent system rather than an idealised object-oriented system has resulted in some limitations on the types and numbers of experiments that can be run. Further experimentation would be suggested on an idealised O-O implementation of this system backed by a proper simulation, in order to be more certain of the results of the mixed-heuristic scheduling approach.

9.4.1.2 Scheduling Heuristics

There are six core rules used by the scheduler, as created from the tuples {operations/time}, and {completed/remaining/slack}, *i.e.* operations completed, operations remaining, operations slack, time completed, time remaining, and time slack. Each of these rules can have a positive or negative weight (corresponding to "most" versus "least") or a weight of zero, in which case it is completely inactive.

The pure heuristics used in experiment Stage 1 and Stage 2 include these six core scheduling rules and their inverses, used by the scheduler as given in Table 9.3. The pure heuristics also include two common heuristics created by equally combining one pair of scheduling rules and their inverses (*i.e.* SPT and LPT), and a third heuristic that involves a lack of scheduling rules (*i.e.* first come first served), as given in Table 9.4.

Table 9.3. Pure heuristics from individual scheduling rules

Precedence of each job in the (re)scheduling task is ordered based upon:	
1. MOC	7. MTC
2. LOC	8. LTC
3. MOR	9. MTR
4. LOR	10. LTR
5. MOS	11. MTS
6. LOS	12. LTS

Table 9.4. Other basic heuristics

Precedence of each job in the (re)scheduling task is ordered based upon:	
13. LPT	16. MTC combined with MTR
14. SPT	17. LTC combined with LTR
15. FCFS	18. no scheduling rules active

The mixed heuristics may be any linear combination of the six core scheduling rules, determined randomly in Stages 1 & 2 or using an evolutionary algorithm in

Stage 3. Note that heuristics that are opposites, such as MTR and LTR, can never be combined together, as they are inverses of the same core scheduling rule (*i.e.* MTR would result from a positive weight for *time remaining*, while LTR would result from a negative weight for *time remaining*).

9.4.1.3 Mixed-heuristic Scheduling Examples

Let us consider an example of a mixed-heuristic scheduler with weights for the six core rules as follows: {−1, −0.6, −0.2, 0.2, 0.6, 1.0}. Assume that the *inputs* to these rules for a particular job are as follows: {3, 1, 4, 60, 120, 210}, which correspond to a job having 3 operations completed, 1 operation remaining, 4 operations slack (*i.e.* delays in its schedule), 1 minute of time completed, 2 minutes of time remaining, and 3.5 minutes of slack time (*i.e.* total time of slack operations) in the *current schedule*. For this job, its scheduling outputs are simply the inputs times their respective rule weights, or: {−3, −0.6, −0.8, 12, 72, 210}. These outputs correspond to the increased precedence due to each rule. The total precedence of this job is simply the sum of these outputs, or: {−3 − 0.6 − 0.8 + 12 + 72 + 210 = 289.6}. All other jobs in the system also have their own unique precedence values; these values are used to order the jobs in preparation for building a *new schedule*, with the highest-precedence jobs choosing their ideal positions in the schedule for each of their operations, given the existence of other operations that have already been placed into the schedule during this round of rescheduling. ASAP jobs perform forward scheduling of their operations, starting with the first operation as early as possible, and so on for each of its operations; JIT jobs would perform backward scheduling of their operations, starting with their last operation as late as possible without being late, and so on for each of its operations.

This entire chain of events is repeated until the schedule converges, up to three times. Many combinations of rules allow convergence in the first (or second) step; however, some rules (such as MTS) are unstable and will not converge (*e.g.* MTS will cause the schedule to keep "flipping" front to back and back to front indefinitely) and so the three-step limit is imposed. In the case of an unstable combination of rules, like MTS, an odd number of steps is appropriate, so that the schedule is shuffled as much as possible in each scheduling step, which fits with the dynamic "Brownian motion" nature desired for the scheduler.

As an aside, it is worth examining the MTS heuristic at this point. The odd number of allowed steps also works well in this case, since the jobs with the poorest time slack characteristics will be on the ends of the sequence that is being flipped: when time advances such that an operation becomes locked into place in the schedule just prior to executing, jobs on both ends of that sequence are equally likely to be the ones having their jobs stick, so that the poorest time slack performers generally get processed first.

Let us consider another example of the SPT (shortest processing time) heuristic in Table 9.4, which is described as a combination of the LTC (least time completed) and LTR (least time remaining) rules. Why is this so? The job with the shortest (or least) processing time is the same as the job with the least time completed plus the least time remaining, since time completed plus time remaining equals processing time. An illustrative calculation is given below.

$$SPT = -1 \times (\text{Processing Time})$$
$$= -1 \times (\text{Time Completed} + \text{Time Remaining})$$
$$= (-1 \times \text{Time Completed}) + (-1 \times \text{Time Remaining})$$
$$= LTC + LTR$$

9.4.1.4 Performance Metrics

A wide array of possible performance metrics exist with which to evaluate the effectiveness of each experiment's chosen scheduling heuristic. A select group of these metrics are measured in each stage, with a simple average of these metrics acting as the objective function to optimise using the evolutionary algorithm in Stage 3.

The performance metrics collected include:

- Job Slack (minimum, average, and maximum) – the downtime between the processing of a job's various operations,
- Resource Slack (minimum, average, and maximum) – the downtime between a resource's processing of its various jobs' operations,
- Makespan – the time between the start of the first job and end of the last job to be processed in the shop in a particular scheduling period,
- Job Flowtime (minimum, average, and maximum) – the time between when a job first enters the shop and when it has been completed, and
- Schedule Efficiency – the ratio between the total time all resources spend actively processing operations and the total time required for an operator to attend the machine (including slack time).

Performance metrics that relate to customer satisfaction or dissatisfaction include Flowtime, and Job Slack. Performance metrics that relate to overall factory costs include the Makespan. Performance metrics that relate to the efficiency of the job-shop scheduler include the Job Slack (*i.e.* the job's perspective), the Resource Slack (*i.e.* the resource's perspective, which may run counter to the job's perspective), and the overall Efficiency. Note that other common metrics such as Earliness or Lateness are not considered here because the current system implementation only performs As Soon As Possible (ASAP) scheduling, though such metrics would be appropriate for ultimate implementation of Just In Time (JIT) scheduling in which jobs have a specified due date.

Of these performance metrics, the authors feel that the least redundant and most relevant and useful set consists of the Average Flowtime (measure of customer satisfaction), the Maximum Flowtime (measure of maximum customer dissatisfaction), the Makespan (measure of the total production costs), and the overall Efficiency (measure of the efficiency of the scheduler). Theoretical best possible cases are used to normalise the first three of these four metrics between zero and one (Efficiency already being a normalised measure), taken as follows: the theoretical best Makespan being the sum total of job lengths divided by the number of resources, the theoretical best Average Flowtime being the average job length, and the theoretical best Maximum Flowtime being the maximum job length.

So, the normalised metric is calculated as the theoretical best metric divided by the actual metric, resulting in a value between zero and one, with one being the best case and zero being the bounding asymptote for the worst case.

The average of the four normalised performance measures is also taken as a Combined Performance measure, to measure the overall effectiveness of a particular scheduling heuristic given certain experimental conditions. This combined metric, which also may vary between zero and one, is used to evaluate all stages, but is especially used in Stage 3 to rank the various evolved scheduling engines that are essentially performing multi-variable optimisations of their respective systems.

9.4.1.5 Randomly Generated Mixed Heuristics

The deterministic and stochastic experimental stages described in the remainder of this section include results from 15 "pure" heuristic schedulers. These stages also include results from 15 "randomly generated" mixed-heuristic schedulers, which were generated to get a feel for how well the mixed-heuristic schedulers perform, without having any prior assumption as to how exactly the heuristics should be "mixed", meaning what weights should be chosen for each rule (although more appropriate weights are evolved in Stage 3, later). For these randomly generated mixed-heuristic schedulers, the weighting of each of the six core rules (which can vary between −1.0 and +1.0) were randomly generated using a known random seed (so that experiments could be repeated) and the *Java.util.Random* class.

To summarise, the randomly generated mixed-heuristic schedulers in Stages 1 and 2 are an attempt to get a feel for how well the mixed-heuristic scheduling approach performs, given the worst case of naively (indeed, *randomly*) chosen weightings for the mixture of heuristic rules.

9.4.2 Stochastic Scenario (Stage 2) Results

The second experimental stage covers the stochastic scenarios, or *Plant Scenarios 2 and 3* from the benchmark. Under Plant Scenario 2, operations may complete in stochastic time, while under Plant Scenario 3, non-bottleneck resources may fail in a stochastic manner.

Each of the 15 possible pure heuristics and the 15 randomly generated heuristics is applied in turn for scheduling one experiment, *i.e.* a complete run of the system from dispatching through scheduling, until all operations have been processed.

One main objective of this experimental stage is to confirm that these stochastic scenarios yield reasonable results, in preparation for the third stage. The other main objective of this stage is to allow comparison of the quality of results obtained in the deterministic scenario versus the stochastic scenarios, between the various scheduling heuristics. This stage allows some measurement of whether this entire system as implemented has met the goals of holonic manufacturing, especially the ability to handle random disturbances.

It is worth repeating that absolutely nothing within the system changes under a stochastic scenario. All of the agent interactions remain exactly the same, as does the scheduler operation. It is the dynamic nature of the scheduling algorithm, and the form of the schedulers themselves that intrinsically encompasses the ability to handle any stochastic events in the job shop.

For the Stage 2 experiments, the 30 schedulers (15 pure heuristics and 15 random mixed heuristics) are compared against each other in Makespan, Schedule Efficiency, Mean Flowtime, and Maximum Flowtime, as well as the Combined Performance Metric. For brevity, we focus on the Combined Performance Metric results here, as summarised in Figures 9.3 and 9.4 (*i.e.* stochastic completion times and machine failures scenarios, respectively). A range of results is presented for each scheduler, since 30 experiments were performed for each of the schedulers, with the graphs again showing ±2 standard deviations about the mean in each measure for each scheduler, said ranges being presented side by side, with one performance measure compared within each figure.

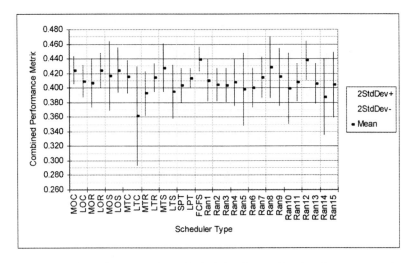

Figure 9.3. Stage 2A comparison of schedulers by combined performance metric

In Stage 2A (stochastic operation completion times), each operation has a 1/3 chance of its completion time being delayed. This delay is defined randomly by a *TRIA(1, 2, 3)* function, with a minimum of one, average of two, and maximum of three minutes, and having a triangular distribution. The random delay is generated with the *Java.util.Random* class using a known random seed for each set of experiments for a scheduler, so that the same conditions can be repeated for each scheduler tested. This flat distribution random number is then adjusted to obtain the desired triangular distribution and the desired range of possible values.

When compared with our preliminary experiments with the deterministic scenario, the top four schedulers remain the same as in the deterministic scenario for the Combined Performance metric: Random 12 on top, MTS and Random 8 next, and FCFS following. For a dynamic scheduler of the form described, it would seem that good performance in a deterministic scenario is also a predictor of good performance in a stochastic operation completion time scenario. This is a very encouraging result.

From simple calculations based on the distributions used for the completion times in Scenario 2A, the normalised metrics should average a 9.6 ±0.4% decrease from the deterministic scenario to the stochastic completion-times scenario. In fact, the average measured change in the Combined Performance metric was only an 8.7 ±0.4% decrease from the deterministic scenario to the stochastic scenario. This is also a very encouraging result, which would seem to indicate that the system presented herein is very able to respond effectively to stochastic operation completion times in the job shop.

In stage 2B, a simple machine-failure scenario was chosen to simplify analysis. Resource No. 1, a non-bottleneck resource, encounters an error condition and stops processing at approximately 20% of the way through the schedule for approximately 10% of the schedule length, based on the average Makespan encountered in the deterministic scenario. Since the average Makespan is 201.7 minutes in the deterministic scenario, the resource error condition is started at 40 minutes into the experiment and continues for 20 minutes. Note that the average job duration is 25.2 minutes, so the sum total duration of all 20 jobs is 504 minutes. The performance of the 30 schedulers for the Combined Performance Metrics is shown in Figure 9.4.

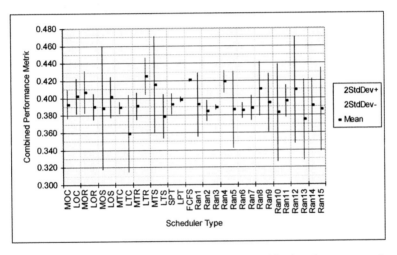

Figure 9.4. Stage 2B comparison of schedulers by combined performance metric

The best-case scenario one could reasonably expect is for only the one resource to be affected, and for the remaining resources to continue processing operations without interruption. In this case, if one considers the job being processed during the resource failure to be extended in length by 20 minutes, then the average job (or operation) length would increase by 20 parts in 504 to 104.0% of the deterministic scenario.   Likewise, the metrics under consideration could also be expected to rise to 104% of the deterministic scenario, so the normalised metrics could be expected to drop to the reciprocal, or 96.2% of the deterministic scenario – a decrease of 3.8 per cent.

The worst-case scenario one could reasonably expect is for all resources to be affected by the one resource failure, due to the blocking of those resources by operations dependent on the completion of the operation trapped on the resource experiencing the failure. In a similar manner, the metrics under consideration could be expected to rise to 115.9% of the deterministic scenario, or the normalised metrics to drop to 86.3% of the deterministic scenario – a decrease of 13.7 ±0.4%.

The actual average decrease in the Combined Performance metric was 12.0 ±0.4%, or very nearly the worst possible case. The reason for this is simple. Due to the need to lock an operation to scheduling somewhat before the current time in the schedule, operations that are dependent on the stopped operation on the failed resource do get stuck on their resources, thereby blocking other operations that would otherwise be available for processing. This is also due in part to the fact that no special measures are taken in any part of the scheduling activities to account for resources that are experiencing error conditions.

This problem is also easily solved. It is necessary for a resource agent to estimate its expected failure duration at some point, either automatically, based upon error codes from the controller and past history, or manually via an estimate entered in by the operator or maintenance personnel shortly after the onset of the failure. Knowledge of this failure duration would allow a resource agent to block out the affected portion of the schedule, thereby allowing (re)scheduling to continue as normal around this schedule blockage. The operations that would otherwise be blocking their resources would immediately be forced to occupy a later position in the schedule, thereby allowing other available operations to be processed and so continue to utilise the shop floor to the fullest extent possible. As a side benefit of these activities, the schedule would remain current (though still just a best estimate, as usual), and so estimated delivery times would also remain current (to the best possible estimate given current shop-floor information).

Under these circumstances, the system could be expected to perform well – similar to the quality of the system's performance in Stage 2A. Also, it should be noted that no other special steps would need to be taken to ensure that the system responds effectively to machine failures (*i.e.* no algorithm modifications, additional rules, separate optimising decisions, *etc.*), and so the system remains a very clean and pleasing solution to the various requirements of real-world dynamic job-shop scheduling. Though the results of the Stage 2B experiments themselves were not good, the performance that could be expected of the system with this slight upgrade in capabilities is actually very encouraging.

9.4.3 Evolving the Mixed-heuristic Scheduler

The third and final experimental stage involves using an evolutionary algorithm to evolve a mixed-heuristics scheduler for the deterministic scenario.

The main objective of this stage is to attempt to evolve a mixed-heuristic scheduler that performs better than standard pure heuristic schedulers under given circumstances, thereby showing that evolutionary tuning of a mixed-heuristics scheduler can, in principle, discover or learn effective scheduling strategies for particular system states, and therefore could be expected to transition between these evolving strategies as the system state itself changes from one scheduling period to the next (*i.e.* days, weeks, months). This system of continuously evolving mixed-heuristic schedulers could then be considered to be a learning system that performs towards optimally for the job-shop scheduling problem, which is a prior-stated goal of such a holonic (and *intelligent*) scheduling system.

9.4.3.1 Results

First, the ability of the system to evolve a good mixed-heuristic scheduler under the deterministic scenario is presented, examining the improvement and convergence of the population under each of the performance metrics, and under the Combined Performance Metric. Figures 9.5 and 9.6 present these results.

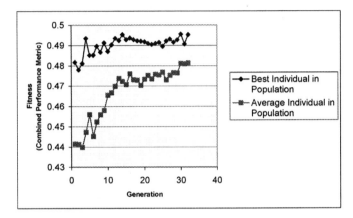

Figure 9.5. Convergence of evolutionary algorithm

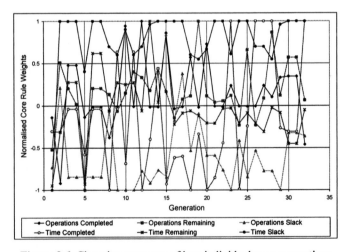

Figure 9.6. Changing genotype of best individual over generations

Next, the top three performing evolved mixed-heuristic schedulers (plus one extra) are compared against the original 30 schedulers (15 pure heuristics and 15 random mixed heuristics) in Makespan, Schedule Efficiency, Mean Flowtime, and Maximum Flowtime, as well as the Combined Performance Metric. Figure 9.7 presents these comparisons of Stage 3 results under the deterministic scenario with the previous deterministic scenario results.

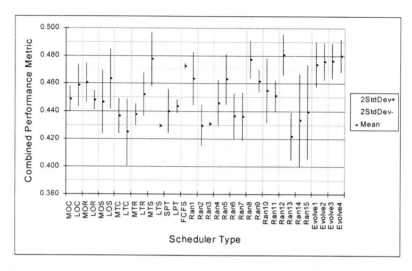

Figure 9.7. Comparing evolved mixed-heuristic schedulers with stage 1A schedulers

9.4.3.2 Evolutionary Terms and Selected Values for Parameters

In the evolutionary scenario, each *individual* corresponds to a particular mixed-heuristic scheduler. An individual's *genotype* (or *chromosome*) is the set of six real values between −1.0 and +1.0 that correspond to the weights for each of the core scheduling rules within the scheduler, with one *gene* corresponding to one of these weights. An individual's *phenotype* is the resulting scheduler, which will have some quality of performance in each of the collected metrics for the one experiment run. An individual's *fitness* corresponds to the value of its Combined Performance Metric. The *population* is the collection of all individual schedulers/experiments.

In an evolutionary algorithm, the mutation rate is the rate/probability at which a gene is randomly adjusted. Each individual mutation may involve flipping one bit of a gene in a genetic algorithm, or taking a real-valued gene and adding (or multiplying by) a random value in an evolutionary strategies algorithm. Mutation allows the genetic makeup of a population to increase in variety. The crossover rate is the rate at which (or probability of) a chromosome crosses another chromosome, exchanging all genes in between the crossover points. Crossover allows individuals to swap genes, which can bring two complementary genetic encodings together in the same individual.

Back [9.39] reports that the following parameters have been found to result in the best online performance of a genetic algorithm: population size of 20–30 individuals, crossover rate of 0.75–0.95 per individual, mutation rate of 0.005–0.01 per bit, and two crossovers points used when performing each crossover operation.

Given the tendency for variance between experiments using identical schedulers, it was decided that more individuals (*i.e.* more schedulers tested) would be an advantage, so the upper end for population size was chosen. Since the particulars of the implementation allowed 32 individuals to be processed as quickly as 30 individuals, a population size of 32 individuals was selected.

Although two crossover points have been found to be very effective when performing crossover operations [9.39], other numbers of crossover points work as well. It was decided to express this as a crossover rate per gene, with an average of two crossover points desired if a crossover occurs, but with some variability allowed. There are six genes in an individual, and the number of crossover points should average two, so the chromosome crossover rate was chosen to be two over six, or 0.333 per gene. This will result in an average number of crossover points of two per chromosome, but still allow some single crossovers, some triple crossovers, *etc.* Also, the individual crossover rate was selected as 0.95 per individual (*i.e.* 95% of individuals have some crossover, with a crossover point rate of 33.3% per gene).

A mutation rate of 0.01 per bit has been found to work well for genetic algorithms that use a binary encoding scheme for variables [9.39]. However, the variables in the evolutionary experiments reported here – *i.e.* the weights of the mixed-heuristic schedulers – are real-valued numbers, and so some conversion must be made. If the real-valued variables had instead been encoded as 8-bit binary values, there would be 48 bits in total, and a mutation rate of 0.01 per bit would result in approximately 0.5 mutations per chromosome. So, the mutation rate was chosen to be one-half over six genes, or 1/12 (*i.e.* 0.0833) mutations per gene.

Also, the implicit mutation range for a binary-encoded variable is approximately half of the total possible value for the variable, since the largest change involves flipping the most significant bit, a change of approximately one half of the largest value. For example, an 8-bit binary number can encode values from 0 to 255, while flipping the most significant bit will change the binary number by 128. So, the possible mutation range was taken to be –0.5 to +0.5, given that the possible values of each gene range from –1.0 to +1.0.

Finally, fitness proportionate selection with roulette-wheel sampling [9.39] was chosen since it is an often-used and very straightforward selection method and, for this simple proof-of-concept experiment, more exotic methods were not necessary. However, to speed up convergence, the common variation known as elitism was also used [9.17], with the four most fit individuals from the previous generation being selected for inclusion in the next generation.

9.4.3.3 Evolving a Mixed-heuristic Scheduler

In any real online evolution of mixed-heuristic scheduler weights, only a limited number of generations would be practically possible, without spending larger amounts of money on computers (or computer cycles). So, when the evolution experiment was seeming to converge at 32 generations, the experiment was stopped. The improvement of the average individual and the best individual of the population over 32 generations is shown in Figure 9.7, with an individual's fitness taken to be its Combined Performance Metric.

Although the algorithm does converge, the difference between the best individual near the beginning of the run and the best individual near the end of the run does not seem to be very large – only about three or four per cent; however, one must remember that a randomly chosen mixed-heuristic scheduler actually has a reasonably good chance of performing well (like Random 12 and Random 8 in Stage

1A – 13% of the random schedulers tested). So, given these circumstances, this evolutionary "tuning" is more about finding which random schedulers are likely to perform well in a given scenario, than being about showing huge improvement while the algorithm converges. It is perhaps more important to note that the best individual found near the end of the run has a fitness more than 12% higher than the average individual found near the beginning of the run.

Figure 9.6 shows the changing genotype of each generation's best individual; a vertical slice through the figure corresponds to one individual, or the change in relative strength of the core rule weights can be tracked horizontally. The values of the rule weights were normalised for clarity such that the absolute value of the largest weight in each individual is one, since only the relative size of the weights compared with each other is important, not any absolute values of the weights. One can see that in Time Slack, the weights tend to be positive (*i.e.* Most Time Slack) and large in magnitude; conversely, the Operations Slack tend to be negative (*i.e.* Least Operations Slack) and large in magnitude. This may seem like a contradiction, but need not be. In fact, the exact way that the weighted heuristics interact to generate a schedule is not precisely known and, in fact, attempting to understand these interactions precisely somewhat misses the point. Each situation will be different, and will rely on different subtleties of interaction, which is why an attempt is being made to evolve weights that are applicable to a certain situation, and/or to have these weights evolve over time to follow the changing situation and states of the manufacturing system.

Operations Completed, Operations Remaining and Time Remaining tend to have small values (positive or negative) – *i.e.* a smaller contribution to the scheduler – while Time Completed does not seem to follow any particular pattern in this deterministic scenario.

9.4.3.4 Testing the Best Evolved Mixed-heuristic Schedulers

The three best individuals from the evolutionary tuning experiment (plus one extra) were run in a standard deterministic scenario, with 30 repetitions for each, and the results compared with the results from the deterministic scenario reported in Stage 1A. Evolve 1 and Evolve 2 are the best individuals from generations 30 and 32, respectively (generation 31 has the same best individual as generation 30). Evolve 3 is a rounding of Evolve 2 weights to the nearest 0.5, or {–0.5, 0, –0.5, 1, 0, 1}, for interests sake. Evolve 4 is the best individual from an earlier peak in the population fitness at generation 14.

The Combined Performance Metrics of these evolved mixed schedulers are compared with the other deterministic scenario schedulers from Stage 1A in Figure 9.7. One can see that these evolved schedulers perform comparably to the best schedulers from Stage 1A (and, referring back to Stage 1B, they also perform comparably to the best schedulers from that stage).

The evolutionary algorithm has indeed tuned several mixed-heuristic schedulers such that they perform well in the Combined Performance Metric in a deterministic scenario. Given how straightforward this experiment was to run, there is no reason to doubt that the evolutionary algorithm can also be used to tune schedulers that work well in a scenario with different characteristics, such as having different numbers and/or types of jobs, or having stochastic completion times or machine failures. These are encouraging results.

9.4.3.5 Issues with the Evolving Schedulers

It is not known whether the evolutionary tuning could keep up to the changing shop conditions from one scheduling period to the next in real time, or whether conditions from one scheduling period to the next would be similar enough for this type of online-tuning approach to work. On the other hand, one could certainly tune schedulers that each had good performance for one of these shop-condition scenarios, and then dynamically match the scheduler to the job conditions on the fly; this would be a form of "meta controller", similar to the meta controller discussed in [9.40].

Within one scheduling period, the behaviour of a scheduler may be good in some circumstances, but not in others. Should the general circumstances change, and a particular evolved scheduler be known to perform better under the new circumstances within this scheduling period, one could switch to the new scheduler (*i.e.* set of core rule weights) on the fly, which would also be a form of meta control like that described above.

Over many scheduling periods, maintaining a diverse population is also a key concern in this active evolutionary scenario, when attempting to allow a population of schedulers to evolve continually and indefinitely. If one's population does not maintain some diversity, the system will be very slow to evolve, and therefore slow to respond as conditions change over scheduling periods. There are various techniques discussed in the literature regarding maintaining population diversity (*e.g.* maintaining separate sub-populations, introducing some randomly generated individuals into each new generation) [9.38].

9.5 Conclusions

How does one know that the system is actually obtaining "towards optimal" solutions, as desired in this research? Evolved solutions did improve over generations, so it can definitely be said to be improving, and therefore moving towards the optimal. The question of how close one actually gets to the true optimum was not examined in this research, and is left for future work. Either alternative solutions could be applied to the benchmark as implemented, or one could extend this research to implement the full benchmark, and then compare alternative solutions against these full benchmark results. In the latter case, the results here could be compared against results obtained by other researchers implementing the benchmark, such as in [9.41]. Work on the benchmarking project presented in [9.5] is progressing, and it is expected that other researchers will have relevant results to compare against before long.

The best results in this chapter do not approach 100% of the theoretical best, as calculated for the various performance metrics. These theoretical bests would only be achievable in perfect conditions and for most circumstances (*i.e.* job-release times, types of jobs, *etc.*) it is an impossible goal to reach, in much the same way that it is impossible to reach the theoretical best Carnot Efficiency of the 2nd Law of Thermodynamics in a real heat engine. Again, this work was not specifically compared against other scheduling approaches, and it is unknown how well those approaches perform relative to the theoretical bests presented here, as the scheduling literature does not consider its performance metrics in the manner presented in this chapter. This question could be resolved when making the comparisons against other research as discussed in the previous section.

Some of the result sets (*i.e.* for each individual scheduler) presented in this section have a large variance. As discussed previously, attempting to evolve schedulers that have a lower variance in their result sets could potentially address part of this problem. On the other hand, this "problem" is not really a problem if the *average* performance of a scheduler is acceptable to industry, especially given the other characteristics of the scheduling approach and its general ability to handle a much more real-world problem.

For the evolutionary experiments run in Stage 3, each generation requires on the order of 15 minutes maximum to complete, given one workstation for each MAS experiment. Experiments were actually begun with slight offsets from each other (of approximately 1 minute) to avoid bugs that were encountered regarding the concurrent start up of Java RMI (Remote Method Invocation) processes. (Since this problem was not in any way central to the work conducted in this research, the simple workaround was deemed to be appropriate at this time.) Also, the number of workstations used was less than the number of experiments to run. Each generation required approximately 40 minutes to complete (assuming the experiment workstations were not tampered with during the experiments), or about 24 hours for one complete set of evolutionary experiments.

However, for an actual implementation of such a system, the time required for each generation is and must be the time involved in one scheduling period of a particular shop. One (or more) workstations would execute the MAS that performs scheduling and control of that shop, with plenty of extra processing power available. An appropriate number of additional workstations can execute the experiments that simulate the other individuals in the generation, such that the necessary processing power never exceeds the available processing power for these normal-time experiments running concurrently. For a scheduling period equivalent to that used for the experiments in this chapter, 2 workstations would suffice, although more could and perhaps should be added to more closely approximate "continuous" rescheduling. Note that the processing power and time required to execute the evolutionary algorithm itself is insignificant compared to that required for the MAS execution and job-shop simulation. Note also that re-implementing the MAS as an object-oriented system would allow significant reduction in the required computing power.

Acknowledgment

The authors wish to thank the Natural Sciences and Engineering Research Council for their generous support of this research under grant OGP-019-7339.

References

[9.1]   Parunak, H.V.D., 1993, "Autonomous agent architectures: a non-technical introduction," *Industrial Technology Institute Report*.

[9.2]   Maturana, F.P. and Norrie, D.H., 1996, "Multi-agent Mediator architecture for distributed manufacturing," *Journal of Intelligent Manufacturing*, 7, pp. 257–270.

[9.3]   Valckenaers, P., Van Brussel, H., Bongaerts, L. and Bonneville, F., 1995, "Programming, scheduling and control of flexible assembly systems," *Computers in Industry*, 26, pp. 209–218.

[9.4]   Weiss, G., 1999, *Multiagent Systems*, The MIT Press, Cambridge, USA.

[9.5]   Cavalieri, S., Bongaerts, L., Macchi, M., Taisch, M. and Wyns, J., 1999, "A benchmark framework for manufacturing control," In *Proceedings of the Second International Workshop on Intelligent Manufacturing Systems*, Leuven, pp. 225–236.

[9.6]   Christensen, J., 1994, "Holonic manufacturing systems – initial architecture and standards directions," *First European Conference on Holonic Manufacturing Systems*, Hannover, Germany, pp. 1–10.

[9.7]   Bussman, S., 1998, "An agent-oriented architecture for holonic manufacturing control," In *Proceedings of IMS'98 – ESPRIT Workshop in Intelligent Manufacturing Systems*, Lausanne, Switzerland, pp. 1–12.

[9.8]   Van Brussel, H., Wyns, J., Valckenaers, P., Bongaerts, L. and Peeters, P., 1998, "Reference architecture for holonic manufacturing systems: PROSA," *Computers in Industry*, 37, pp. 255–274.

[9.9]   Zweben, M. and Fox, M., 1994, *Intelligent Scheduling*, Morgan Kaufman, San Francisco, CA.

[9.10] Prosser, P. and Buchanan, I., 1994, "Intelligent scheduling: past, present, and future," *Intelligent Systems Engineering*, 3(2), pp. 67–78.

[9.11] Dilts, D.M., Boyd, N.P. and Whorms, H.H., 1991, "The evolution of control architectures for automated manufacturing systems," *Journal of Manufacturing Systems*, 10(1), pp. 79–93.

[9.12] Duffie, N. and Prabdu, V., 1994, "Real time distributed scheduling of heterarchical manufacturing systems," *Journal of Manufacturing Systems*, 13(2), pp. 94–107.

[9.13] Lin, G. and Solberg, J., 1994, "Autonomous control for open manufacturing systems," In *Computer Control of Flexible Manufacturing Systems*, Joshi, S. and Smith, J. (eds.), Chapman & Hall, London, pp. 169–206.

[9.14] Baker, A.D., 1998, "A survey of factory control algorithms which can be implemented in a multi-agent heterarchy," *Journal of Manufacturing Systems*, 17(4), pp. 297–320.

[9.15] McFarlane, D.C. and Bussman, S., 2000, "Developments in holonic production planning and control," *International Journal of Production Planning and Control*, 11(6), pp. 522–536.

[9.16] Ramos, C., 1996, "A holonic approach for task scheduling in manufacturing systems," In *Proceedings of IEEE Conference on Robotics and Automation*, Minneapolis, pp. 2511–2516.

[9.17] Gou, L., Luh, P. and Kyoya, Y., 1998, "Holonic manufacturing scheduling: architecture, cooperation, mechanism and implementation," *Computers in Industry*, 37(3), pp. 213–231.

[9.18] Sugimura, N., Hiroi, M., Moriwaki, T. and Hozumi, K., 1996, "A study on holonic scheduling for manufacturing system of composite parts," *Japan/USA Symposium on Flexible Manufacturing*, pp. 1407–1410.

[9.19] Marcus, A., Kis, T., Vancza, and Monostori, L., 1996, "A market approach to holonic manufacturing," *Annals of CIRP*, 45(1), pp. 433–436.

[9.20] Heikkila, T., Jarviluoma, M. and Juntunen, T., 1997, "Holonic control for manufacturing systems: design of a manufacturing robot cell," *Integrated Computer Aided Engineering*, **4**, pp. 202–218.

[9.21] Ng, A., Yeung, R. and Cheung, E., 1996, "HSCS – The design of a holonic shopfloor control system," In *Proceedings of the IEEE Conference on Emerging Technologies and Factory Automation*, pp. 179–185.

[9.22] Agre, J., Elsley, G., McFarlane, D., Cheng, J. and Gunn, B., 1994, "Holonic control of cooling control system," In *Proceedings of Rensslaers Manufacturing Conference*, New York, pp. 134–141.

[9.23] McFarlane, D., Marett, B., Elsley, G. and Jarvis, D., 1995, "Application of holonic methodologies to problem diagnosis in a steel rod mill," In *Proceedings of the IEEE Conference on Systems, Man and Cybernetics*, Vancouver, Canada, **1**, pp. 940–945.

[9.24] Brown, J. and McCarragher, B., 1998, "Maintenance resource allocation using decentralised cooperative control," *Internal Report, Australian National University*, Canberra, Australia.

[9.25] Bongaerts, L., Van Brussel, H., Valckenaers, P. and Peeters, P., 1997, "Reactive scheduling in holonic manufacturing systems: architecture, dynamic model and cooperation strategy," *Proceedings of Advanced Summer Institute of the Network of Excellence on Intelligent Control and Integrated Manufacturing Systems*, Budapest.

[9.26] Biswas, G., Sugato, B. and Saad, A., 1995, "Holonic planning and scheduling for assembly tasks," *Internal Report CIS-95-01*, Centre for Intelligent Systems, Vanderbilt University, Nashville, TN.

[9.27] Maturana, F., Gu, P., Naumann, A. and Norrie, D.H., 1997, "Object-oriented job-shop scheduling using genetic algorithms," *Computers in Industry*, **32**, pp. 281–294.

[9.28] Morton, T.E. and Pentico, D.W., 1993, *Heuristic Scheduling Systems*, John Wiley and Sons, Inc., New York.

[9.29] Sun, J. and Xue, D., 2001, "A dynamic reactive scheduling mechanism for responding to changes of production orders and manufacturing resources," *Computers in Industry*, **46**(2), pp. 189–207.

[9.30] Sauter, J.A., Parunak, H.V.D. and Goic, J., 1999, "ANTS in the supply chain," In *Proceedings of the Workshop on Agents for Electronic Commerce*, Seattle, WA.

[9.31] Valckenaers, P., Van Brussel, H., Bongaerts, L. and Wyns, J., 1997, "Holonic manufacturing systems," *Integrated Computer-Aided Engineering*, **4**, pp. 191–201.

[9.32] French, S., 1982, *Sequencing and Scheduling: An Introduction to the Mathematics of the Job-Shop*, Ellis Horwood Limited, UK.

[9.33] Walker, S.S., Brennan, R.W. and Norrie, D.H., 2001, "Demonstrating emergent intelligence: an evolutionary multi-agent system for job shop scheduling," In *Proceedings of the Evolutionary Computation in Multi-Agent Systems Workshop (ECOMAS), Genetic and Evolutionary Computation Conference (GECCO) 2001*, San Francisco.

[9.34] Brooks, R.A., 1991, "Intelligence without reason," In *Proceedings of the International Joint Conference on Artificial Intelligence*, **1**, pp. 569.

[9.35] Walker, S.S., Brennan, R.W. and Norrie, D.H., 2006, "Experience and reflection on the development of a holonic job shop scheduling system," *International Journal of Computer Applications in Technology*, **26**(1/2), pp. 15–27.

[9.36] Zhang, X. and Norrie, D.H., 1999, "Holonic control at the production and controller levels," In *Proceedings of the 2nd International Workshop on Intelligent Manufacturing Systems (IMS'99)*, Leuven, Belgium, pp. 215–224.

[9.37] Zhang, X. and Norrie, D.H., 1999, "Dynamic reconfiguration of holonic lower level control," In *Proceedings of the Second International Conference on Intelligent Processing and Manufacturing of Materials*, Honolulu, Hawaii, **2**, pp. 887–893.

[9.38] Bauer, A., Bowden, R., Browne, J., Duggan, J. and Lyons, G., 1994, *Shop Floor Control Systems: From Design to Implementation*, Chapman & Hall, London.

[9.39] Back, T., 1996, *Evolutionary Algorithms in Theory and Practice: Evolution Strategies, Evolutionary Programming, Genetic Algorithms*, Oxford University Press, Oxford.

[9.40] Mitchell, M., 1996, *An Introduction to Genetic Algorithms*, The MIT Press, Cambridge, Massachusetts.

[9.41] Brennan, R.W. and Sorensen, C., 2000, "Evaluating machine agent coupling of reactive and planning-based control architectures for manufacturing," In *Proceedings of the IEEE International Conference on Systems, Man and Cybernetics 2000 (SMC2000)*, pp. 1673–1678.

10

Distributed Scheduling in Multiple-factory Production with Machine Maintenance

Felix Tung Sun Chan and Sai Ho Chung

Department of Industrial and Manufacturing Systems Engineering
The University of Hong Kong, Pokfulam Road, Hong Kong, China
Email: ftschan@hkucc.hku.hk

Abstract
In general, the distributed scheduling problem focuses on solving two issues simultaneously: (i) allocation of jobs to suitable factories, and (ii) determination of the corresponding production scheduling in each factory. Its objective is to maximise the system efficiency by finding an optimal plan for a better collaboration among various processes. This makes the distributed scheduling problem more complicated than the classical production scheduling ones. With the addition of alternative production routing, the problems are even more complicated. Conventionally, machines are assumed to be available without interruption during the production scheduling. Maintenance is usually not considered. However, in reality, this assumption is not true in most cases. Maintenance policy always directly affects the machine availability. Consequently, it interrupts the production. In this connection, maintenance should be considered with the distributed scheduling problems. In this chapter, a *genetic algorithm with dominant genes* (GADG) approach is introduced to deal with this problem. The significance and benefits of considering maintenance are demonstrated by simulation runs in an example.

10.1 Introduction

The significance of distributed manufacturing has been recognized by many researchers and industrialists in recent years due to the changes in the mode of today's production environment. Nowadays, markets are frequently shifting. New technologies are continuously emerging and competitors are globally multiplying. To increase international competitiveness and responsiveness to market changes, many companies have changed their production framework from traditional single-factory to multi-factory, by building new factories, merging, and factory acquisition, *etc.* [10.1]. These factories are geographically distributed in different locations. This allows them to be closer to their customers, to employ professionals, to comply with local laws, to focus on a few product types, to produce and market their products more effectively, and respond to market changes more quickly [10.2]–[10.6].

In distributed manufacturing, each factory can be considered as an individual entity and is usually capable of manufacturing a variety of product types. Some may be unique in a particular factory, while some may not. They have different efficiencies and are subject to different constraints, for example, machine advances, labour skills and education levels, labour cost, government policy, tax, nearby suppliers, and transportation facilities, *etc.* In addition, they are usually subject to different production peak seasons. As a result, different factories have different operating cost, production lead time, customer service levels, constraints, *etc.*, and this induces the question of how to allocate a job to a suitable factory, and the scheduling problem in it.

The distributed manufacturing problem can be considered as a DS (distributed scheduling) problem, which consists of a set of processes, some of which are subject to time constraints, and are executed in different locations. DS has gained an increasing importance and is applied in a wide range of areas, from multimedia to industrial control, and extensive efforts have been invested in solving various open research issues [10.7]–[10.10]. Basically, there are two types of DS problems in a multi-factory environment. In the first type, all the factories belong to the same company. They operate cooperatively to maximise the social welfare, which is the sum of the factories' profits. They are willing to take individual losses to benefit the others, maximising the revenue of the company as a whole. Another type is called a virtual enterprise. A number of different individual companies join together to form a multi-factory production network, in which these companies can operate more economically than operating individually. However, in such a network, each individual company usually focuses on self-benefits. They plan to maximise their own profit and do not care much about the others within the network [10.11]. In this chapter, the first type will be discussed.

In general, DS problems in distributed manufacturing can be divided into two main issues: (i) allocation of jobs to suitable factories, and (ii) determination of the production scheduling in each factory [10.12]. In detail, job allocation is to match jobs to suitable/optimal work places (factory). Since there are many suitable places that a job can be assigned to and executed, the first question is to determine how to assign them among the multiple alternatives. This assignment is known as job allocation or global scheduling. Then, the next problem is to determine the production schedule of the allocated jobs in each factory. In general, each job has a number of operations, and each operation may usually be processed by two or more machines. Since each machine may have a number of waiting operations and possesses different production capabilities, decisions of production schedule highly depend on the urgency of the jobs and the availability of the machines at the moment of scheduling.

The main purpose of DS is to maximise the system reliability and the resources utilisation through collaboration among different entities. It enhances the utilisation of manpower, factories, machines, and raw materials, *etc.* [10.13]. Companies can achieve better quality, lower production and distribution cost, and reduce the risk of uncertainties [10.14]. The entities have to efficiently share their available resources in order to assign and schedule tasks among themselves. This task-scheduling problem consists of defining a schedule that can meet all timing and logical constraints of the tasks being scheduled, and in general, it has been classified as NP-complete [10.15].

DS problems are much more complicated than the classical single-factory ones because they involve not only the scheduling problems in each factory, but also the problems in an upper level of how to allocate the jobs to suitable factories [10.16]. In fact, these issues are highly interrelated and should be considered simultaneously, since once a job is allocated to a factory and processed, it is usually not possible or uneconomical to transfer this work-in-process part to another factory for the remaining operations. Moreover, the production scheduling(s) in the factories have to depend on the jobs allocated. Inappropriate allocation of jobs may result in inefficient utilisation of resources, overloading or idling machine capacity, long production lead time, and unreliable due-date assignment and achievement (earliness and tardiness).

To enhance supply-chain performance, many researchers have proposed to concurrently optimise the production activity with other related activities, such as production planning and inventory control, distribution and logistics, production and distribution, and inventory and distribution, *etc.* [10.17]–[10.21]. Some literature reviewed that concurrently planning the production capability, production capacity, production cost, rough production scheduling, and job due date can provide goods and service to the customers at a lower cost, and achieve a higher customer service level [10.22][10.23]. However, there is a lack of papers studying distributed scheduling problems, especially those involving alternative production routings.

Flexible manufacturing systems (FMS) have been widely implemented in many factories because they are capable of responding faster to various changes such as demand and product mix. Flexible machines are equipped with tool magazines, which can accommodate up to 160 different tools [10.24]. Depending on the number and variety of tools equipped, each machine is capable of performing a number of different operations. In general, FMS consists of a set of machines, storage racks, and an automated material handling system with no manual intervention during the operational phase, providing flexibility of part routing. Since FMS can provide great flexibility in the job-shop floors and raise the efficiency of large volume production, the production scheduling and control in FMS can be very complex as the number of jobs, operations, parts, tooling and machines increase.

The advantages of flexibility in FMS are created due to the use of versatile and/or redundant machines and these in turn enable alternative routing in the system [10.25]. The introduction of alternative routing makes it possible to better balance the machine workload and achieve higher system robustness in case of machine failure. In order to achieve higher system productivity and reliability, the selection of the correct/optimal scheduling policy to control the system is the most crucial task. In the meantime, it is the most difficult task, especially when these systems are operating in a highly dynamic environment, where product mix and overall system objectives are changing rapidly.

At the production level, machine maintenance is inevitable. Maintenance policy influences the machine availability, and its utilisation ratio. It also influences the production rate, and product quality. The purpose of maintenance management is to reduce the effect of breakdown and maximise the facility availability at the minimum cost [10.18]–[10.20]. A commonly used indicator called machine age is an important measurement in maintenance. It reflects the expected/estimated inspection time, repairing time, production rate, product quality, and failure rate, *etc*. After each time of maintenance, the machine age has to be adjusted, depending on the type of maintenance carried out. A new set of inspection time, repairing time, production rate, and product quality will be generated [10.21][10.22]. If the production schedule obtained from DS does not consider maintenance, the planning determined will be seriously interrupted because of the mismatch among various processes. Consequently, the system reliability will be damaged and the purpose of DS will not be achieved.

In this chapter, an optimisation algorithm named *genetic algorithm with dominated genes* (GADG) is proposed to deal with the DS problems in multi-factory FMS production with the consideration of machine maintenance. GADG implements the idea of adaptive strategy. A new crossover mechanism termed dominated gene crossover will be introduced to enhance the optimisation ability and eliminate the determination processes of crossover rate. To prevent the saturation of solution pool, a saturation operator will also be introduced to monitor the similarity of the chromosomes. The chapter is divided into the following sections. Section 10.2 gives a literature review. Section 10.3 presents the DS problem and the objective function. Section 10.4 introduces the optimisation methodology of the proposed GADG. An example is studied and discussed in Section 10.5 for verification of the reliability of the proposed GADG, and Section 10.6 concludes the chapter.

10.2 Literature Review

Job-shop scheduling problems have been studied by many researchers and practitioners for many years since it is one of the most crucial tasks. A good schedule can increase the efficiency of a manufacturing system. However, because of the inherent complexity of the problem and the governed constraints, determining a good schedule is a difficult and time-consuming activity. It usually involves a heavy computational effort. The computational effort grows exponentially with the increment of the problem size. Application of pure mathematical optimisation approaches to determine an optimal solution may not be efficient in practice even in classical scheduling problems [10.26]–[10.28]. The assignment of portions of the time resources to the tasks is known to be defined as a task schedule. The task-scheduling problem consists of defining a schedule that can meet all timing and logical constraints of the tasks being scheduled. This problem, in its general form, has been shown to be NP-complete [10.15].

Heuristic approaches, which can obtain a near-optimal solution in a relatively short period, are more appreciated and practical. Many different heuristic methodologies have been proposed, for example, dispatching rules such as first-in-first-out, shortest processing time, and critical ratio, *etc*. Although there are many conventional sequencing rules such as the shortest processing time rule, each rule shows different performances according to the state of a shop floor. In recent years, some other heuristic methodologies have been widely adopted, such as branch and bound, hill climbing, simulated annealing, Tabu search, and genetic algorithm (GA). Among different heuristic approaches, GA is widely recognised as an appropriate and efficient approach in scheduling [10.29]–[10.32].

GA is a promising tool for solving real-world problems [10.33]. Aytug *et al.* [10.33] provided a review of papers using GA to solve operation problems and classified them according to the problem nature. Cheung *et al.* [10.34] gave a detailed tutorial survey on papers using GA for solving classical job-shop scheduling problems (JSP) in their Part I survey. In Part II, they reviewed papers using hybrid GA for tackling JSP [10.35]. Jain and ElMaraghy [10.36] proposed a GA to solve single process plan scheduling (SPPS) problems. Cavalieri and Gaiardelli [10.37] applied a hybrid GA, which combines GA with the dispatching rule (earliest due date), to solve multi-objective scheduling problems. Sakawa [10.38] combined GA with fuzzy logic to model the uncertainties of production lead time and order due date when scheduling problems. Wadhwa and Chopra [10.39] proposed a GA-based technique to obtain reasonably good schedules for a dynamic reconfigurable production system, which comprises multiple production lines that are capable of being reconfigured into two or three independent production lines or recombined into single lines. The results indicated that its performance is better than the traditional standard scheduling rules, such as *shortest processing time* and *earliest due date*. Mori and Tseng [10.40] compared the performance of their GA with stochastic construction method (STOCOM) in multi-mode resource-constrained project scheduling. They proposed an order-crossover approach and applied mutation to alter modes of operations. Ghedjati [10.41] proposed GA with a heuristic mixing method, which merges the crossover operation with a specific heuristic design when solving job-shop scheduling problems with several unrelated parallel machines.

GA is recognised as an appropriate and efficient approach to solve scheduling problems by many researchers. However, to maximise the performance of a genetic search, an optimal set of genetic parameters has to be determined, such as population size, crossover rate, mutation rate, generation gap, scaling window, and selection strategy [10.40][10.42][10.43]. Different genetic parameter settings in different problems will lead to different performance since the setting of each parameter has its function. For instance, a high crossover rate can increase global searching, while a low one favours fine local searching. Determining an optimal set of parameters is critical to the performance, but it is difficult to achieve due to the large number of possible combinations. This process is time consuming and is not practical in real situations.

In traditional GA, the crossover rate and mutation rate are usually fixed throughout the whole evolutions and between each pair of parents, for example, one-point, two-point, or multi-point crossovers. Michalewicz [10.43] stated that each of the crossovers is particularly useful for some classes of problems but may perform poorly in other problems. In addition, the crossover rate defines the number of genes to be selected in the traditional crossover operation, and the selection process is totally arbitrary. The selected genes for crossover may not be able to represent the critical structure of the chromosome. In this chapter, the function of the proposed DG is to identify and record the best genes and structure in each chromosome. During crossover, only those DG will undergo crossover to ensure that the best genes and structure will be selected. Other researchers also stated that by changing the genetic parameters (such as population size, crossover rate, mutation rate, generation gap, scaling window, and selection strategy), will the performance of GA be different [10.36][10.42].

To strategically strengthen the genetic search in different phases of evolution, during the genetic evolution, the genetic parameters, such as population size, crossover rate, and mutation rate will change strategically [10.31][10.37][10.43]. Michalewicz [10.43] proposed a non-uniform mutation that allows the operator to search through the solution space uniformly in the beginning phases to prevent prematurity of the solution pool, and then to locally alter for fine tuning. Cavaliefi and Gaiardelli [10.37] adopted a dynamic population size approach to replace the old chromosomes with new ones to maintain the diversity of the solution pool. However, applying such an approach leads to the problem of how to define the initial and maturity phases.

In the literature regarding DS problem solving, many researchers have proposed various heuristic approaches. DiNatale and Stankovic [10.44] applied a simulated-annealing algorithm to distributed static systems, in which tasks are periodic and have arbitrary deadlines, precedence, and exclusion constraints. They presented a general framework consisting of an abstract architecture model and a general programming model. Sandholm [10.11] discussed the issues that arise in extending automated contracting to operate among self-interested agents in distributed scheduling systems. They presented a levelled commitment contracting protocol that allows self-interested agents to efficiently accommodate future events by having the possibility of unilaterally recommitting from a contract based on local reasoning. Vincent and Stephen [10.10] applied wasp-like agents for distributed coordination. These agents use a model of wasp task allocation behaviour to determine which new jobs should be accepted into the machine's queue, and when to bid and when not to bid for arriving jobs. They benchmark the performance of their system on a real problem, which has to assign trucks to paint booths in a simulated vehicle paint shop. There are many other heuristic approaches that can be found in [10.45]–[10.47].

Recently, Jia et al. [10.13][10.48] proposed a modified GA to solve multi-factory distributed scheduling problems. They proposed an encoding of chromosome, crossover mechanism, and two mutation mechanisms. Their modified GA has been compared with other classical single-factory scheduling problems and heuristic approaches, and it obtained satisfactory results. However, the problem sizes they tested are small, and the algorithm can only handle fixed production routing. Chan et al. [10.16] further enhanced the GA approach by introducing a concept named dominant genes (DG), which improves the optimisation results and minimises the deviation of the solution obtained. However, it was again designed for fixed production routing.

In a practical manufacturing environment, FMS is usually involved. Alternative production routings are inevitable. Studies in FMS scheduling problems have been published by many researchers in the last few decades [10.49]–[10.58]. Distributed scheduling problems involving FMS are more complicated, since the FMS scheduling problem in a single factory is already NP-hard [10.15][10.41]. It is now concurrently performing the scheduling tasks on more than one FMS factory and the associated job allocation. For alternative production routing, Chan et al. [10.59] proposed a new encoding mechanism. They also proposed a DG approach that demonstrated its ability to enhance the optimisation reliability. However, to the authors' knowledge up to this time, there is a lack of papers that take account of machine maintenance in DS.

10.3 Problem Background

Figure 10.1 shows an outline of a typical distributed scheduling network. It generally consists of a number of factories ($F = 1, 2, ..., l$), which are geographically distributed in different locations. Each factory has different numbers of machines/ work centres (H_l) and can produce various product types with different production efficiency, and cost, *etc.*

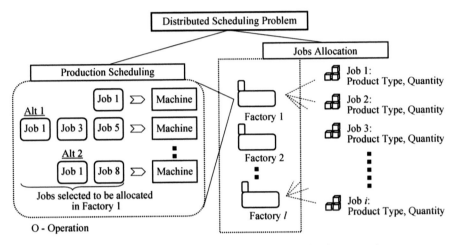

Figure 10.1. Outline of a multi-factory FMS production network

When the system receives a number of jobs ($N = 1, 2, ..., i$), first it has to deal with the job-allocation problem. Allocating a job to a factory among alternative choices is not an easy task, since it has to benefit/optimise the objective function, for example, minimisation of the total system operation cost, and makespan are commonly used by many research groups. In this chapter, minimisation of the makespan will be used. In order to optimise it, an estimated completion time for the jobs has to be calculated. This means the production scheduling of the jobs is required. In general, each job has up to N_i number of operations. Since each of which can be performed on more than one suitable machine (but not all) and each machine in different factories bears different processing time (T_{ijhl}) and numbers of awaiting operations, *etc.*, the solution of job allocation will directly influence the completion time of each job and finally the makespan of the system. Therefore, the production-scheduling problem has to be simultaneously considered.

In practice, machine maintenance is inevitable. After a certain operating period, machines are usually required to be shut down for maintenance or repair. The estimated inspection/repair time can be represented on a graph, which is a common practice, obtained empirically as shown in Figure 10.2. Whether the time is long or short depends on the corresponding machine age (A), which represents the total cumulated processing time of operations. If the machine age reaches the maximum machine age (*M*), it means that maintenance has to be carried out immediately after the completion of its current operation. After maintenance, the machine age may be reset to 0, depending on the type of maintenance performed. In general, different graphs are used to represent different types of machines as shown in Figures 10.3 and 10.4. In this chapter, the graph shown in Figure 10.2 will be used. Since machines will become unavailable during maintenance and this will therefore influence the completion time of jobs and the production scheduling, scheduling of machine maintenance should be considered simultaneously with job allocation and production scheduling.

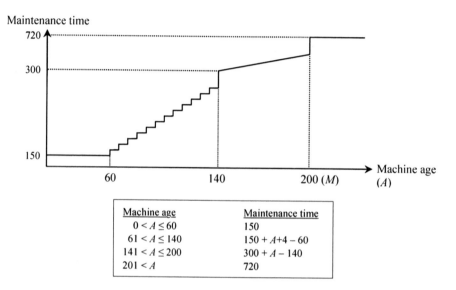

Machine age	Maintenance time
$0 < A \leq 60$	150
$61 < A \leq 140$	$150 + A + 4 - 60$
$141 < A \leq 200$	$300 + A - 140$
$201 < A$	720

Figure 10.2. Maintenance time related to machine age

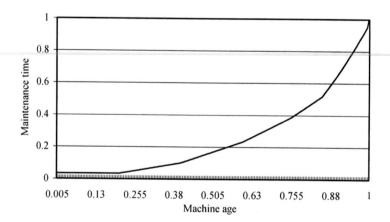

Figure 10.3. A sample repair time for some electric, mechanic, and computer failures

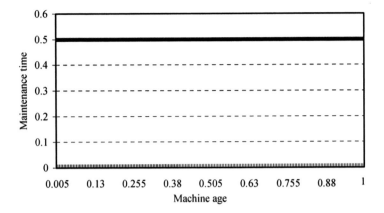

Figure 10.4. A sample repair time to replace an old component

Similar to classical scheduling problems, it is generally assumed that each machine can only handle one operation at a time. Each operation will be completed before another operation is loaded on a machine. Each job will be operated as its order quantity (a batch) without splitting. Note that once a job is allocated to a factory, all of its operations will be processed in it. The problem is expressed in the following notation:

χ_{il} Binary integer, defined as 1 if job i is allocated to factory l; otherwise 0.

δ_{ijkhl} Binary integer, defined as 1 if operation j of job i occupied time slot k on machine h in factory l; otherwise 0.

γ_{ijhl} Binary integer, defined as 1 if machine h in factory l will be maintained after operation j of job l; otherwise 0.

F Number of factories.

N Number of jobs.

N_i Number of operations of job i.

H_l Number of machines in factory l.

D_{li} Travel time between factory l and job i.

K Time horizon under consideration.

T_{ijhl} Processing time of operation j of job i on machine h in factory l.

A_{hl} Age of machine h in factory l.

M Maximum machine age.

S_{ij} Starting time of operation j of job i.

E_{ij} Ending time of operation j of job i.

C_i Completion time of job i.

In the problem, it is assumed that F, N, N_i, H_l, and T_{ijhl} are given. The decision variables are χ_{il}, δ_{ijkhl}, and γ_{ijhl}. With the solution χ_{il}, δ_{ijkhl}, and γ_{ijhl} obtained, the value of S_{ij}, E_{ij}, and C_i can be calculated. The objective is to minimise the makespan of jobs.

$$Objective\ Z : \min\ (\max\{\,C_i\,\}) \tag{10.1}$$

Completion time (C_i) of job i equals the summation of the completion time of the last operation (N_i) of job i and the travel time between factory l and job i, as defined by Equation (10.2).

$$C_i = E_{iN_i} + \sum_{li} D_{li} \chi_{il} \tag{10.2}$$

The problem is subject to the following constraints:

Precedence constraint:

$$S_{ij} \geq E_{i(j-1)} \qquad\qquad (i = 1, 2, ..., N; j = 2, 3, ..., N_i) \tag{10.3}$$

Processing time constraints:

$$E_{ij} - S_{ij} = \sum_{hl} \chi_{il} T_{ijhl} \qquad\qquad (i = 1, 2, ..., N; j = 1, 2, ..., N_i) \tag{10.4}$$

$$\sum_{khl} \delta_{ijkhl} \geq \sum_{hl} \chi_{il} T_{ijhl} \qquad\qquad (i = 1, 2, ..., N; j = 1, 2, ..., N_i) \tag{10.5}$$

Operation constraint:

$$\sum_{khl} \delta_{ijkhl} \geq 1 \qquad\qquad (i = 1, 2, ..., N; j = 1, 2, ..., N_i) \tag{10.6}$$

Processing operation constraint:

$$\sum_{hl} \delta_{ijkhl} \leq 1 \qquad (i = 1, 2, ..., N; j = 1, 2, ..., N_i; k = 1, 2, ..., K) \tag{10.7}$$

Machine capacity constraint:

$$\sum_{ij} \delta_{ijkhl} \leq 1 \qquad (k = 1, 2, ..., K; h = 1, 2, ..., H_l; l = 1, 2, ..., F) \tag{10.8}$$

Factory constraint:

$$\sum_{l} \chi_{il} = 1 \qquad\qquad (i = 1, 2, ..., N) \tag{10.9}$$

In the above constraints, Constraint (10.3) defines that each operation can only start upon the completion of its preceding operation. Constraint (10.4) defines that once an operation starts, it will be finished without interruption. Constraint (10.5) indicates that the allocated time slot equals the required operation time. Constraint (10.6) forces each operation to be carried out on one machine throughout the horizon. Constraint (10.7) forces each operation to be only carried out on one machine at each time unit, and Constraint (10.8) forces each machine to carry out only one operation at each time unit. Constraint (10.9) forces all the operations of a job to be finished in the same factory.

10.4 Optimisation Methodology: Genetic Algorithm with Dominant Genes

GA is robust in finding the global or very strong local optimal. However, GA requires a set of predefined genetic parameters, which directly control the ability of the genetic search. However, to determine an optimal set of such parameters is extremely time consuming because of the large number of combinations and each combination requires a certain amount of running time for testing. For this reason, GADG is introduced. In this approach, the crossover rate and mutation rate are not required to be predefined. It does not require the crossover rate and mutation rate to be determined, so that it can reduce the computational time markedly. Meanwhile, it can improve the performance of the genetic search.

10.4.1 Dominant Genes

In GA, the function of a chromosome is to simulate a potential solution to the problem studied. It consists of a series of genes and each gene represents a decision variable. When designing for scheduling problems, many researchers conventionally utilise the position of the genes to model the scheduling priority of jobs. For example, although two chromosomes can have the same value for each decision variable, if the structures of the chromosomes (the position of the genes) are different, the production scheduling may be totally different.

The idea of dominant genes (DG) is to identify and record the best genes in each chromosome, and its corresponding structure. If any changes in the genes, either in the values or positions, can increase the fitness value (compared with its parent), those genes are marked as DGs. Figure 10.5 shows an example of a chromosome, in which the second, third, and seventh genes are DGs. These genes, structured in this order and with each value, can increase the fitness value for that particular chromosome. During the crossover operation, the whole set of DGs will be selected. Note that this set of DGs may not strengthen the other chromosomes. The purpose of this approach is to prevent the loss of any identified potential critical structures and give more chance for them to grow stronger.

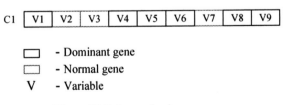

Figure 10.5. A sample chromosome

Figure 10.6 demonstrates the idea of the DG. Assuming that, we have to find an optimal solution as shown below. The solution pool consists of 4 chromosomes (C1–C4) and each of them has the same value in the variables, but different structure and with some genes randomly assigned as seed DGs. The fitness value is calculated by the total number of genes that match those in the optimal chromosome, *i.e.* in C1, the fifth gene (value: V5) in position 5 matches the one in the optimal. Therefore, its fitness value is 1.

Position	1	2	3	4	5	6	7	8	9
Optimal	V1	V2	V3	V4	V5	V6	V7	V8	V9

Initial Pool

										Fitness Value
C1	V9	V8	V7	V6	V5	V4	V3	V2	V1	1
C2	V4	V3	V5	V1	V6	V2	V7	V8	V9	3
C3	V1	V3	V5	V4	V9	V8	V7	V6	V2	3
C4	V4	V2	V8	V9	V5	V6	V7	V1	V3	3

Situation A
Evolution 1 (C2×C4)

										Fitness Value
O1	V4	V2	V3	V1	V5	V6	V7	V8	V9	7
O2	V4	V3	V2	V1	V5	V6	V7	V8	V9	5

Evolution 2 (O1×C3)

										Fitness Value
O3	V1	V2	V3	V4	V5	V6	V7	V8	V9	9
O4	V1	V2	V3	V4	V5	V6	V7	V8	V9	9

Situation B
Evolution 1 (C1×C2)

										Fitness Value
O1	V6	V5	V4	V3	V2	V1	V7	V8	V9	3
O2	V4	V5	V6	V7	V8	V9	V3	V2	V1	0

Figure 10.6. An outline demonstrating the idea of DG

To undergo evolution, a number of chromosomes will be randomly selected as parents (P) and grouped into pairs for crossover operation. For example, C2 and C4 are paired in Situation A. Their potential critical structures are preserved and combined to generate offspring (O1 and O2), *i.e.* C2 generates O1 and C4 generates O2 (the detailed crossover methodology will be discussed in Section 10.4.3). After that, O1 and C3 are selected to generate O3 and O4, which are similar to the optimal solution.

Note that the DG may not necessarily represent the critical structure, such as the last three genes in C1 and the second gene in C2. In addition, such as in Situation B, the offspring generated after crossover may not be stronger than its parents. O2 with the fitness value 0 is weaker than parent C2 with the fitness value 3. In this case, the inherited DG will be denoted as normal genes. On the contrary, the inherited DG will still be DG such as those inherited from C2 in O1.

10.4.2 Encoding of Chromosome

A chromosome is designed to model the DS problem in multi-factory FMS production with the consideration of machine maintenance. It consists of $\sum_i N_i$ genes, and each gene consists of six parameters, representing Factory, Machine, Job, Operation, Scheduling of maintenance, and Domination (FMJOSD). The S parameter represents that the machine has to be repaired immediately after the completion of its current operation, and it will be denoted as 1, otherwise 0. The D parameter will be denoted as 1 if the gene is a DG, otherwise 0. Figure 10.7 demonstrates the solution of allocating 3 jobs (each job has up to 3 operations) into 2 factories, in which the first gene shows that O3 of J3 is allocated to M2 of F1, and after this operation, M2 will be maintained; it is also denoted as a DG.

123311–121210–111100–222110–111300–212300–232200–133111–113200

Figure 10.7. A sample of chromosome encoding

Figure 10.8 also shows the decoding of the chromosome. The production priority of jobs on machines is defined by the ordering, from the highest priority on the left to the lowest on the right. Any operation can only start after its preceding operation is complete. Therefore, the first and second genes cannot be processed. The third gene (J1O1) will be processed. Then it will come back to the second gene (J1O2). A detailed production schedule is shown in Figure 10.8. In an FMS environment, assuming that, O3 of J3 can also be performed on M3, the second gene can be represented as (3330) as shown in Figure 10.9.

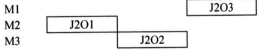

Production priority in Factory 1

M1	J1O1	J3O2	J1O3		
M2		J1O2	Maintenance	J3O3	Maintenance
M3	J3O1	Maintenance			

Production priority in Factory 2

M1			J2O3
M2	J2O1		
M3		J2O2	

*Assuming that each operation and maintenance takes 1 unit of time

Figure 10.8. Another sample of chromosome decoding

1333̲1̲1̲–121201–111100–2221̲1̲0–111300–212300–232200–133101–113200

Figure 10.9. A sample of encoding of alternative routing

10.4.3 Dominant Genes Crossover

The purpose of the crossover operation is to import into a chromosome the structure of another chromosome, attempting to increase its own fitness value. During crossover, the whole set of DGs will be selected. This prevents the loss of the critical structure and gives more chance to test whether this setting of genes can contribute to other chromosomes. In the initial pool, some genes will be randomly assigned as seed DGs in each chromosome. During the evolution process, each pair of selected parents generates a pair of offspring. DG crossover can be divided into two cases. If there are no DG conflicting at the same position of the two parents, and if there are no identical jobs dominating in both parents, it will be classified as Case A, as shown in Figure 10.10, otherwise Case B, as shown in Figure 10.11.

Case A

Crossover in Case A is carried out in three steps. Figure 10.10 shows a sample crossover in which P1 generates O1 and P2 generates O2.

Step 1: O1 reserves DG from P1, and the values of these genes in P1 will be temporarily marked as zero. This indicates that these genes are passed to the offspring and prevents the repetition of genes. Then, O1 inherits DG from P2, and the genes with the same values of J and O in P1 will be temporarily marked as zero. Similarly, this practice is to avoid repetition.

Step 2: Copy the remaining genes with the same number of J equal to the DG from P2 to replace those in P1.

Step 3: Copy the non-zero genes from P1 to the non-zero genes in O1. Reset the genes of P1 to their original values.

P1: **123311**–121210–111100–222110–111300–212300–232200–**113111**–113200
P2: 222300–211100–221300–232100–**212201**–231210–113100–133200–123300

Step
1   O1: **123311**–000000–000000–000000–**212201**–000000–000000–**113111**–000000
2   P1: 000000–121210–111100–*222100*–111300–*212300*–000000–000000–113200
3   O1: **123311**–121210–111100–222110–**212201**–212300–111300–**113111**–113200

Step
1   O2: **123311**–000000–000000–000000–**212201**–000000–000000–**113111**–000000
2   P2: 222300–211100–221300–232100–000000–231210–000000–*133200*–000000
3   O2: 123310–222300–211100–221300–**212201**–232100–231210–113110–133200

Figure 10.10. Dominant gene crossover of Case A

Similar steps will be carried out for O2 except Step 1, where O2 reserves DG from P2, then inherits from P1. In Step 2, the remaining genes will be replaced from P1 to P2, and the non-empty genes will be copied from P2 to O2.

One of the advantages of this crossover mechanism is that the whole set of recognised best genes will undergo crossover. In addition, these newly changed genes will be tested to see whether they can increase the fitness value. If they make a contribution to the chromosome, they will be identified, recorded, and inherited to its offspring. Assuming that if the fitness value of O1 is better than P1, the inherited DG from P2 will retain their dominance. However, if it is weaker than its parents, like O2 is weaker than P2, the inherited DG from P1 will become normal genes.

Case B

In this case, DGs conflict between the two parents either at the same location(s) or job(s), as shown in Figure 10.11. A selection is required to test which sets of DGs can contribute more to the offspring. The checking will be done by setting one set of DGs as normal genes, and then the other to satisfy the criteria of Case A in O1A and O1B, respectively. For example, P1 and P2 in Figure 10.11 will be converted to the forms in Figures 10.12 and 10.13 to generate two offspring and the stronger one will be O1. Note that the authors decided to compare the whole set of DGs instead of individually comparing each pair of conflicting DGs, because this can prevent the dramatic change in the chromosome structure.

P1: **123311**–121210–111100–222110–111300–212300–232300–**113111**–133200
P2: **222301**–211100–221300–232100–212200–231210–113100–133200–123300

P3: **123311**–121210–111100–222110–111300–212300–232300–**113111**–133200
P4: 222300–211100–221300–232100–212200–231210–113100–133200–**123301**

Figure 10.11. Dominant genes conflicting at the same location between P1 and P2 and jobs between P3 and P4 (Case B)

P1: **123311**–121210–111100–222110–111300–212300–232300–**113111**–133200
P2: 222300–211100–221300–232100–212200–231210–113100–133200–123300

Figure 10.12. P1 and P2 in Figure 10.11 converted to generate O1A

P1: 123310–121210–111100–222110–111300–212300–232300–113110–133200
P2: **222301**–211100–221300–232100–212200–231210–113100–133200–123300

Figure 10.13. P1 and P2 in Figure 10.11 converted to generate O1B

10.4.4 Mutation Operator

There are two types of mutation. In Mutation 1, a pair of genes will be randomly selected and swapped, as shown in Figure 10.14. This mutation aims to reschedule the priority of job operations.

In Mutation 2, some genes will be randomly selected and mutated either in the *F*, *M*, or *S* parameter, as shown in Figure 10.15. The number of mutated genes is governed by a predefined mutation rate(s). The purpose of this mutation is to increase the diversity of the chromosomes. In both mutation operators, if O1 is stronger than P1, the mutated gene will become a DG.

P1: **123311**–*121210*–111100–222110–*111300*–212300–232300–**113111**–133200
O1: **123311**–*111300*–111100–222110–*121210*–212300–232300–**113111**–133200

Figure 10.14. A sample procedure of Mutation 1

P1: **123311**–*121210*–111100–222110–111300–212300–232300–**113111**–133200
O1: **123311**–*131210*–111100–222110–111300–212300–232300–**113111**–133200

Figure 10.15. A sample procedure of Mutation 2

10.4.5 Elitist Strategy

To prevent the loss of the best chromosome during successive generations, the best chromosome will be identified and recorded. If the best chromosome is lost or becomes weaker after evolution, it will be inserted back into the mating pool for the next evolution.

10.4.6 Prevention of Prematurity and Local Searching

To prevent prematurity, and the algorithm from searching around at a local space, an adaptive mutation rate is applied. A high rate favours a global search, while a low rate favours a fine local search [10.37][10.43]. In this chapter, a one-point rate is basically adopted throughout the whole genetic evolution process. The reason for applying this rate is to strengthen the local optimal search. In addition, this proportion is larger than the others because fine local tuning usually requires a longer searching time. In order to switch from the local search to a global search, the mutation rate of the next evolution in the process will increase when the similarity value is larger than the predefined threshold, so that the algorithm will not be trapped at a local point. The sudden increase in the rate is to strengthen the global optimal search by increasing the diversity of the solution pool. After this change, the mutation rate of the following evolution will then return to the one-point rate and it will remain until there is another change.

Similarity checking is applied to measure the maturity of the solution pool. When a solution pool becomes mature, which implies many genes are similar, the effectiveness of crossover is usually reduced because of the lower chances of forming a new chromosome structure. If the similarity is evaluated to be larger than a threshold, a high mutation rate will be applied to increase the diversity of the solution pool. Similarity checking measures the number of identical genes in the same column. Its purpose is to increase the diversity of crossover. During crossover, the selected gene(s) will crossover with other genes within the same column only, for example column 1 (C1) in Figure 10.16. If identical genes saturate in a column, the chances of forming a new chromosome structure through crossover will be reduced. The similarity is evaluated by Equation (10.10).

$$Similarity = Same / Comparison \qquad (10.10)$$

Same   – the number of identical pairs
Comparison – the total number of comparisons among the genes in the same
                    column

Figure 10.16 shows an example of the proposed similarity checking for four chromosomes. Each column is individually evaluated. In C1, the gene in P1 will be compared with the genes in P2, P3, and P4 with a total of 3 comparisons, among which, 2 comparisons show as identical (P1 = P2, and P1 = P3). Next, P2 is compared with P3 and P4, then, P3 with P4. As a result, the number of identical pairs (*Same*) equals 3 and *Comparison* equals 6. The similarity is evaluated to be (3/6 = 0.5), which is equal to the probability of selecting 2 identical genes to crossover in the column. In this chapter, when any one of the columns' similarity exceeds the threshold (0.25), 25% of the identical genes in that column will be randomly selected to perform the mutation operation using Mutation 1. The reason for using this low mutation rate (25%) is to prevent the dramatic change of chromosome structure.

```
      C1      C2      C3     C4     C5     C6     C7     C8     C9
P1: 123311–121210–111100–222110–111300–212300–232300–113111–133200
P2: 123311–231210–211100–212300–211300–232300–222110–133200–113111
P3: 123311–121210–111100–222110–111300–212300–232300–113111–133200
P4: 223311–231210–231100–222110–221300–212300–313111–232300–333200
```

Figure 10.16. A sample of similarity checking

10.5 Example

Consider a problem with 2 factories (F1 and F2) and 10 jobs. Table 10.1 shows the problem parameters with processing time of operations (O) on different machines (M), and Table 10.2 shows the travelling distance between factories and customers. Assuming that the processing time of the operations in F1 and F2 is the same, the number of evolutions adopted is 5000 with the solution pool size of 100, and each set will be run for 50 times individually to measure the deviation of the results.

The significance of considering maintenance during DS will be demonstrated by two approaches. In Approach 1 (A1), maintenance will not be considered during DS, while in Approach 2 (A2), it will. Table 10.3 summarises the results of 50 trails. It shows the average makespan (Avg.), the standard deviation (Std. dev.), and the minimum makespan (MIN) obtained from the 50 trails. From A1, when machine maintenance is not required, the average makespan obtained is 510 units of time. However, if each machine is forced to stop for maintenance after its concurrent operation when the machine has been operating for longer than 200 units of time (A ≥ M), the average makespan obtained becomes 1780 units of time. However, if maintenance is considered with DS (A2), the average makespan can be shortened to 1280 units of time, which has a 28% improvement.

Table 10.1. Problem parameters (production lead time) of a sample problem

Job 1 O	M	Process time	Job 2 O	M	Process time	Job 3 O	M	Process time	Job 4 O	M	Process time	Job 5 O	M	Process time
1	1	7	1	1	8	1	1	10	1	2	9	1	1	10
	3	4		2	12		2	15		3	5		3	15
2	2	3	2	3	4		3	8	2	1	6	2	2	7
3	1	3	3	1	7	2	2	2		3	2		3	14
	3	6		2	14		3	6	3	2	7	3	1	5
4	1	2	4	1	8	3	1	2		3	12		2	8
	2	4		3	4		3	4	4	1	9	4	1	4
						4	1	6		2	6		2	6
							2	3		3	3		3	8

Job 6 O	M	Process time	Job 7 O	M	Process time	Job 8 O	M	Process time	Job 9 O	M	Process time	Job 10 O	M	Process time
1	1	7	1	1	8	1	1	10	1	2	9	1	1	10
	3	4		2	12		2	15		3	5		3	15
2	2	3	2	3	4		3	8	2	1	6	2	2	7
3	1	3	3	1	7	2	2	2		3	2		3	14
	3	6		2	14		3	6	3	2	7	3	1	5
4	1	2	4	1	8	3	1	2		3	12		2	8
	2	4		3	4		3	4	4	1	9	4	1	4
						4	1	6		2	6		2	6
							2	3		3	3		3	8

Table 10.2. Problem parameters (transportation lead time) of a sample problem

Travel time Factory	Job									
	1	2	3	4	5	6	7	8	9	10
1	10	10	10	10	10	10	10	10	10	10
2	20	20	20	20	20	20	20	20	20	20

Table 10.3. Optimisation results for Set 1 and Set 2

		Avg.	Std. dev.	MIN
A1	No maintenance	510	12.87	490
	Maintenance	1780	21.83	1920
A2	Maintenance	1280	13.14	1220

Figure 10.17 shows one of the scheduling samples obtained from A1 with the makespan of 490 units of time, and the corresponding scheduling after maintenance is 1920 units of time, as shown in Figure 10.18. Figure 10.19 shows a sample scheduling obtained from A2 with only 1220 units of time and the encoding of the chromosome is (13$\underline{1}$300 21$\underline{2}$301 23$\underline{6}$101 13$\underline{8}$100 11$\underline{7}$101 23$\underline{4}$201 12$\underline{9}$100 22$\underline{4}$301 21$\underline{10}$100 21$\underline{2}$101 13$\underline{9}$200 11$\underline{5}$100 22$\underline{10}$200 12$\underline{8}$201 23$\underline{4}$100 13$\underline{8}$301 23$\underline{3}$100 11$\underline{8}$401 12$\underline{9}$301 13$\underline{5}$201 11$\underline{1}$100 13$\underline{7}$210 22$\underline{6}$200 23$\underline{2}$201 11$\underline{7}$300 12$\underline{1}$200 21$\underline{6}$300 22$\underline{3}$211

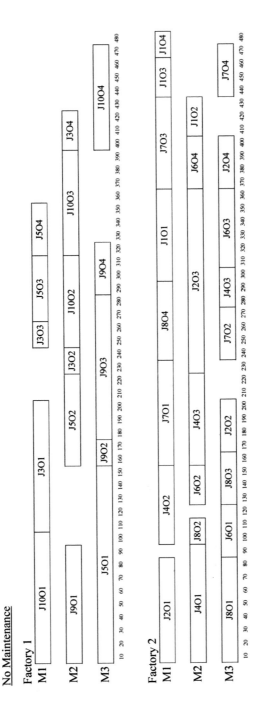

Figure 10.17. Result of distributed scheduling without consideration of maintenance (A1)

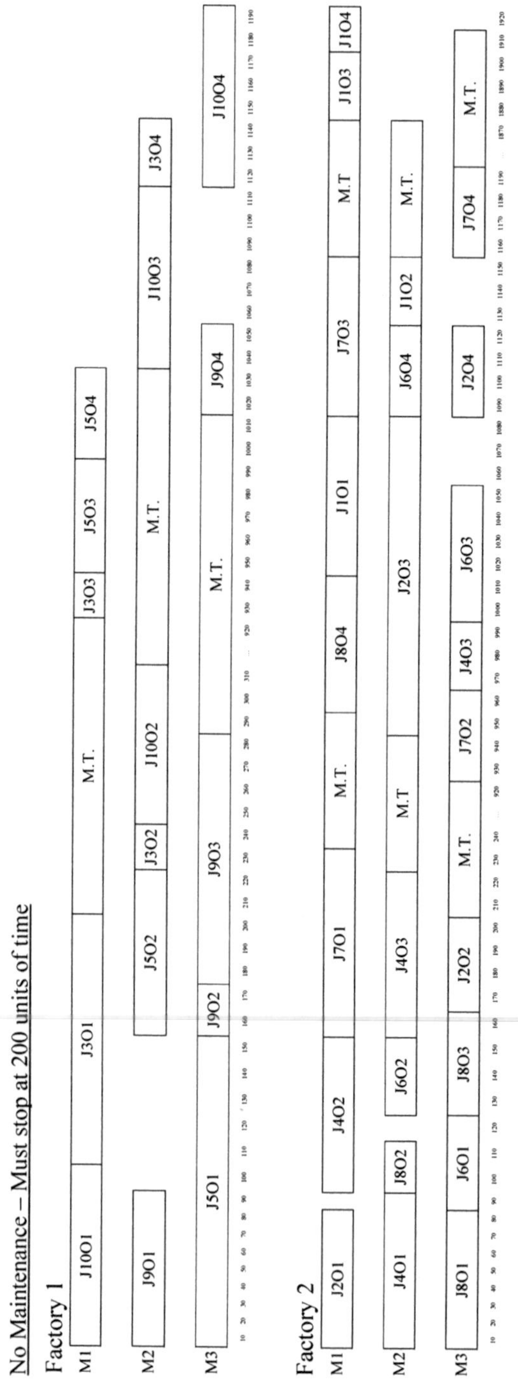

No Maintenance – Must stop at 200 units of time

Figure 10.18. Distributed scheduling of Figure 10.17 after consideration of the maintenance factor

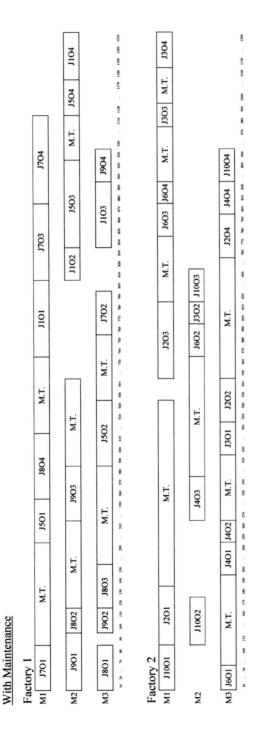

Figure 10.19. Result of distributed scheduling with consideration of maintenance (A2)

11*7*401 23*2*400 21*6*401 12*5*301 21*3*301 21*3*400 22*10*300 13*9*400 23*4*400 12*5*400 12*1*401 23*10*400). The jobs are indicated in italics and underlined. This comparison demonstrates that adequate scheduling maintenance during DS can shorten the makespan by improving the machine utilisation ratio, such as in Figure 10.18, the total maintenance time is 6480 (9×720) units of time, while Figure 10.19 is only 3060   (170+320+200+160+200+230+230+340+160+150+150+230+150+160+210) units of time. The machine utilisation ratio has a 53% improvement.

10.6 Conclusions

The significance of DS in the multi-factory FMS problem has been recognised and studied by many researchers. However, there is a lack of papers taking machine maintenance into consideration. In practice, machine maintenance is inevitable, and it can interrupt the production. Therefore, considering machine maintenance is necessary. In solving the scheduling problems, GA has been widely applied. However, in order to optimise the ability of the genetic search, it is known that an optimal set of genetic parameters has to be determined, but this process is time consuming. Especially in the traditional crossover mechanism, a number of genes will be randomly selected, and governed by a predefined crossover rate(s). However, it is usually difficult to ensure that the important part of the chromosome structure can be selected and inherited to its offspring. To deal with these problems, GADG is introduced. In the new approach, the DG identifies and records the best genes in each chromosome, and the corresponding structure. The significance of considering maintenance during DS has been demonstrated by two approaches, where A1 does not consider maintenance, while A2 does. The comparison of the results demonstrates that considering maintenance during DS can shorten the makespan by improving the machine utilisation. The average makespan obtained from A2 is shorter. In addition, the machine utilisation is improved.

References

[10.1]   Shen, W. and Norrie, D.H., 1999, "Agent-based systems for intelligent manufacturing: a state-of-the-art survey," *International Journal Knowledge and Information Systems*, 1(2), pp. 129–156.

[10.2]   Schniederjans, M.J., 1999, *International Facility Acquisition and Location Analysis*, Quorum Books, Westport.

[10.3]   Sule, D.R., 2001, *Logistics of Facility Location and Allocation*, Marcel Dekker, Inc. New York, NY.

[10.4]   Drezner, Z., 1995, *Facility Location: A Survey of Applications and Methods*, Springer-Verlag Inc., New York.

[10.5]   Jayaraman, V., 1998, "Transportation, facility location and inventory issues in distribution network design," *International Journal of Operations and Production Management*, 18(5), pp. 471–494.

[10.6]   Dhaenens-Flipo, G. and Finke, G., 2001, "An integrated model for an industrial production-distribution problem," *IIE Transactions*, 33(9), pp. 705–715.

[10.7]   Stankovic, J., 1996, "Strategic directions in real-time and embedded systems," *ACM Computing Surveys*, 28(4), pp. 751–763.

[10.8]   Kim, K.H., Bae, J.W., Song, J.Y. and Lee, H.Y., 1996, "A distributed scheduling and shop floor control method," *Computer Industrial Engineering*, 31(3/4), pp. 583–586.

[10.9]   Wang, H.H. and Wu, Z.M., 2003, "The application of Adaptive Genetic Algorithms in FMS dynamic rescheduling," *International Journal of Computer Integrated Manufacturing*, 16(6), pp. 382–397.

[10.10] Vincent, A.C. and Stephen, F.S., 2004, "Wasp-like agents for distributed factory coordination," *Autonomous Agents and Multi-Agent Systems*, 8, pp. 237–266.

[10.11] Sandholm, T.W., 2000, "Automated contracting in distributed manufacturing among independent companies," *Journal of Intelligent Manufacturing*", 11, pp. 271–283.

[10.12] Barroso, A.M., Leite, J.C.B. and Loques, O.G., 2002, "Treating uncertainty in distributed scheduling," *The Journal of Systems and Software*, 63, pp. 129–136.

[10.13] Jia, H.Z., Nee, A.Y.C., Fuh, J.Y.H. and Zhang, Y.F., 2003, "A modified genetic algorithm for distributed scheduling problems," *Journal of Intelligent Manufacturing*, 14, pp. 351–362.

[10.14] Ballou, R.H., 1998, *Business Logistics Management*, Prentice Hall, Upper Saddle River, New Jersey.

[10.15] Garey, M.R. and Johnson, D.S., 1979, *Computers and Interactability: A Guide to the Theory of NP-completeness*, Freeman and Co., San Francisco, CA.

[10.16] Chan, F.T.S., Chung, S.H., and Chan, P.L.Y., 2005, "An adaptive genetic algorithm for distributed production and scheduling problems with alternative production routines," *Expert Systems with Application*, 29(2), pp. 261–371.

[10.17] Cohen, M.A. and Lee, H.L., 1998, "Strategic analysis of integrated production-distribution systems: models and methods," *Operations Research*, 36(2), pp. 216–228.

[10.18] Beamon, B.M., 1999, "Supply chain design and analysis: models and methods," *International Journal of Production Economics*, 55(1), pp. 281–294.

[10.19] Lee, Y.H., Kim, S.H. and Moon, C., 2002, "Production-distribution planning in supply chain using a hybrid approach," *Production Planning and Control*, 13(1), pp. 35–46.

[10.20] Thomas, D.J. and Griffin, P.M., 1996, "Coordinated supply chain management," *European Journal of Operational Research*, 94(1), pp. 1–15.

[10.21] Chan, F.T.S. and Chung, S.H., 2005, "Multi-criteria genetic optimization for due date assigned distribution network problems," *Decision Support Systems*, 39, pp. 661–675.

[10.22] Willis, A.K., 1996, "Customer Delight and Demand Management: Can they be integrated?" *Hospital Material Management Quarterly*, 18(2), pp. 58–65.

[10.23] Wu, S.H.., Fuh, J.Y.H. and Nee, A.Y.C., 2002, "Concurrent process planning and scheduling in distributed virtual manufacturing," *IIE Transactions*, 34(1), pp. 77–89.

[10.24] Siwamogsatham, T. and Saygin, C., 2004, "Auction-based distributed scheduling and control scheme for flexible manufacturing systems," *International Journal of Production Research*, 42(3), pp. 542–572.

[10.25] Byrne, M.D. and Chutima, P., 1997, "Real-time operational control of an FMS with full routing flexibility," *International Journal of Production Economics*, 51(1–2), pp. 109–113.

[10.26] Abdinnour-Helm, S., 1999, "Network design in supply chain management," *International Journal of Agile Management Systems*, 1(2), pp. 99–106.

[10.27] Glover, F., 1986, "Future paths for integer programming links to artificial intelligence," *Computers and Operations Research*, 13(5), pp. 533–589.

[10.28] Glover, F., 1989, "Tabu search part I", *ORSA Journal on Computing*, 1(3), pp. 190–206.

[10.29] Goldberg, D.E., 1989, *Genetic Algorithms in Search, Optimization and Machine Learning*, Reading, MA: Addison-Wesley.

[10.30] Vignaux, G.A. and Michalewicz, Z., 1991, "A genetic algorithm for the linear transportation problem," *IEEE Transactions on Systems, Man, and Cybernetics,* **21**(2), pp. 445–452.

[10.31] González, E.L. and Fernández, M.A.R., 2000, "Genetic optimization of a fuzzy distribution model," *International Journal of Physical Distribution and Logistics Management,* **30**(7/8), pp. 681–696.

[10.32] Al-Hakin, L., 2001, "An analogue genetic algorithm for solving job shop scheduling problems," *International Journal of Production Research,* **39**(7), pp. 1537–1548.

[10.33] Aytug, H., Khouja, M. and Vergara, F.E., 2003, "Use of genetic algorithms to solve production and operations management problems: a review," *International Journal of Production Research,* **41**(17), pp. 3955–4009.

[10.34] Cheung, R., Gen, M. and Tsujimura, Y., 1996, "A tutorial survey of job-shop scheduling problems using genetic algorithms – I," *Computers and Industrial Engineering,* **30**(4), pp. 983–997.

[10.35] Cheung, R., Gen, M. and Tsujimura, Y., 1999, "A tutorial survey of job-shop scheduling problems using genetic algorithms – II," *Computers and Industrial Engineering,* **37**(1), pp. 51–55.

[10.36] Jain, A.K. and ElMaraghy, H.A., 1997, "Single process plan scheduling with genetic algorithm," *Production Planning and Control,* **8**(4), pp. 363–376.

[10.37] Cavalieri, S. and Gaiardelli, P., 1998, "Hybrid genetic algorithms for a multiple-objective scheduling problem," *Journal of Intelligent Manufacturing,* **9**, pp. 361–367.

[10.38] Sakawa, M., 2002, *Genetic Algorithms and Fuzzy Multiobjective Optimization,* Kluwer Academic Publishers, pp. 188–222.

[10.39] Wadhwa, S. and Chopra, A., 2000, "A Genetic Algorithm application: dynamic reconfiguration in agile manufacturing systems," *Studies in Informatics and Control,* **9**(4), "http://www.ici.ro/ici/revista/sic2000_4/index.html".

[10.40] Mori, M. and Tseng, C.C., 1997, "Genetic algorithms for multi-mode resource constrained project scheduling problem," *European Journal of Operational Research,* **100**(1), pp. 134–141.

[10.41] Ghedjati, F., 1999, "Genetic algorithms for the job-shop scheduling problem with unrelated parallel constraints: Heuristics mixing method machines and precedence," *Computers and Industrial Engineering,* **73**, pp. 39–42.

[10.42] Grefenstette, J.J., 1986, "Optimization of control parameters for Genetic Algorithms", *IEEE Transactions on Systems, Man, and Cybernetics,* **16**(1), pp. 122–128.

[10.43] Michalewicz, Z., 1996, *Genetic Algorithms + Data Structures = Evolution Programs,* Springer-Verlag, Berlin, Heidelberg, New York.

[10.44] DiNatale, M. and Stankovic, J.A., 1995, "Applicability of simulated annealing methods to real-time scheduling and jitter control," In *16th IEEE Real-time Systems Symposium,* Pisa, Italy.

[10.45] Barroso, A.M., Torreao, J.R.A., Leite, J.C.B., Loques, O.G., and Fraga, J.S., 1997, "A new technique for task allocation in real-time distributed systems," In: *Proceedings of the 7th Brazilian Symposium of Fault Tolerant Computers,* Campina Grande, Brazil, pp. 269–278.

[10.46] Santos, J., Ferro, E., Orozco, J. and Cayssials, R., 1997, "A heuristic approach to the multi-task-multiprocessor assignment problem using the empty-slots method and rate-monotonic scheduling," *Journal of Real-time Systems,* **13**, pp. 167–199.

[10.47] Tindell, K.W., Burns, A. and Wellings, A.J., 1992, "Allocating hard realtime tasks: an NP-hard problem made easy," *Journal of Real-time Systems,* **4**, pp. 145–165.

[10.48] Jia, H.Z., Fuh, J.Y.H., Nee, A.Y.C. and Zheng, Y.F., 2002, "Web-based multi-functional scheduling system for a distributed manufacturing environment," *Concurrent Engineering: Research and Applications,* **10**(1), pp. 27–39.

[10.49] Erschler, J., Roubellat, F. and Thuriot, C., 1985, "Steady state scheduling of a flexible manufacturing system with periodic releasing and flow time constraints," *Annals of Operations Research*, **3**, pp. 333–353.

[10.50] Shriskandarajah, C. and Ladet, P., 1986, "Some no-wait shops scheduling problems," *European Journal of Operational Research*, **24**, pp. 424–445.

[10.51] Langston, M.A., 1987, "Interstage transportation planning in the deterministic flow-shop environment," *Operations Research*, **35**(4), pp. 556–564.

[10.52] Wittrock, R.J., 1998, "An adaptable scheduling algorithm for flexible flow lines," *Operations Research*, **36**, pp. 445–453.

[10.53] Shriskandarajah, C. and Sethi, S.P., 1989, "Scheduling algorithms for flexible flow shop: worst and average case performance," *European Journal of Operational Research*, **43**, pp. 143–160.

[10.54] Ghosh, S. and Gaimon, C., 1992, "Routing flexibility and production scheduling in a flexible manufacturing environment," *European Journal of Operational Research*, **60**, pp. 344–364.

[10.55] Stecke, K.E., 1992, "Planning and scheduling approaches to operate a particular FMS," *European Journal of Operational Research*, **61**, pp. 273–291.

[10.56] Caprihan, R. and Wadhwa, C.S., 1997, "Impact of routing flexibility on the performance of an FMS – a simulation study," *International Journal of Flexible Manufacturing Systems*, **9**, pp. 273–298.

[10.57] Lee, C.Y. and Vairktarakis, G., 1998, "Performance comparison of some classes of flexible flow shops and job shops," *International Journal of Flexible Manufacturing Systems*, **10**, pp. 379–405.

[10.58] Jawahar, N., Aravindan, P. and Ponnambalam, S.G., 1998, "A genetic algorithm for scheduling flexible manufacturing systems," *International Journal of Advanced Manufacturing Technology*, **14**, pp. 588–607.

[10.59] Chan, F.T.S., Chung, S.H. and Chan, P.L.Y., 2006, "Application of genetic algorithms with dominant genes in distributed scheduling problem in FMS," *International Journal of Production Research*, **44**(3), pp. 523–543.

11

Resource Scheduling for a Virtual CIM System

Sev Nagalingam, Grier Lin and Dongsheng Wang

Centre for Advanced Manufacturing Research
University of South Australia, Mawson Lakes, SA 5096, Australia
Emails: sev.nagalingam@unisa.edu.au, grier.lin@unisa.edu.au,
dongsheng.wang@postgrads.unisa.edu.au

Abstract
The present global market, which is highly competitive, demands high-quality products, reduced cost and shorter delivery times. In order to lead the market, as a minimum requirement, manufacturers should have a capability to satisfy these global demands and be proactive. Globally distributed enterprises can gain leverage in leading the market due to their potential to utilise the distributed resources, and the capability to have collaborative expertise to meet and satisfy the diverse customer requirements. In addition, optimal utilisation of globally distributed resources is a necessity for success in globally distributed business enterprises. With the developments taking place in computer-integrated manufacturing (CIM) and its related technologies, CIM can be further expanded as a globally distributed CIM to overcome the above issues. A network of interconnected CIM systems, which are globally distributed, is regarded as a virtual CIM (VCIM) in this chapter. The VCIM concept was proposed as a solution to satisfy the emerging technological application of virtual enterprises. This chapter focuses on the issues that need to be addressed when resource scheduling is considered for collaborating enterprises that have adopted the VCIM system and it provides a simple approach for resource scheduling with the use of an agent-based architecture.

11.1 Introduction

The economic environment of many countries is under threat from emerging competitors. At the same time, sophisticated customers demand high quality products with shorter delivery times and a diverse number of customised products. In addition, today these customers can reach various manufactures across the world and choose their preferred ones, due to the advancements in information and communication technology (ICT) and sophisticated ICT tools that are available today at affordable price and with increased reliability. Similarly, in this same environment, manufacturing enterprises face competition across the globe, encounter challenges in an open market, and struggle to extend their businesses to reach customers in every part of the world. However, an enterprise that relies on traditional manufacturing systems cannot satisfy the needs of globally distributed customers and match the capability of the competitors, since these traditional systems are only deployed within an enterprise and this enterprise does not have the capability to maximise the potential and strength that is available in different parts of the world. Hence, proactive enterprises seek the application of intelligent and integrated manufacturing systems in order to meet these needs and be the winners in this competitive market.

Therefore, a more flexible and comprehensive methodology is necessary to overcome the distance barriers, facility-sharing problems and communication obstacles. This need has led to the concept of *virtual computer-integrated manufacturing* (Virtual CIM) [11.1], which is a network of interconnected and globally distributed *computer-integrated manufacturing* (CIM) systems across geographical boundaries, whereas CIM is an integration of localised manufacturing facilities.

11.2 VCIM System

Virtual CIM (VCIM) is an evolved concept of CIM and has the vision to integrate all activities in an enterprise or network of enterprises to share resources and management objectives in a holistic and cohesive manner to work as a seamless global CIM system. This VCIM concept has been conceived to move the traditional CIM systems to the next level of virtual enterprises to meet today's needs. A VCIM system enables a consistent framework to be built for marketing, planning, design and production, maintenance, administration, financial management and other activities, no matter where they occur (local or remote) or in whatever environment (hardware and software) the components are placed.

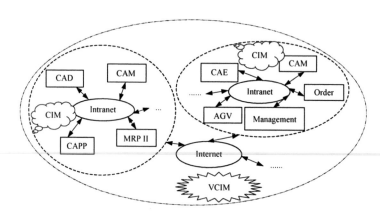

Figure 11.1. Conceptual application of a VCIM system

The application of VCIM is described in Figure 11.1. In a traditional CIM system, components are connected in local networks to communicate, share and exchange information with each other. However, today companies are expanding their business boundaries locally and internationally, and merging or cooperating with other companies across geographical demarcations. As a result, departments of a company or several companies that are involved in the same project may be located far away from each other even in different continents. Consequently, it is almost impossible to deploy a proper and dedicated information technology (IT) infrastructure to support a cooperative environment and it is very difficult to efficiently utilise all of the available facilities and share common data in a real-time environment among all the partners.

To represent the evolving process of CIM, and to reflect the need for a VCIM to meet the current global market and environmental conditions, a new virtual CIM wheel, as shown in Figure 11.2, has been developed at the Centre for Advanced Manufacturing Research (CAMR) of the University of South Australia [11.2]. This new wheel stresses the importance in strategic and integrated management of implementing VCIM across globally distributed enterprises, while enhancing the CIM wheel (also known as the CASA/SME manufacturing enterprise wheel) developed by the Society of Manufacturing Engineers (SME) in 1992.

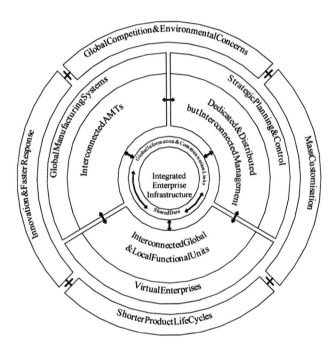

Figure 11.2. Virtual CIM wheel to represent the global market conditions

The concept of the VCIM wheel was explained as follows [11.2]:

- The outer circle represents the present world situation and depicts the characteristics such as global competition, environmental concerns, mass customisation to satisfy the variety of customer requirements, shorter product lifecycles, requirement for innovative products and need for a faster response.
- The second circle represents global systems and concepts needed to address the situation.

- The third circle explains briefly how the concepts and systems can be realised.
- The fourth circle represents the need for global information and communication links, and the need for the shared data among the systems.
- The inner circle represents the outcome of the CIM as a globally integrated enterprise by means of an integrated architecture.

11.2.1 VCIM Issues

Technologies to implement VCIM are becoming available from various research projects and ICT tools that are available today. However, in order to have an effective application, the research on VCIM should be further strengthened and the interaction of ICT tools with a VCIM system should be investigated. The significance of the VCIM is that all distributed manufacturing resources can work together as a united and interconnected system. The final objective of the research project at CAMR is to integrate the distributed manufacturing resources across the globe and let them work as a concurrent, cohesive and cooperative manufacturing system. Only with the concurrency and cooperating features can a VCIM system quickly respond to changing market requirements, reduce costs and offer high-quality products. Some key issues on VCIM that have to be resolved to make the VCIM happen are: how to organise manufacturing resources for information and material flow in a VCIM system in order to produce a specific product, how to interconnect the resources on a virtual environment to share activities, and how to let the distributed but autonomous resources work together as a united system.

However, many of the VCIM issues are similar to a virtual enterprise issues. A virtual enterprise, also known as an extended enterprise [11.3] is a new terminology denoting an activity where enterprises work in a collaborative manner towards a common goal across national boundaries. Crossing global boundaries established by distance, time, language, political and economical systems, and satisfying new rules of economics and the emergence of new consumer and institutional markets are some of the complex challenges faced by today's virtual enterprises [11.4].

A virtual enterprise's functionality can be classified into internal and external activities. The external activities are mainly involved in electronic commerce (e-commerce), which leverages a network-based sales channel to enhance marketing, purchasing and selling of products and services. It then evolves to electronic business (e-business), thereby improving business performance by use of information technologies to connect suppliers and customers at all links along the value chain. Traditional CIM is the internal technological hub for a single manufacturing enterprise and focuses on improving internal process efficiency and effectiveness with integration of computer-assisted processes and enterprise resource planning. It supports current enterprise strategy and is committed to optimising internal operation. On the other hand, e-business extends each enterprise's internal resources into the external environment and focuses on product promotion and external, cross-enterprise interaction. The different features of internal and external environments generate a gap in technological capabilities, which cannot be fulfilled by a traditional CIM system, due to the fact that the merging of both businesses and Internet technologies is incomplete and its exact nature is unclear. Therefore, it is vital for virtual enterprises to have balanced internal and external functionality and appropriate linkage between internal and external environments. In addition, many technical and managerial issues engulf collaboration of virtual enterprises. Some of the issues are:

- Structuring the level of cooperation of entities and configuring different enterprise architectures for a virtual enterprise.
- Identifying methods to build an adaptable, dynamically reconfigurable and open architecture for the distributed functional entities, and providing seamless integration of various management and manufacturing systems of partners/ members in real time.
- Interconnecting partners' hardware/software across geographical boundaries and diverse operational environments, and resolving conflict across these different systems by using suitable interconnection standards and protocols.
- Determining communication among the partners/members of the virtual enterprise, and capturing and managing knowledge across various enterprises for optimal output and to have synergy in operations.
- Identifying the nature of control systems, whether it is a hierarchy or oligarchy or distributed control, and appropriately delegating authority and control among partners.
- Overcoming scalability issues in software management and integration, and resolving legacy software interfacing issues.
- Generating virtual scheduling and execution for dynamic environment, and holistically integrating supply chain and logistics management of the distributed enterprises.
- Facilitating optimal material and information flow across various distributed entities in a virtual enterprise.
- Obtaining the running status of the distributed functional entities in real time and avoiding security breaches on the system integrity, while having an open system.
- Assessing the capability of different entities across the whole virtual enterprise, quantifying the value added to the product/project, and distributing the cost and profit accordingly.
- Accommodating country- and organisation-specific legal, technical, administrative and marketing requirements/issues within a virtual enterprise framework.

Various researchers are pursuing solutions to many of these issues and as stated earlier many of the virtual enterprise related issues are still valid for a VCIM system. However, in order to have an efficient and effective VCIM system, an architecture is paramount to provide seamless integration of various management and manufacturing systems in real time, and to enable the VCIM system in having a capability to respond dynamically and autonomously to the changing real-time situations. Since, in a virtual enterprise, activities of various enterprises are grouped together to reach rationalised integrated operations across many functional units [11.5], whereas VCIM aims to interlink all activities and functional units of distributed CIM systems through information integration and an architecture in a cohesive manner to work as a seamless global CIM system.

A VCIM system needs to be dynamic, reconfigurable and comprehensive, while still having an open architecture [11.1]. This requirement can overcome some of the above-mentioned issues that are related to integrating and cooperating of entities within virtual enterprises. The objective of the VCIM system is to dynamically respond, report, and even predict various statuses and processes of corresponding cooperative projects to achieve optimal results. The reconfigurable feature can be achieved by configuring related components of the system as plug-ins and enabling them to collaborate with different partners during different projects and development stages. "Comprehensive" denotes the scale of the integration of the system, which varies in line with the characteristics of cooperative projects and the size of the enterprises. However, the trend is to include as many related activities as possible into the VCIM system to pursue overall optimisation. As a result, VCIM can provide a mechanism to bring e-commerce, e-business, supply chain, CIM, and enterprise resource planning systems together to form an open and fully integrated virtual enterprise. Then, information, material and energy flows in an enterprise can be managed more efficiently.

11.2.2 Need for a VCIM Architecture

Today, almost all organisations are making use of ICT tools to maximise their productivity and improve the efficiency. In addition, many organisations around the world are offering services online to build up a stronger e-commerce environment. Many large companies are investing heavily in the corporate deployment of advanced ICT systems with an objective to realize real-time information flow and to enable accurate and responsive decisions. Researchers, designers, engineers, purchasers and consumers have connected their personal computers to the World Wide Web (WWW) where they announce innovative offers and look for new opportunities. By using the WWW, goods and services can be sold anywhere and everywhere. The traditional reliance on physical goods production for governing a business has been shifted. Moreover, market dynamics is driving manufacturing companies to integrate their strategies relating to people, processes and technologies for improving their overall agility to the market. Hence, the broader, deeper and faster sharing of manufacturing knowledge is generally considered as a critical competitive resource to serve the customers in the most efficient and effective manner. Therefore, a VCIM system should have an effective and efficient architecture to facilitate this effort.

Applying the VCIM concept in today's enterprises means establishing real-time connection among components of a VCIM system, where most have inconsistent data format among different participants' enterprises. The developments in Internet, object-oriented technology and agent-based systems provides a required means to bring distributed units of an organisation or even different companies together and to cooperate with each other to complete multidisciplinary and comprehensive projects through the framework of VCIM.

As stated in Section 11.2, it is almost impossible to deploy a proper IT infrastructure for cooperating environments. One feasible solution to the distributed and cooperative situation is to set up a common database to manage the availability of facilities and data that can be used in a common project. For instance, data on manufacturing equipment and information of application software packages such as computer-aided design (CAD), computer-aided manufacturing (CAM), computer-aided processing planning (CAPP), computer-aided engineering (CAE) and others can be stored in a common database. Users can obtain a proper facility by querying the database with their requirements. However, two drawbacks exist in this approach. First, if the required IT infrastructure (network connection either cabled or wireless) is not available between the corresponding departments, it may not be possible to use a centrally located database and access it through the network. Second, even if the central database can be developed and applied, the management and manipulation of the shared facilities are neither dynamic nor automatic, which means the database cannot dynamically report whether a specific facility is working or inactive or waiting for requests, nor automatically start the facility, then the system cannot be used in a real-time environment. In these types of systems, users need to find out first whether a facility can complete certain tasks. Then, they need to apply other mechanisms to determine the availability of the facility and initiate operations remotely or on the local system. These drawbacks will impact the efficiency and effectiveness of using expensive but available software and hardware, and adversely influence the cooperation among partners.

As the agent-based technology provides a better means of conceptualising and implementing an application system [11.6] and many real-world systems are naturally distributed by spatial, functional, or temporal differences, a multi-agent system can easily accommodate the distributed environments [11.7] such as VCIM. It provides a simpler and more understandable approach to study and implement the VCIM system. Jennings and Wooldridge [11.6] defined that an intelligent agent is a computer system situated in some environment and is capable of flexible autonomous action in order to meet its design objectives. Earlier, in a landmark article [11.8], they described that the properties of intelligent agents include *Autonomy*, *Social ability*, *Reactivity* and *Pro-activeness*. By using these properties, an integrated manufacturing environment can be formed when distributed manufacturing functional entities are encapsulated as intelligent agents and connected together as an agent community. In the meantime, Shen and Norrie [11.9] in their extensive review of agent-based systems for manufacturing and related applications stated that the intelligent agent technology has been applied in many areas including manufacturing enterprise integration, supply-chain management, manufacturing planning, scheduling and control, materials handling, and holonic manufacturing systems.

With this agent-based environment, the agents of distributed manufacturing resources can negotiate with each other to reach efficient resource planning and scheduling based on the real-time working status information of the distributed resources. In addition, manufacturing resources in an enterprise can be encapsulated as intelligent agents and connected at different levels, such as an enterprise level or a shop-floor level. By connecting the agents at shop-floor levels as sub multi-agent systems, the manufacturing resources can be integrated in an agent-based environment with a multi-level structure. With this integration environment, the complex task of an enterprise resource planning can be subdivided as a number of simple tasks at the individual levels. Accordingly, the use of intelligent agent technology can potentially offer a practical approach for real-time distributed manufacturing resources integration and scheduling. Unlike this agent-based approach, there are many other approaches for resource scheduling, such as heuristics, constraint-propagation techniques, simulated annealing and Tabu search, all of which have been described by Shen [11.10] as traditional approaches. These approaches are usually based on search strategies and use simplified theoretical models with centralised computation through a single node.

Various researchers [11.11]–[11.22] have attempted to develop agent-based architectures for distributed manufacturing resource integration and scheduling. Although their efforts have brought many valuable solutions in particular application areas, these approaches have some shortcomings, and will not be able to support a global integrated manufacturing system in the form of VCIM due to the nature of the VCIM concept and its ultimate aim to provide a cohesive interaction among all interlinked VCIM components. Although, these approaches provided solutions for scheduling manufacturing resources using agent negotiation, most of these approaches did not consider optimised resource allocation when multiple manufacturing resources provide proposals for a single task request. Some of these research projects are briefly reviewed in this section.

In 1991, Sycara *et al.* [11.11] proposed a multi-agent system to schedule facilities in multiple shop floors. This system designed intelligent agents with capabilities to perform the scheduling of a set of jobs among a set of resources. A few years later, towards integrated resource scheduling, Sikora and Shaw [11.12] presented a multi-agent system. In this system, components such as decision software modules, and process controllers in the information system were converted as intelligent agents. This system also developed coordination mechanisms for agents to schedule operations and achieve the desirable system performances. At the same time, Kouiss *et al.* [11.13] implemented a multi-agent based system for dynamic job-shop scheduling. This system dedicated work centres as resource agents, which could perform local facility planning by selecting suitable dispatching rules. This architecture also integrated a supervisory agent to control those resource agents. Meanwhile, Barbuceanu and Fox [11.14] introduced an agent network to simulate a supply chain. In that system, intelligent agents were integrated to perform one or more supply-chain functions, and a coordination language was developed to resolve coordination issues by the researchers.

Later, towards shop-floor resource integration and scheduling, Parunak *et al.* [11.15] developed an agent-based architecture that was named AARIA. This architecture modelled shop-floor facilities as a series of unit processes, parts, and resources. The unit processes represented operations. Parts denoted inputs (materials) and outputs (products) for a unit process. Resources represented various manufacturing resources such as machines, and handling devices. This AARIA architecture encapsulated those facilities as intelligent agents and used heuristic techniques such as the simulation technologies to perform distributed resource scheduling.

MetaMorph [11.16] is a well-known agent-based architecture that was developed to integrate manufacturing resources for intelligent manufacturing applications. This approach encapsulated distributed resources as intelligent agents and grouped them according to the functionalities such as design, marketing, scheduling and material supply. In addition, each group contained a coordinator named *Mediator* to connect with the agents in that group. An enterprise Mediator was also included to connect with Mediators of each group. By using this approach a Mediator-centric hierarchical architecture is formed and with this architecture manufacturing resources of multiple enterprises can be integrated according to distinctive enterprise management layers. At the same time, some architectures [11.17][11.18] included agents with special functionalities for resource scheduling. However, these architectures would not enable a product order to be rapidly responded in order to meet today's industry needs such as satisfying a customer's required due date/time, or delivering to the desired destination.

In 2003, Karageorgos et al. [11.19] developed an agent-based system for the optimisation of logistics and production planning. This system used a holonic viewpoint to consider business networks as a temporary holon, where service providers were considered as sub holons. This system considered an enterprise holarchy by extending the holonic concept to lower-level resources in an enterprise and formed an agent-based hierarchical integration environment by mapping intelligent agents in this holarchy. This system also used a wholesaler agent to accept customers' query and used a directory facilitator (DF), which was provided by a FIPA compatible platform, to find agents that represented partner enterprises. Then, the enterprise agents further decomposed requests as basic products and requested proposals from those agents that represented manufacturing facilities through agent negotiation, which utilised a nested contract net protocol. In addition, these enterprise agents were also responsible for selecting and sending proposals information to the wholesaler for customers to make decisions on the proposed plan. However, their approach did not include strategies to perform optimised resource allocation in the planning process. At the same time, Odrey and Mejia [11.20] built a multi-agent architecture. This architecture incorporated production agents, error-recovery agents, and a structure of a mediator agent towards the construction of a reconfigurable system. It also used an approach that was based on a Petri net to perform process planning and error recovering in a multi-agent system environment.

Later, around 2004, Jia et al. [11.21] developed an agent-based system for coordinated product development and manufacturing. This system integrated a managing agent and many functional agents together. In this system, the managing agent connected the functional agents that might be located at different regions. With this integrated environment, a product lifecycle could be optimised by comprehensively considering production procedures that ranged from product design to final manufacturing in an entire production process. Recently, Nahm and Ishikawa [11.22] proposed an agent-based architecture to support inter-enterprise resource integration and collaboration. Their architecture included a hybrid agent architecture to enable agents to exhibit hybrid behaviour and interaction. It also included a hybrid network architecture to help build large-scale distributed manufacturing systems. As a starting point, this project developed an attribute unit agent, a design interface module agent, a service wrapper agent, and a design module agent to form a collaborative product development environment.

As indicated earlier, some of these previous research results assisted us in our search for a solution in identifying/developing an applicable architecture for the VCIM concept. However, an agent-based VCIM architecture for resource integration and scheduling that would be more applicable to a network of small and medium enterprises has to be developed, as none was able to provide all the required features to effectively implement a VCIM system. Our developed architecture includes many *Facilitator* agents (a coordinator in this system is called a Facilitator). These Facilitator agents are designed with the capability to coordinate activities of all manufacturing resource agents and at the same time be a DF. Each Facilitator agent uses a "backward network algorithm for the shortest path" to perform alternative optimisation for resource scheduling. With this VCIM architecture, the Facilitator agents can process information flow across the agent community in a parallel manner and dynamically schedule the distributed resources in order to rapidly respond to customer product requests.

11.2.3 An Agent-based VCIM Architecture

To implement an agent-based VCIM system, three categories of agents have been identified [11.23][11.24]. These agents include Facilitator agents, Customer agents, and Resource agents. Facilitator agents are designed to act as coordinators to route the information flow across the VCIM agent community. Customer agents are designed to provide interfaces for customers to participate in the VCIM system. Finally, Resource agents are designed as agent interfaces to encapsulate distributed manufacturing functional entities and connect them with the agent community. The functionalities and responsibilities of these agents are described as follows:

- Facilitator agent: Users need to send request messages to this agent to participate in VCIM and get a response from it. This agent decomposes the request message from the users as a set of subtasks and sends subtask requests to related Resource agents. This agent receives proposal messages for subtasks from Resource agents, combines the proposals as production schedules, and seeks an optimised candidate schedule. Resource agents register their information such as functionalities, name, and address in a database of this agent. To have these capabilities, message formats for agent communication, and strategies for manufacturing resources planning and scheduling are stored as a knowledge base of this agent.
- Customer agent: Users input request message consisting of information such as product name, required due date/time, delivery place and others by using a graphical user interface (GUI) of this agent. This agent gets the responses from the Facilitator agent and displays the results to the users.
- Resource agent: This Resource agent is responsible for further decomposing tasks and for local resource planning and optimisation. To have these capabilities, agent communication actions and functional records such as the time and cost for this agent to complete a subtask are identified and stored in the knowledge base and database of this agent. Each Resource agent maintains an independent working schedule and renews this schedule when a new task is confirmed.

By connecting a Facilitator agent, a Customer agent and many Resource agents through the Internet, these agents can form a basic multi-agent VCIM system as depicted in Figure 11.3. However, a real VCIM system includes many Facilitator agents, Customer agents and Resource agents, while the functionalities of Resource agents may include design, manufacture, assembly, delivery, material supply and others.

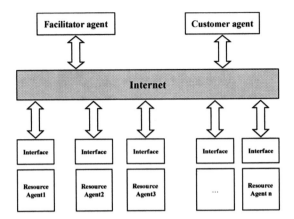

Figure 11.3. A multi-agent VCIM system model

In our VCIM concept, a Resource agent can include a manufacturing organisation or a unit that has an authority to plan internal resources to meet a customer order. With this definition, a Resource agent can be any of the following or a combination of: an assembly workshop, a manufacturing system, a group of delivery facility, a material provider, a design department, a branch company or a small enterprise. This Resource agent usually has a group of functionalities that should be registered with the Facilitator agent. This Resource agent may include many functional units, and a functional unit may include many functional entities. In addition, when manufacturing resources at lower levels are encapsulated as sub multi-agent systems, a Resource agent hierarchy for VCIM systems can be formed. However, our current work focuses on integrating and scheduling the Resource agents at the VCIM level and the sub-functionalities (or sub-agents) will be considered in our future work. Based on these approaches, we formulated an agent-based architecture with a three-layered structure as shown in Figure 11.4 to support the VCIM system implementation. In this structure, as shown in the figure, all Customer agents, Facilitator agents and Resource agents are accommodated in three layers such as a Customer layer, a Facilitator layer, and a Resource layer.

Generally, each manufacturing organisation has its own benefits, partner relationships, resource-selection criteria, and other elements related to manufacturing planning. Our agent-based VCIM architecture uses these realistic issues related to the benefits of an organisation, as a particular knowledge for partner resource selection, and dictates this knowledge with the Facilitator agents. We consider a VCIM system is formed when an organisation has built a Customer agent to provide a user interface and a Facilitator agent to coordinate the Resource agents shared by partner organisations. However, our approach allows an organisation to implement multiple Facilitator agents that may have different knowledge on resource planning depending on the local issues and situation, and multiple Customer agents at different areas for users to invoke VCIM services. By means of these approaches, when many organisations join together as a network of organisations to collaboratively work to achieve the competitive edge by building Facilitator agents and Customer agents for a VCIM system, an agent-based architecture with a three-layered structure is formed.

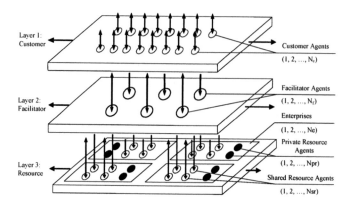

Figure 11.4. Three-layered structure of the agent-based VCIM architecture

With this proposed three-layered architecture, the VCIM systems can potentially achieve much reliability, efficiency and needed flexibility. In this architecture, all agents (except sub-agents) are connected through the Internet and new Facilitator agents, Customer agents, and Resource agents are permitted to connect with the existing structure when and where necessary (sub-agents can either be connected by the Internet or an Intranet). This feature enables a VCIM system to have the necessary flexibility in extending information processing capability, customers' convenience, and manufacturing capability of the partner organisations. In addition, each Facilitator agent has the capability to communicate and negotiate with all other Customer agents and Resource agents. Therefore, when an error occurs or information is blocked in one facilitator agent, a Customer agent can select and use other Facilitator agents to invoke the VCIM services. This alternative selection of the Facilitator agent to process a task allows the information to flow in a more reliable manner than a centrally controlled system that can have a possible communication bottleneck.

11.2.4 A Java Implementation Environment for a Multi-agent VCIM System

In order to improve the agent interoperability, a Java environment is used to implement the intelligent agents and to organise the agents as a multi-agent VCIM system. This environment includes two Java-implemented software tools: Java Agent DEvelopment Framework (JADE) [11.25] and Jess [11.26].

JADE enables multi-agent application development through the Internet and can act as a middleware. The framework of JADE is fully implemented in the Java programming language and complied with the specifications of the Foundation for Intelligent Physical Agents (FIPA) [11.27], where FIPA is an IEEE Computer Society standards organisation, originally formed as a Swiss-based organisation in

1996 and later accepted by the IEEE as its standards committee. FIPA acts as the standards organisation for agents and multi-agent systems and promotes agent-based technology and interoperability of its standards with other technologies [11.27]. Meanwhile, Jess [11.26], which is a Java expert system shell, acts as an intelligent functional module to facilitate JADE agents to automatically perform information analysis. Based on this agent development environment, construction approaches of VCIM agents are developed in this research project.

Using the JADE to implement VCIM agents provides many common capabilities. These capabilities are:

- Software behaviours for message operations such as sending, receiving, saving, storing and deleting messages related to agent communication.
- Software services to generate real-time agent information such as agent ID, agent address and agent activation status.
- Communication interface to support information exchange, thereby maintaining a message queue to control both incoming and outgoing messages.
- Agent GUI to provide menu commands for users to operate communication messages.

Besides these common capabilities, to perform the requested functionalities of VCIM agents, special software behaviours and functional modules are developed with the JADE agent structure in our research project. Some of the developments include:

- A Jess functional module to provide expert-system components such as inference engine, knowledge base and database. This module also assists in analysing the production order messages to perform the requested task decomposition.
- Software behaviours such as selecting Resource agents and requesting subtask proposals, combining proposals as production schedules and performing optimisation of resource allocation, and processing messages with respect to agent communication in a Facilitator agent.
- In a Resource agent, enabling software behaviours to perform local resource planning, providing a proposal message to satisfy a requested subtask, recording new tasks when receiving acceptance messages or cancelling temporary proposals information when receiving refusal messages, and submitting registration information of Resource agents.
- GUI for the customers to assist send, receive and confirm product order message.
- Output module for customers to save, view or print information of production schedules.

Based on the analysis of functional modules and software behaviours, three categories of agents that were defined in this research project for the implementation of a VCIM system were constructed [11.28]. Figure 11.5 depicts configurations and the connection pattern of those VCIM agents based on the JADE framework. In the configuration of these three categories of agents, similar functional modules are designed to support the consistent operation in the same architecture. These similar components include a GUI for each agent, a communication interface, and a message-queue manager. However, due to the different functionalities, each type of agent has different components in their configuration: The Facilitator agent includes a Jess module with components of an inference engine, a knowledge base, and a database; The Customer agent is designed for the customer's application, therefore, it does not have a Jess module, but has an output module to view or print the proposal documents; The Resource agent includes a Jess module to analyse the incoming request message and an output module to view and print the working schedule.

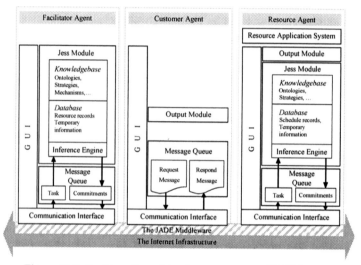

Figure 11.5. Configuration of the three categories of VCIM agents

Besides this agent configuration, we also use the JADE specifications to identify the agent communication. JADE uses the FIPA Agent Communication Language (ACL) to support agent communication. The standard ACL message is formatted with three layers: communication layer, message-identification layer, and content layer. However, we use a VCIM message, which has a simple variation from the standard ACL message and each layer is identified with information that is relevant to the VCIM resource scheduling process. In the VCIM message, the communication layer only includes names of the Sender and the Receiver agents. The message identification layer uses ten communicative actions to describe the message action and indicates the data structure for decoding. The action includes terms such as: a Request message to ask for a production proposal; a Proposal message to respond to a request; an Accept-proposal message to accept a proposal; a Reject-proposal message to reject a proposal; a Query-ref message to ask for the customers' opinion on a production schedule of the proposal; an Agree message to agree with a query message; a Refuse message to refuse a query; a Confirm message to confirm an action that has been identified; an Inform message to register a new agent; a Failure message to indicate an action that has not been completed. Finally, the content layer comprises the message content. This message includes the customer's requirements for a particular product. Nevertheless, this message format is to be reviewed and improved at the next stage of VCIM development.

11.3 Resource Scheduling with the VCIM Architecture

With our proposed three-layered VCIM architecture, as a preliminary approach a "backward network algorithm for the shortest path" is used to perform the optimisation of distributed manufacturing resource scheduling for defined product requests. This resource scheduling approach is explained in the following section.

11.3.1 Resource Scheduling in a VCIM System

In order to initiate the VCIM services to manufacture a product, a customer first uses the GUI of a Customer agent to submit a product-request message to a selected Facilitator agent. Then, the Facilitator agent that received this request decomposes the requested task into interrelated subtasks, according to the registered functionalities of Resource agents and the built-in knowledge base of the Facilitator agent for this requested product. Our approach considers that the functionalities of Resource agents at the VCIM level include many working procedures of a production process, such as product design, material supply, manufacturing, assembly, test, package and warehousing. Then, this Facilitator agent sends request messages to Resource agents that are considered to have a capability to complete a subtask. In this approach, the Facilitator agent performs the requested procedure by matching the subtask and the Resource agent's functionality in its database. Similarly, when receiving a request message, the Resource agent checks its functional records in its database to determine a required processing time and cost, then checks its working schedule to allocate available time segments and form a proposal message that includes information such as start time, finish time and cost to perform the requested subtask. This approach includes cost and time in the proposal message, as they are the main factors that influence partner resource selection. However, other factors, such as quality, friendship, credit and delivery reliability, which Su et al. [11.29] and Chu et al. [11.30] have discussed for resource selection in virtual enterprises, will be considered at our next stage of development by using a multiple criteria approach. In addition, the Facilitator agent also sends request messages for proposals for transportation services, since the subtasks to manufacture a product may be performed by Resource agents at different places among the collaborating organisations and transportation procedures are necessary to link the material/parts flow.

Combining the proposals of subtasks and transportation procedures results in many candidate production schedules with different start time and cost, since manufacturing resources in a VCIM system may have different cost structure, time frame and availability. Among these candidate schedules, our approach aims to achieve an optimised schedule that has the lowest cost as the primary criteria and the shortest production time as the secondary criteria. In order to achieve this objective, the Facilitator agent uses a "backward network algorithm for the shortest path" to perform the resource alternative optimisation in the scheduling process. With this approach, when the optimised schedule has been reached, the Facilitator agent will send this schedule as a proposal for the customer to review through the Customer agent. If the customer accepts the schedule and sends back a confirmation message, the Facilitator agent sends confirmation messages to all selected Resource agents and the Resource agents renew their working schedule by adding this new task. Otherwise, if the customer refuses the schedule, the Facilitator agent cancels all temporary proposal messages for the current product request. However, customers can refine their requirements and submit a new request. Currently, we are investigating additional features, such as allowing a customer to modify the product schedule or to request that certain features be produced at pre-selected locations by pre-selected Resource agents of the customer's choice.

11.3.2 VCIM Resource Scheduling Process

The resource scheduling process in a VCIM system starts when a customer-requested product order activates a Facilitator agent to perform the task decomposition and ends when this agent achieves an optimised production schedule. To describe the resource allocation process explicitly, a simple product "PRODUCTA", which is considered as having eight sequential operations (as shown in Figure 11.6), is utilised. The eight different manufacturing operations that are considered for this "PRODUCTA" to illustrate the VCIM resource scheduling process are: a material supply, a part manufacturing, two standard part supplies, a sub assembly, a general assembly, a testing and a packaging procedure.

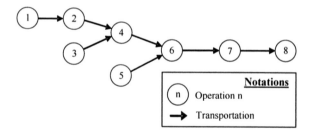

Figure 11.6. Task decomposition and subtask sequence for a product

The Contract Net protocol (CNP) was first proposed by Smith [11.31] in 1980, which has been widely used in agent-based application systems to facilitate distribution of subtasks among agents for agent negotiation. Although there are some modified versions such as the extended contract net protocol [11.32] and nested contract net protocol [11.19], CNP is usually used when an agent wants to get a response for a task. When a principle agent sends a request message to one or many agents and gets replies, it awards a contract to the best offer according to a selection criterion determined by the principal agent. Our agent-based VCIM architecture uses CNP as the basic protocol to support the resource scheduling. This application involves a series of agent negotiation activities: when a customer submits a request message to a Facilitator agent for a product, the Facilitator agent sends subtask messages to Resource agents for proposals based on that request.

Then, Resource agents send proposals to the Facilitator agent and the Facilitator agent sends confirmation messages to award contracts to selected Resource agents. Proposal messages are generalised through agent negotiation based on the real-time information of distributed manufacturing resources.

The following notations are used to formulate the resources scheduling process and to explain the intermediate stages in this process.

O	=	A subtask that includes manufacturing operations such as {Operation 1, Operation 2, ..., Operation 8}
T	=	A transportation subtask corresponding to the subtasks in O that includes procedures such as {Transportation 1, Transportation 2, ..., Transportation 8}
m_i	=	Total number of Resource agents to perform Operation i, where, $i = 1$ to 8.
N_t	=	Total number of Transportation agents
RA_{ij}	=	jth Resource agent to perform Operation i, where, $i = 1$ to 8, $j = 1$ to m_i
TA_k	=	kth Transportation agent, where, $k = 1$ to N_t
Loc_{ij}	=	Location of RA_{ij}
S	=	Start time in a proposal for a manufacturing operation
S^t	=	Start time in a proposal for a transportation procedure
E	=	End time in a proposal for a manufacturing operation
E^t	=	End time in a proposal for a transportation procedure
C	=	Manufacturing operation cost
C^t	=	Transportation procedure cost
TC	=	Total cost of an intermediate production schedule
TS	=	Start time of an intermediate production schedule
R^T		Required finish time for a transportation agent
CT	=	Current time/date
DT	=	Customer requested due date/time
Flc	=	Customer requested delivery destination

In order to complete a product request, there are nine steps in the resource allocation process. They are depicted in Figure 11.7 and described below:

- *Step 1: Select a subtask by using a backward sequence.* The Facilitator agent uses a backward sequence to select a subtask and subtasks are selected among Operation i, where $i = 1$ to 8. Therefore, for this sample product "PRODUCTA", Operation 8, which is the final subtask, is first selected.
- *Step 2: Select Resource agents to perform the subtask that is selected in Step 1.* In this process, the Facilitator agent identifies Resource agents to perform the subtasks selected in Step 1 by matching Resource agents' functionality to subtasks. Since Resource agents register their functionalities with a Facilitator agent and subtasks are also stored according to Resource agents' functionalities in the database of Facilitator agents. Considering Operation 8, Resource agents are selected among the RA_{8j}, where $j = 1$ to m_8.

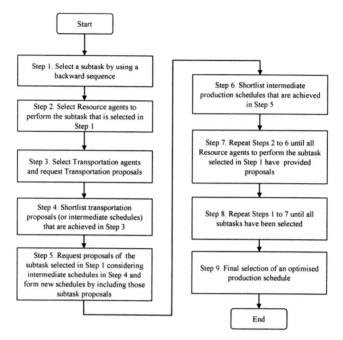

Figure 11.7. Flow chart for the VCIM resource allocation process

- *Step 3: Select Transportation agents and request Transportation proposals.*
  The Facilitator agent identifies suitable Transportation agents and requests proposals for a transportation procedure. The Facilitator agent performs this process until all Transportation agents have provided proposals related to the subtasks that are selected in Step 1. Considering Operation 8, the Facilitator agent identifies the Transportation agents among the TA_k, where $k = 1$ to N_t, to perform Transportation 8. A Transportation agent TA_1 is first selected. Then, the Facilitator agent forms a message to request proposals of Transportation 8 from TA_1. To form the request messages, the Facilitator agent includes Loc_{81}, the location of RA_{81}, as the start address. Since Transportation 8 is the final procedure in a production process, in the request message, the Facilitator agent includes Flc as the end address and includes DT as the required finish time. After that, the Facilitator agent sends the request message with information of start address, end address and required finish time to TA_1. Then, TA_1 replies with a proposal to perform Transportation 8 with information of start time S^t, end time E^t and cost C^t. In this approach, a Facilitator agent considers a proposal is valid if the proposal satisfies $E^t \leq DT$, and $S^t \geq CT$, where DT is the required due time in the request message and CT is the current time. Therefore, this backward-sequence algorithm to select subtasks and request subtask proposals enables candidate production schedules to satisfy the customer required due time and delivery destination. This process is repeated by the Facilitator agent for the remaining Transportation agents among TA_k, where $k = 2$ to N_t and sends messages to those agents to request proposals to perform Transportation 8.

Then, each Transportation agent replies with proposals to perform Transportation 8. These transportation proposals need to satisfy $E^t \leq DT$, and $S^t \geq CT$. In addition, as Transportation 8 is the final production procedure, the proposals to perform Transportation 8 form the initial intermediate production schedules. Therefore, the start time and total cost of these initial intermediate schedules are allocated as $TS = S^t$, and $TC = C^t$. At this stage, since multiple Transportation agents can provide proposals, there are multiple intermediate schedules. The costs of these intermediate schedules are calculated by adding together the cost of the process by the particular Resource agent and the transportation cost for the next destination.

- *Step 4: Shortlist transportation proposals (or intermediate schedules) that are achieved in Step 3.* In this process, the Facilitator agent that is processing the product order shortlists intermediate schedules that are achieved in Step 3. For the shortlisting process, the Facilitator agent selects an intermediate schedule that has the lowest cost and moves it to a selected schedule list. If multiple schedules have the lowest cost, a schedule with the latest start time is moved to the selected schedule list. Then the Facilitator agent deletes all schedules that have a start time earlier than the selected one. In the remaining intermediate schedules, the Facilitator agent repeats this selection process until no schedule is in the waiting list. This shortlisting process records intermediate schedules that have a lower cost or a higher cost but with a later start time. This approach deletes redundant intermediate schedules, as it reduces agent negotiation activities in the subsequent step of requesting next-stage subtask proposals. Shortlisting the intermediate candidate schedules assists in reaching a feasible schedule quickly by reducing related agent negotiation activities and related computation. As Facilitator agents request proposals and later schedule the resources through agent negotiation, some of the intermediate candidate schedules that are feasible in the initial calculation may become infeasible at the next step of calculation. For example, if the due date/time is too close to the current date/time, some of the initial schedules cannot reach a potential Resource agent to execute the next subtask because of the tightly required time period. To illustrate this shortlisting process, a flow chart is given in Figure 11.8.

- *Step 5: Request proposals of the subtask selected in Step 1 considering intermediate schedules in Step 4 and form new schedules by including those subtask proposals.* In order to perform this step, the Facilitator agent sends request messages to Resource agents for the subtask proposals. The request messages are formed by including the start time of transportation proposals that are recorded in Step 4 as the required finish time for the subtasks that have been selected in Step 1 and the corresponding Resource agents that were identified in Step 2. Considering each of the intermediate schedules recorded in Step 4, the Facilitator agent forms a request message, sends this message to ask for a subtask proposal, and combines the new proposal with that related to intermediate schedules. By including the start time of each proposal that is recorded in Step 4 to perform Transportation 8, as the required finish time R^T, the Facilitator agents form request messages and send these messages to RA_{81} to request proposals to perform Operation 8.

Then, RA_{81} replies with a proposal to perform Operation 8 with the information of the start time, end time and cost. These proposals need to satisfy $E \leq R^T$, and $S \geq CT$. When achieving these proposals, total cost TC and start time TS of intermediate schedules are allocated as TC = TC1 + C, and TS = S, where TC1 is the total cost of an original intermediate schedule. With this approach, the lead time between the procedures to perform Operation 8 and Transportation 8 is minimised.

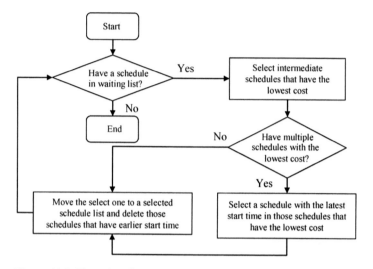

Figure 11.8. Flow chart for shortlisting intermediate production schedules

- *Step 6: Shortlist intermediate production schedules that are achieved in Step 5.* For this step, considering the intermediate schedules that are recorded in Step 5, the Facilitator agent takes similar approaches to those of Step 4 to shortlist the intermediate schedules by comparing the total cost and the start time of intermediate schedules. This shortlisting process is repeated in order to further reduce the redundant intermediate schedules resulted from the proposals of the Resource agents. However, in this process, intermediate schedules that have a lower cost or a higher cost but with a later start time are recorded.
- *Step 7: Repeat Steps 2 to 6 until all Resource agents to perform the subtask selected in Step 1 have provided proposals.* At this stage, the Facilitator agent repeats Step 2 to Step 6 until all Resource agents that are selected in Step 1, have provided proposals. Considering Operation 8, the Facilitator agent repeats Step 2 to Step 6 until all Resource agents RA_{8j}, where $j = 2$ to m_8 have provided proposals. This resource scheduling process forms and records a series of intermediate production schedules. At this stage, the intermediate schedules include proposals for Operation 8 and Transportation 8.
- *Step 8: Repeat Steps 1 to 7 until all subtasks have been selected.* The Facilitator agent repeats Step 1 to Step 7 to process the remaining subtasks. According to a backward sequence, Operation 7 is selected as the next

subtask. Then, the Facilitator agent repeats Step 2 to Step 7 until all Resource agents RA_{7j}, where j =1 to m_7, to perform Operation 7 have provided proposals. With this resource scheduling process, new intermediate schedules are formed and recorded by including proposals of Operation 7 and Transportation 7. However, as Transportation 7 links Operation 7 and Operation 8, when requesting proposals of Transportation 7, locations of RA_{7j}, where j =1 to m_7, are considered as start addresses. In addition, the required finish time and end address are identified from locations of Resource agents and start time to perform Operation 8, in intermediate schedules. This approach enables the lead time between Transportation 7 and Operation 8 to be minimised. This Step 8 of the resource allocation process in VCIM resource scheduling continues until all remaining subtasks operations have been selected and the final production schedules have been achieved.

- *Step 9: Final selection of an optimized production schedule.* When all the final production schedules are achieved, the Facilitator agent performs a final selection of an optimised production schedule. In this step, the Facilitator agent selects an optimised schedule that has the lowest cost as the primary criteria and shortest time as the secondary criteria. In this selection process, as shown in Figure 11.9, the Facilitator agent selects a schedule that has the lowest cost in the final production schedules; if multiple final schedules have the lowest cost, the Facilitator agent selects a schedule that has the latest start time in those schedules with the lowest cost; if multiple schedules have both the lowest cost and the same start time, the Facilitator agent selects a schedule with more Resource agents that are selected from a priority Resource agent domain. A priority Resource agent domain is to be formed by considering credit-related factors for partner resource selection and it is used by the Facilitator agent to further reduce the search of the suitable resources and to reduce the agent negotiation activities. Furthermore, when a Resource agent has reached a required credit level, it will be allocated into the priority domain. This selection approach conforms to the real enterprise cooperation scenario, as Berger *et al.* [11.33] described that the enterprises are using long-term partnerships to achieve the benefits provided by multiple sourcing. For the final schedule, the total cost of the final production schedule is calculated by adding together the selected resource agents' costs and the transportation agents' costs. The start time of the final production schedule is considered as the one with the earliest start time.

To support VCIM resource scheduling, a Facilitator agent needs a capability to decompose a product order related tasks into subtasks. To achieve this capability, as shown earlier in Figure 11.5, a Jess module is included in the agent structure of the Facilitator. This Jess module includes knowledge base, database and inference engine and in this module, heuristic knowledge of human experts is identified as a set of rule-based knowledge by specifying a set of actions considering a group of conditions for each individual product and capability. With this rule-based knowledge, a Jess inference engine can match a collection of facts with given conditions of the rules, and when a rule is initiated, actions of that rule are offered as inference solutions. Therefore, by including product order information as conditions and subtask information as actions in a rule-based knowledge, when a customer submits a product order that matches given conditions of a rule, that rule is initiated and task decomposition is achieved. These mechanisms of a Jess module enable a Facilitator agent to achieve task decomposition when receiving a customer's product order. However, to expand the capability of these agents, more information is to be added by considering all the products that can be manufactured by the collaborating enterprises.

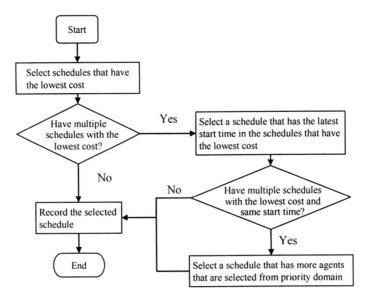

Figure 11.9. Flow chart for selecting the finalised production schedule

The product order information consists of sender agent, receiver agent, required product name, delivery address, and due date/time. Subtask information includes production procedures in a production process considering a specific product order. Subtask information also indicates the sequence and relationships of the production procedures. Although a Facilitator agent is currently imbedded with predefined knowledge for decomposing each product within the capability for the VCIM system, our research project endeavours to imbed the agent community with more artificial intelligence to include many variations in each product profile that can be manufactured by the collaborating partners.

In order to support VCIM resource scheduling, Resource agents need a capability to perform local resource planning and to submit subtask proposals to the Facilitator agents. Resource agents achieve this objective by determining the required cost and processing time from the subtask request of the Facilitator agent and by matching its capability information, and by considering the local lead time and manufacturing process issues by checking its database first. Later, the Resource agent allocates a time segment that is closest to the required finish time by checking its working schedule, and forms a proposal message by including information of the start time, end time and required cost. This message is then sent as the proposal to the Facilitator agent.

Currently, the manufacturing managers for the enterprise, where the Resource agents are located, are allowed to update the lead time, capacity and capability issues for the Resource agents. Similarly, managers who own the transportation systems, update and validate the Transportation agents considering their local issues. However, research is in progress to inbuild this knowledge without human intervention by considering dynamic resource planning issues of the collaborating enterprises [11.34][11.35].

11.4 Conclusions

To assist manufacturing industries in the current global market, researchers are seeking solutions towards intelligent manufacturing and the application of virtual computer-integrated manufacturing (VCIM) systems. Today, we are at the threshold of utilising innovative and improved computer-related and communication technologies for the betterment of manufacturing industries, compared to the situation that prevailed a decade ago. The successful future of the manufacturing industry is inextricably involved in the efficient and effective utilisation of intelligent equipments and its components in the form of VCIM. The research in VCIM and related technologies, and its application in manufacturing industries are progressing towards a better future. Current developments in VCIM illustrate that these systems can be extended beyond the boundaries of the manufacturing industries to improve the future for mankind.

In order to illustrate the concept of VCIM resource scheduling, a simple resource scheduling approach for a product is given in this chapter. In this approach, three categories of agents, which include Facilitator agents to coordinate the agent community, Customer agents to provide user interfaces to participate in VCIM activities, and Resource agents to provide manufacturing functionalities, are identified. With these agents, an agent-based architecture with a three-layered structure, which can accommodate all these agents for a VCIM system no matter where the agents are located and to which enterprise these agents belong, is proposed. With this architecture, each participating organisation can have one or more Facilitator agents to dictate their particular knowledge of resource planning taking into consideration their own benefits, while each Facilitator agent can potentially coordinate activities of all Resource agents through the Internet. This feature enables many Facilitator agents to process information flow across the VCIM environment in a parallel manner for multiple product requests, and has the potential to be more reliable and efficient to support the distributed resource integration than centrally controlled architectures by a single coordinator agent. In addition, this illustrated approach permits new agents, whatever the category, to be connected when and where necessary. This feature enables the VCIM system to have more flexibility, information processing capability, and manufacturing capability to quickly respond to the customer requirements.

In this approach, proposal messages for subtasks to manufacture a requested product are generated through agent negotiation, which uses real-time working status information of the resources that are considered. The Facilitator agent uses a backward sequence to select subtasks, request proposals, and uses a network algorithm for the shortest path to reach a feasible schedule. With this approach, all the candidate production schedules for the subtasks satisfy the customer-required due date/time and delivering destination. Meanwhile, this approach achieves a schedule that has the lowest cost as the primary criteria and the shortest production time as the secondary criteria. In addition, this approach enables shortlisting of the intermediate schedules to narrow the resource search scope by using a priority Resource agent domain, which also helps to reduce the agent negotiation activities and increase the speed in reaching a realistic schedule. By using these proposed approaches to form the agent-based VCIM architecture, a prototype system has been developed and tested at the University of South Australia's Centre for Advanced Manufacturing Research. The test results demonstrated that this approach for resource integration and scheduling with the agent-based VCIM architecture is viable. Based on the previous work, we are now working on extending the prototype system and investigating more potential issues for a VCIM system implementation in a real industrial application. With successful results, it is anticipated that the VCIM implementation technology could be further improved, be practical, and be easily adoptable by organisations, particularly small and medium manufacturing enterprises.

In addition, this research of VCIM systems should be further strengthened towards developing optimised VCIM systems, with more imbedded artificial intelligence to control the scarce resources we have today, and to meet the competitive and agility requirements of present and future business environments. Furthermore, various developments that have been achieved in manufacturing, computation, communication, and other related technologies needs to be integrated into the exiting systems in a cohesive manner to provide a complete and intelligent solution to industries. This cohesive integration will help industries move forward into the next decade with confidence and competitive ability.

VCIM will help improve the competitiveness of many industries. The principle of VCIM can be applied to a diverse section of enterprises, including agriculture, biomedical engineering, building and construction industries, environmental monitoring and control, medical laboratories, pharmacology, energy-related industries, water resource monitoring, recycling and control, and even in applications that enhance wildlife.

References

[11.1] Lin, G.C.I., Nagalingam, S.V. and Zhou, J., 2000, "Virtual CIM for globalised manufacturing," In *Proceedings* of *International Conference on Engineering and Technology Sciences*, pp. 319–327.

[11.2] Nagalingam, S.V. and Lin, G.C.I., 1999, "Latest developments in CIM," *Robotics and Computer-Integrated Manufacturing*, 15(6), pp. 423–430.

[11.3] Rembold, U., Reithofer, W. and Janusz, B., 1998, "The role of models in future enterprises," *Annual Reviews in Control*, 22, pp. 78–83.

[11.4] Center, J.W. and Thompsen, J.A., 1996, "The virtual enterprise framework and toolbox," In *IEEE International Engineering Management Conference on Managing*

Virtual Enterprises, pp. 117–123.

[11.5]   Rembold, U., Lueth, T. and Ogaswara, T., 1994, "Intelligent manufacturing components for future CIM systems," *A Post Print Volume from the IFAC (International Federation of Automatic Control) Workshop*, Vienna, pp. 1–10.

[11.6]   Jennings, N.R. and Wooldridge, M.J., 1998, "Applications of intelligent agents," *Agent Technology: Foundations, Applications, and Markets*, Jennings, N.R. and Wooldridge, M.J. (eds.), Springer Verlag, pp. 3–28.

[11.7]   Barber, K.S., Liu, T.H. and Han, D.C., 1999, "Agent-oriented design," *Multi-Agent System Engineering*, LNAI 1647, Garijo, F.J. and Boman, M. (eds.), Springer Verlag, Heidelberg, Germany, pp. 28–40.

[11.8]   Wooldridge, M. and Jennings, N.R., 1995, "Intelligent agents: theory and practice," *Knowledge Engineering Review*, **10**(2), pp. 115–152.

[11.9]   Shen, W. and Norrie, D.H., 1999, "Agent-based systems for intelligent manufacturing: a state-of-the-art survey," *Knowledge and Information Systems*, **1**(2), pp. 129–156.

[11.10] Shen, W., 2002, "Distributed manufacturing scheduling using intelligent agents," *IEEE Intelligent Systems and Their Applications*, **17**(1), pp. 88–94.

[11.11] Sycara, K.P., Roth, S.F., Sadeh, N. and Fox, M.S., 1991, "Resource allocation in distributed factory scheduling," *IEEE Expert*, **6**(1), pp. 29–40.

[11.12] Sikora, R. and Shaw, M.J., 1997, "Coordination mechanisms for multi-agent manufacturing systems: applications to integrated manufacturing scheduling," *IEEE Transactions on Engineering Management*, **44**(2), pp. 175–187.

[11.13] Kouiss, K., Pierreval, H. and Mebarki, N., 1997, "Using multi-agent architecture in FMS for dynamic scheduling," *Journal of Intelligent Manufacturing*, **8**(1), pp. 41–47.

[11.14] Barbuceanu, M. and Fox, M.S., 1997, "Integrating communicative action, conversations and decision theory to coordinate agents," In *Proceedings of the 1st International Conference on Autonomous Agents*, pp. 49–58.

[11.15] Parunak, H.V.D., Baker, A.D. and Cleark, S.J., 1998, "The AARIA agent architecture: from manufacturing requirements to agent-based system design," In *Proceedings of Workshop on Agent-Based Manufacturing of the 2nd International Conference on Autonomous Agents*, Minneapolis, pp. 136–145.

[11.16] Maturana, F., Shen. W. and Norrie, D.H., 1999, "MetaMorph: an adaptive agent-based architecture for intelligent manufacturing," *International Journal of Production Research*, **37**(10), pp. 2159–2173.

[11.17] Peng, Y., Finin, T., Labrou, Y., Cost, S., Chu, B., Long, J., Tolone, W.J. and Boughannam, A., 1999, "An agent-based approach for manufacturing integration – the CIIMPLEX experience," *Applied Artificial Intelligence*, **13**(1–2), pp. 39–64.

[11.18] Lazansky, J., Stepankova, O., Marik, V. and Pechouchek, M., 2001, "Application of the multi-agent approach in production planning and modeling," *Engineering Applications of Artificial Intelligence*, **14**(3), pp. 369–376.

[11.19] Karageorgos, A., Mehandjiev, N., Weichhart, G. and Hammerle, A., 2003, "Agent-based optimisation of logistics and production planning," *Engineering Applications of Artificial Intelligence*, **16**(4), pp. 335–348.

[11.20] Odrey, N.G. and Mejia, G., 2003, "A re-configurable multi-agent system architecture for error recovery in production systems," *Robotics and Computer Integrated Manufacturing*, **19**(1–2), pp. 35–43.

[11.21] Jia, H.Z., Ong, S.K, Fuh, J.Y.H, Zhang, Y.F. and Nee, A.Y.C., 2004, "An adaptive and upgradable agent-based system for coordinated product development and manufacture," *Robotic and Computer Integrated Manufacturing*, **20**(2), pp. 79–90.

[11.22] Nahm, Y.E. and Ishikawa, H., 2005, "A hybrid multi-agent system architecture for enterprise integration using computer networks," *Robotics and Computer Integrated Manufacturing*, **21**(3), pp. 217–234.

[11.23] Wang, D., Nagalingam, S.V. and Lin, G.C.I., 2004, "Development of a parallel processing multi-agent architecture for a virtual CIM system," *International Journal of Production Research*, **42**(17), pp. 3765–3785.

[11.24] Wang, D., Nagalingam, S.V. and Lin, G.C.I., 2004, "A parallel processing multi-agent architecture for virtual CIM system," In *Proceedings of the 14th International Conference on Flexible Automation and Intelligent Manufacturing*, pp. 1178–1185.

[11.25] http://sharon.cselt.it/projects/jade

[11.26] http://herzberg.ca.sandia.gov/jess/index.shtml

[11.27] http://www.fipa.org/about/index.html

[11.28] Wang, D., Nagalingam, S.V. and Lin, G.C.I., 2003, "Implementation approaches for a multi-agent virtual CIM system," In *Proceedings of the 9th International Conference on Manufacturing Excellence*, CD-ROM.

[11.29] Su, P., Wu, N. and Yu, Z., 2003, "Resource selection for distributed manufacturing in agile manufacturing," In *Proceedings of IEEE International Conference on Systems, Man and Cybernetics*, **2**, pp. 1578–1582.

[11.30] Chu, X.N., Tso, S.K., Zhang, W.J. and Li, Q., 2000, "Partners selection for virtual enterprises," In *Proceedings of IEEE 3rd World Congress on Intelligent Control and Automation*, **1**, pp. 164–168.

[11.31] Smith, R.G., 1980, "The contract net protocol, high level communication and control in a distributed problem solver," *IEEE Transactions on Computers*, **29**(12), pp. 1104–1113.

[11.32] Fischer, K., Muller, J.P., Pischel, M. and Scheer, D., 1995, "A model for cooperative transportation scheduling," In *Proceedings of the 1st International Conference on Multi-agent Systems*, pp. 109–116.

[11.33] Berger, P.D., Gerstenfeld, A. and Zeng, A.Z., 2004, "How many suppliers are best? A decision-analysis approach," *Omega*, **32**(1), pp. 9–15.

[11.34] Lin, H.-W., Nagalingam, S.V., Chiu, M. and Lin, G.C.I., 2004, "Development of a dynamic task rescheduling approach for small and medium manufacturing enterprises," In *Proceedings of the 8th International Conference on Manufacturing and Management*, CD-ROM.

[11.35] Lin, H.-W., Nagalingam, S.V. and Chiu, M., 2005, "Development of a collaboration decision-making model for a network SMEs," In *Proceedings of the 18th International Conference on Production Research*, CD-ROM.

12

A Unified Model-based Integration of Process Planning and Scheduling

Weidong Li[1], S.K. Ong[2] and A.Y.C. Nee[2]

[1] School of Applied Science, Cranfield University
Cranfield, Bedfordshire, MK43 0AL, UK
Email: w.li@cranfield.ac.uk

[2] Department of Mechanical Engineering, National University of Singapore
9 Engineering Drive 1, 117576, Singapore
Emails: mpeongsk@nus.edu.sg, mpeneeyc@nus.edu.sg

Abstract
To increase the flexibility and responsiveness of a job shop in the more competitive market, process planning and scheduling systems have been actively developed and deployed. It is ideal to integrate the two systems more tightly to achieve the global optimisation of product development and manufacturing. In this chapter, a unified representation model and a simulated annealing-based approach have been developed to facilitate the integration and optimisation process. Three methods, namely, processing flexibility, operation sequencing flexibility and scheduling flexibility, have been used to effectively explore the vast search space to support the optimisation approach. Performance criteria, including makespan, the balanced level of machine utilisation, job tardiness and manufacturing cost, have been defined in the optimisation approach to address the various practical requirements. Case studies under different working conditions have been conducted to show the merits and characteristics of the developed approaches.

12.1 Introduction

Manufacturing planning usually involves various decision-making stages. Traditionally, these stages are organised in a sequential manner, *i.e.* the current stage is based on the result of the previous stage, and the newly obtained result is used as the input for the next stage. This strategy is reasonable when the objectives of these stages are independent. However, when the variables of these stages are coupled or the objectives are contradictory, the entire decision-making process will become very complex. A typical example is in process planning and scheduling, which are two important functions when planning the manufacturing operations for a batch of parts. These functions are usually arranged to run consecutively. In process planning, manufacturing resources (*e.g.* machines and tools) in a shop floor are assigned for each individual part to ensure the application of good manufacturing practice and maintain the consistency of the desired functional specifications of the part. In the following scheduling, it is necessary to decide how and when to assign the manufacturing resources for the entire batch of parts. In a complex situation, it is difficult to produce a satisfactory or even acceptable result through the consecutive execution of the two functions. A fundamental reason is that there are many conflicting and/or contradictory considerations and objectives in the functions. In process planning, it is assumed that all the relevant machines and tools are available when planning for parts. The main objectives are to achieve the minimal cost and to meet the manufacturing specifications of each part. In scheduling, these parts need to compete for machines to meet temporal and/or resource constraints, and therefore, not all the generated process plans for the parts might be schedulable according to the time and resource feasibility of a job shop.

Based on the above observation, it is necessary to develop a strategy with respect to both process planning and scheduling more cooperatively. In this chapter, a unified model-based integration scheme of process planning and scheduling has been developed to address various objectives and enable greater dynamic interactions and information sharing between the two functions. A simulated annealing (SA)-based method has been developed to optimise the integration problem. Three methods, namely, processing flexibility, operation sequencing flexibility and scheduling flexibility, have been proposed to effectively explore the vast search space to support the optimisation approach. Performance criteria, such as makespan, balanced level of machine utilisation, job tardiness and manufacturing cost, have been defined in the optimisation approach to address the various practical requirements. To show the characteristics of the developed approach, some case studies under different working conditions have been conducted.

12.2 Recently Related Works

In recent years, different works have been developed to consider process planning and scheduling in a closer way. These works can be generally classified into three categories: the iterative approach, the concurrent approach and the agent-based approach.

The iterative approach is similar to the basic sequential arrangement of process planning and scheduling, while improvements have been made to enhance the interactions and integration between them. Multiple process plans for each part with identical or similar manufacturing cost are first generated. A schedule is then determined through choosing a suitable process plan of each part from their alternative sets according to the current resource constraints of a job shop and scheduling performance criteria [12.1]–[12.6]. The advantage of the approach is that a set of process plans for each part with the optimised manufacturing cost are generated for the scheduling module to choose from so that the objectives of both process planning and scheduling can be achieved. In some further improved works, process planning and scheduling are both in dynamic adjustment until satisfactory performance criteria can be reached [12.7].

In the concurrent approach, some unified optimisation models have been developed [12.8]–[12.12]. For example, both process planning and scheduling have been represented in linear mixed-integer programming [12.8]. The concurrent approach can effectively alleviate the empirically iterative process, while the unified models create much larger search spaces. To facilitate the search, intelligent evolutionary algorithms and heuristic rules have been recently employed to generate optimised solutions with the satisfaction of constraints efficiently. An optimisation model was developed to combine the considerations of process planning and scheduling, such as the production cost, the tardiness time, the setup cost, and the early finish time [12.11][12.12]. Based on this, a Tabu search (TS)-based method [12.11] and an improved hybrid genetic algorithm (GA)-based method [12.12] have been designed to optimise planning and scheduling simultaneously. Other GA-based methods include [12.9].

Recently, the agent-based approach has been under active investigation. Process planning and scheduling have been encapsulated as agents, and the agent framework provides a distributed intelligent solution for the multiple agents to work together through negotiation, coordination and cooperation [12.13]–[12.15]. The recent relevant works are surveyed in [12.13]. The major research issues include: (1) how to encapsulate functional systems as agents effectively, (2) how to construct a better multi-agent architecture, (3) how to design a reasonable negotiation mechanism, and (4) how to empower and enhance agents with learning capabilities to improve the performance of a multi-agent system, considering learning is one of the key techniques for an agent.

In sophisticated situations, more careful considerations are needed. For process planning and scheduling, different criteria are used to address specific practical cases. From the process planning perspective, the lowest manufacturing cost is usually a desired target, while the scheduling usually needs to look for the most balanced utilisation of machines, the minimum number of tardy jobs, the shortest makespan, *etc.* To meet the various requirements in practical situations, further improvement is required on optimisation algorithms to make them more adaptive to accommodate diverse objectives for users to choose from. Complex manufacturing constraints, such as operation precedence constraints and manufacturing resource constraints, also need to be considered carefully. It is imperative to develop a more generic method to represent the exploration space with constraints. A unified model-based method needs to be naturally utilised in various heuristic or evolutionary algorithms to support the manipulation operators, such as the crossover, mutation, shifting and swapping of operations, *etc.*

12.3 A Unified Model to Integrate Process Planning and Scheduling

A process plan for a part can be represented by a series of machining operations, applicable resources for the operations, setup plans, operation sequence, *etc.* A setup can be generally defined as a group of operations that are manufactured on a single machine with the same fixture (here, a setup is simplified as the tool approach direction (TAD)). Based on the generated process plans of parts, the scheduling task is to assign the parts and their machining operations to specific machines to be executed in different time slots, targeting at a good shop-floor performance, such as the shortest makespan, the most balanced machine utilisation, and the least total tardiness, *etc.*

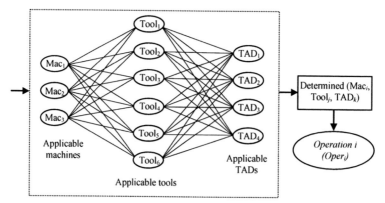

(a) Processing flexibility to generate alternative process plans

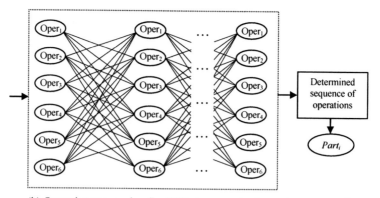

(b) Operation sequencing flexibility to generate alternative process plans

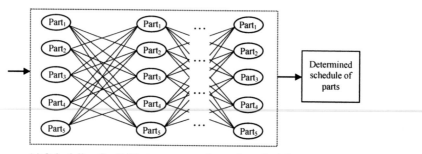

(c) Scheduling flexibility to generate alternative schedules

Figure 12.1. Three strategies to generate alternative process plans or schedules

Intelligent algorithms are imperative to look for desired optimised results effectively according to process planning and scheduling criteria. Three strategies have been developed here: processing flexibility, operation sequencing flexibility and scheduling flexibility. Usually, a part can be manufactured through different process plans, and a batch of parts can be manufactured through different schedules. Processing flexibility refers to the possibility of performing an operation on alternative machines with alternative TADs or tools. Operation sequencing flexibility corresponds to the possibility of interchanging the sequence in which the operations are performed. Scheduling flexibility relates to the possibility of arranging different schedules to manufacture the parts and the operations. For a shop floor with multiple available manufacturing resources, a group of parts can generate a vast search space for determining process plans and schedules.

In Figure 12.1, the above three strategies to generate alternative process plans or schedules are illustrated. For the processing flexibility strategy shown in Figure 12.1(a), an alternative process plan of a part can be generated if a different combination of machine, tool and TAD is chosen for the part. With the application of the operation sequencing strategy shown in Figure 12.1(b), different sequences of operations can be arranged in a part to produce alternative process plans. The scheduling flexibility strategy is illustrated in Figure 12.1(c), in which a new arrangement of the part sequence is chosen to bring about a new schedule.

Each operation of a process plan for a part can be represented in a class *Operation*. Based on this class, a process plan can be represented in a class *Process_Plan*. The details of these two classes are given in Figure 12.2 and Table 12.1. The *Process_Plan* class consists of a series of operations. It assumes that a process plan of a part has *n* operations for manufacturing the part, and its sequence is the operation sequence of the process plan. The *Operation* class consists of three types of data: the indices for the operation and the part that the operation belongs to, the applicable (machines, tools and TADs) for the operation, and the chosen (machine, tool and TAD) to execute the operation. In the *Process_Plan*, the manufacturing cost for the process plan of a part has been defined in [12.16][12.17] to include the costs from the machine utilisation, the tool utilisation, the setups, the machine changes and the tool changes. A Gantt chart has been popularly used to represent a schedule of a group of parts. In a Gantt chart, the order in which the parts and their operations are carried out is laid out and the linkage/dependencies of the tasks are represented. The *X*-axis of the Gantt chart represents time. Each row in the *Y*-axis represents a machine and the specific arrangement for the operations of the parts on the machine. Each machine is represented as a class *Machine*, which is defined in Table 12.2. The *Machine* class is comprised of a number of time slots, which can be further classified into idle time slots, preparation time slots for machining operations (further including the setup time, the machine change time, or

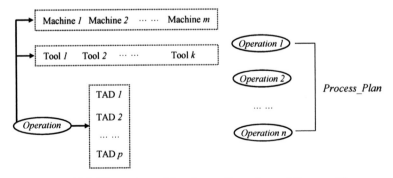

Figure 12.2. Illustrations of the classes *Operation* and *Process_Plan*

Table 12.1. Definitions of *Operation* and *Process_Plan* classes

Operation

Variables	Descriptions
Operation_id	The id of the operation
Part_id	The id of the part that the operation belongs to
Machine_id	The id of a machine to execute the operation
Tool_id	The id of a cutting tool to execute the operation
Setup_id	The id of a setup (TAD) to apply the operation
Machine_list[ ]	The candidate machine list for executing the operation
Tool_list[ ]	The candidate tool list for executing the operation
Setup_list[ ]	The candidate setup (TAD) list for applying the operation
Operation_ parameters	Other machining parameters of the operations
Setup	The number of the setup required to execute the operation. It is defined in Equation (12.1)
Setup_T	The setup time, which is defined in Equation (12.2)
MC	The number of machine changes required to execute the operation, which is defined in Equation (12.3)
MC_T	The machine change time, which is defined in Equation (12.4)
TC	The number of tool changes required to execute the operation, which is defined in Equation (12.5)
TC_T	The tool change time, which is defined in Equation (12.6)
Pre_T	The preparation time required to execute the operation, which is defined in Equation (12.7)
Idle_T	The idle time before the operation is executed, which is defined in Equation (12.8)

Process_Plan

Variables	Descriptions
Operation[n]	Define a process plan *Process_Plan* based on the above class – *Operation. n* is the number of operations in the plan.
TC	Total cost of the process plan, which includes the setup cost, the machine change cost, the tool change cost, the machine utilisation cost and the tool utilisation cost.

Table 12.2. Definition of *Machine* class

Machine

Variables	Descriptions
Machine_id	The id of a machine to execute operations
Operation_list[]	The executed operations on this machine
Total_T	The total time to use the machine, which is defined in Equation (12.10)
Makespan	The makespan for a group of parts, which is defined in Equation (12.11)

the tool change time), and machining time slots of operations. In the class, the starting and ending moments for each operation are indicated to facilitate some performance computations.

Criteria to evaluate the performances of process plans and schedules are defined here. The criteria include the manufacturing cost, the makespan, the job tardiness, and the balanced level of machine utilisation. In the following definitions, it assumes that there are m machines available in a job shop, and there are n operations to be executed on it for a specific machine.

Operation i is denoted as Operation[i]. Machine j is denoted as Machine[j].

$$\Omega_1(X,Y) = \begin{cases} 1 & X \neq Y \\ 0 & X = Y \end{cases}, \text{ and } \Omega_2(X,Y) = \begin{cases} 0 & X = Y = 0 \\ 1 & otherwise \end{cases}.$$

(1) Setup time for Operation[i] – Operation[i].Setup_T:

For the first operation in a machine, there is a setup (Setup). For two consecutive operations, its computation is given below (the first operation needs a setup):

$$Operation[1].Setup = 1, \text{ and}$$

$$Operation[i].Setup = \Omega_2(\Omega_1(Operation[i].Part\_id, Operation[i-1].Part\_id),$$
$$\Omega_1(Operation[i].Setup\_id, Operation[i-1].Setup\_id))$$

$$(i = 2,...n) \tag{12.1}$$

That is, a setup is required for the first operation. The number of setups required for other operations will be determined by the Part_id and Setup_id of an operation and the operation just prior to it.

The setup time for each setup is considered to be the same, and Setup_Index is the time index for each setup. Hence,

$$Operation[i].Setup\_T = Operation[i].Setup * Setup\_Index, (i = 1,...n) \tag{12.2}$$

(2) Machine change time for Operation[i] – Operation[i].MC_T:

A machine change of Operation[i] is denoted as Operation[i].MC, and its relevant computation is given below (the change for the first operation is 1):

$$Operation[1].MC = 1, \text{ and}$$

$$Operation[i].MC = \Omega_1(Operation[i].Part\_id, Operation[i-1].Part\_id),$$

$$(i = 2,...n) \tag{12.3}$$

That is, a machine change is required for the first operation. The number of machine changes required for other operations will be determined by the Part_id of an operation and the operation just prior to it.

The machine change time is considered to be the same for each machine change, and MC_Index is the time index for each machine change. Hence,

$$Operation[i].MC\_T = Operation[i].MC * MC\_Index, (i = 1,...n) \tag{12.4}$$

(3) Tool change time for *Operation*[*i*] – *Operation*[*i*].*TC_T*:

A tool change of *Operation*[*i*] is denoted as *Operation*[*i*].*TC*, and the relevant computation is as follows (the change for the first operation is 1).

$$Operation[1].TC = 1, \text{ and}$$

$$Operation[i].TC = \Omega_1(Operation[i].Tool\_id, Operation[i-1].Tool\_id),$$

$$(i = 2,...n) \tag{12.5}$$

That is, a tool change is required for the first operation. The number of tool changes required for other operations will be determined by the *Tool_id* of an operation and the operation just prior to it.

The tool change time is considered to be the same for each machine change, and *TC_Index* is the time index for each tool change. Hence,

$$Operation[i].TC\_T = Operation[i].TC * TC\_Index, (i = 1,...n) \tag{12.6}$$

(4) Preparation time for *Operation*[*i*] – *Operation*[*i*].*Pre_T*:

The preparation time for an operation consists of the setup time, the machine change time and the tool change time for the operation. That is:

$$Operation[i].Pre\_T = Operation[i]Setup\_T + Operation[i].MC\_T$$

$$+ Operation[i].TC\_T, (i = 1,...n) \tag{12.7}$$

(5) Idle time for *Operation*[*i*] – *Operation*[*i*].*Idle_T*:

Operation[1].*Idle_T* is the start time of the machine to execute *Operation*[1], and

$$Operation[i].Idle\_T = Operation[i].Start\_M - Operation[i-1].Start\_M -$$

$$Operation[i-1].Pre\_T - Operation[i-1].Mac\_T$$

$$(i = 2,...n) \tag{12.8}$$

where *Operation*[*i*].*Start_M* is the start time leading to execution of *Operation*[*i*], and *Operation*[*i*].*Mac_T* is the machining time for *Operation*[*i*].

(6) Total time for *Machine*[*j*] – *Machine*[*j*].*Total_T*:

The total machine occupation time starts from the moment of its utilisation, and the calculation consists of the idle time, preparation time and machining time of all operations. That is:

$$Machine[j].Total\_T = Machine[j].Start\_M$$

$$+ \sum_{i=1}^{n} (Operation[i].Idle\_T + Operation[i].Pre\_T + Operation[i].Mac\_T) \tag{12.9}$$

(7) Makespan:

$$Makespan = \underset{j=1}{\overset{m}{Max}}(Machine[j].Total\_T) \tag{12.10}$$

(8) Balanced level of machine utilisation:

The *standard deviation* concept is introduced here to evaluate the balanced machine utilisation (assuming that there are m machines, and each machine has n operations).

$$Average\_Utilisation = \frac{\sum_{i=1}^{n}(Operation[i].Mac\_T)}{n}, (j=1,..,m) \qquad (12.11)$$

$$\chi = \frac{\sum_{j=1}^{m}(Machine[j].Utilisation)}{m} \qquad (12.12)$$

$$Utilization\_Level = \sqrt{\sum_{j=1}^{m}(Machine[j].Utilisation - \chi)^2} \qquad (12.13)$$

(9) Part tardiness:

The due date of a part is denoted as DD, and the completion moment of the part is denoted as CM. Hence,

$$Part\_Tardiness = \begin{cases} 0 & if\ DD\ is\ later\ than\ CM \\ CM - DD & Otherwise \end{cases} \qquad (12.14)$$

(10) Manufacturing cost for the process plan of a part:

In the authors' previous work, the manufacturing cost associated with the process plan of a part has been defined in terms of machine utilisation, tool utilisation, setup changes, machine changes and tool changes. The relevant computations are elaborated in [12.16][12.17]. Manufacturing constraints and a constraint-handling algorithm for the operations in a process plan have been developed to ensure the manufacturability of the generated process plans. The definitions of the constraints and the constraint-handling algorithm can also be found in [12.16][12.17].

12.4 Simulated Annealing-based Optimisation Approach

One of the major advantages of an SA (simulated annealing) algorithm is that it enables the global minimum of an objective function in a complex search space to be found efficiently. In SA, a parameter called "temperature" is used to guide and control the iterations of the algorithm [12.18]. The temperature begins at a high level and is cooled until equilibrium is reached. The way in which the algorithm avoids being trapped in a local minimum is that it generates and accepts random solutions in which the performance evaluation function has a greater value, *i.e.* the solution has a higher energy. When the temperature is high, the algorithm will be likely to accept a higher-energy solution, while at a very low temperature the algorithm will almost only accept solutions of lower energy (therefore, a finally refined process). Solutions are accepted according to the Boltzmann probability (the definition is in Equation (12.15)).

The structural design of the optimisation algorithm is affected by the chosen performance criteria. To be more flexible to meet the various requirements in practical situations, different performance criteria are considered in the algorithm, including (1) manufacturing cost, (2) makespan, (3) the balanced level of machine utilisation, and (4) part tardiness. Two or more criteria are incorporated as a simultaneous consideration. The criteria are added up with weights as a single criterion.

The major processes of the SA are described as follows:

1. Decide on *"an initial schedule"*. The schedule is based on the process plan for each part. The process plans are generated by a process planning algorithm [12.16][12.17]. A constraint-handling algorithm [12.16][12.17] is applied to the generated plans to adjust them to satisfy the manufacturing constraints.

2. The initial schedule is chosen as *"the current schedule S"*.

3. Determine the start and end temperatures T_0 and T_1. Set *"the current temperature T"* as $T = T_0$.

4. While not yet frozen ($T > T_1$), perform Steps (a)–(c) below:

    (a) Generate *"a temporary schedule S'"* by making some random changes on the current schedule S using several neighbourhood strategies. The neighbourhood strategies are:
        -   *Shift.* This strategy removes an operation from its present position and inserts it at another position in the current schedule.
        -   *Adjacent swapping.* This strategy exchanges two adjacent operations in the current schedule.
        -   *Mutation.* Two types of mutation strategies are applied. The first mutation strategy exchanges two operations chosen randomly in the current schedule. The second mutation strategy randomly selects an operation in the current schedule, and replaces the set of machine, tool, and the TAD used in this operation from the candidate lists.

        After the neighbourhood strategies have been applied, the constraint handling algorithm is applied to S' to adjust it to the feasible domain.

    (b) Set $\Delta = Performance\_Criterion(S') - Performance\_Criterion(S)$.

        The *Performance_Criterion* function is one of the above defined: makespan, the balanced machine utilisation, the part tardiness, the manufacturing cost, or a combined consideration.

        if $\Delta \leq 0$ (downhill move): set $S = S'$.
        else (uphill move):
            choose a random number r from [0, 1]. Set $S = S'$ when

        $$r < e^{-k*\Delta/T} \tag{12.15}$$

    (c) Return to Step (4) after lowering the temperature T as

        $$T = \alpha * T \tag{12.16}$$

        where $0 < \alpha < 1$.

5. If frozen ($T \leq T_1$), the algorithm ends.

The value of the parameter k in Equation (12.15) is determined in the case studies discussed later. To achieve optimal or near-optimal solutions, the cooling process needs to be very slow, and α in Equation (12.16) should be very close to 1.

12.5 Case Studies and Discussions

Some assumptions that are commonly used in most of the previous research studies regarding scheduling are taken for this study.

The assumptions include:

- Parts are independent, and part pre-emption is not allowed.
- The sequence of the operations of each part complies with manufacturing constraints.
- All parts, machines and tools are available at time zero simultaneously.
- Each operation is performed on a single machine, and each machine can only execute one operation at a time.
- The time for a setup is identical and independent of specific operations. The time for a machine change or a tool change follows the same assumption.
- Machines are continuously available for production.
- If a machine or a tool is broken down, or a new part is inserted, the algorithm can restart and generate new process plans and a schedule due to the efficient optimisation performance of the algorithm.

Based on multiple groups of parts, various experiments have been conducted for different conditions and based on different performance criteria to measure the adaptability and robustness of this approach. Eight parts have been used to test the developed approach. For the group of parts, the results of the following three conditions are taken here to demonstrate the performance of the algorithm.

1. The performance criterion is makespan (to achieve the minimum makespan).
2. The performance criterion is manufacturing cost (to achieve the minimum cost).
3. The performance criteria are makespan and the balanced utilisation of several machines. The two criteria are incorporated as a single criterion so as to optimise them simultaneously. The criterion is:

   Performance_Criterion = Makespan + Weight * Utilisation_Level

   where *Weight* is used to adjust the incorporation rate of the two criteria to achieve the best performance.

The optimisation results for the group of the parts are shown in Figure 12.3, Figure 12.4 and Figure 12.5, respectively. All of the results are prone to stabilisation after several hundreds of iterations. Through further trials on other groups of parts, the developed approach has been verified to have stable performance and good optimisation results.

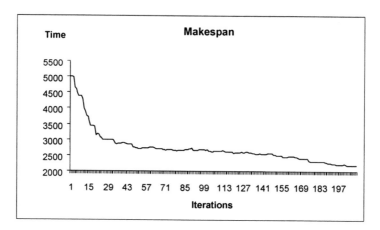

Figure 12.3. The optimisation results of Condition (1) for eight parts

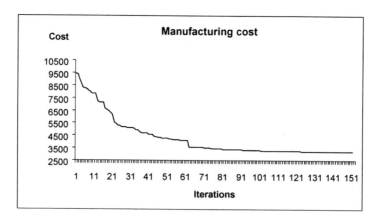

Figure 12.4. The optimisation results of Condition (2) for eight parts

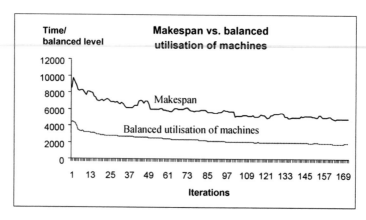

Figure 12.5. The optimisation results of Condition (3) for eight parts

Meanwhile, the algorithm has been compared with other popular heuristic or evolutionary algorithms, including GA, SA and particle swarm optimisation (PSO) for the purposes of benchmarking. A GA is modelled based on natural evolution in that the employed operators are inspired by a natural evolution process. These genetic operators, including selection, crossover and mutation, can be used to manipulate the chromosomes (solutions to a problem) in a population over several generations to improve its fitness function gradually. A PSO algorithm is a recently developed evolutionary algorithm, and it is inspired by the social behaviour of birds flocking or fish schooling [12.19]. In a PSO algorithm, the potential solutions, called particles, fly through the problem space to find the best solution. The optimisation process is controlled by several parameters/functions, such as the fitness function, neighbourhood functions and velocity.

From the results in Figure 12.6, it can be observed that all of the approaches can reach good results, while different characteristics are shown due to the inherent mechanisms of the algorithms. The SA-based approach usually takes a shorter time to find good solutions but it is vigilant to its parameters (such as the starting temperature and the cooling parameter) and the problems to be optimised. The GA- and PSO-based approaches are slow in finding good solutions but they are robust for optimisation problems. Meanwhile, the SA-based approach is much "sharper" to find optimal or near-optimal solutions, and the common shortcoming of the GA- and PSO-based approaches is that they are prone to prematurity in some cases (converge too early and have difficulty in finding the optimal or near-optimal solutions).

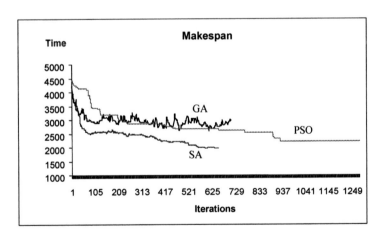

Figure 12.6. The comparisons of SA-, GA- and PSO-based approaches

12.6 Conclusions

An active research direction is to integrate process planning and scheduling more seamlessly to reinforce the whole performance by utilising their complementary roles in product development and manufacturing. In this chapter, a representation model to unify the process planning and scheduling problems has been designed. Based on this, a SA-based approach has been developed to determine the optimised results from the complex search space effectively and efficiently. Three strategies, *i.e.* processing flexibility, operation sequencing flexibility and scheduling flexibility, have been used to support the algorithm to explore the search space extensively. Several commonly used performance criteria in practice, including makespan, the balanced level of machine utilisation, job tardiness and manufacturing cost, have been developed in the algorithm to meet the various practical requirements. Case studies under various working conditions have been used to test the algorithm, and the comparisons have been made for this approach and three modern evolutionary approaches to indicate their characteristics and advantages.

The major characteristics of the chapter include:

- Based on a unified model to incorporate the process planning and scheduling problems, a set of systematic performance criteria have been developed and embedded in the algorithm. The algorithm is enabled to choose one or more specific criteria to address the different practical requirements, and its adaptability is therefore enhanced.
- Processing flexibility, operation sequencing flexibility and scheduling flexibility can represent the extensive search space effectively. Meanwhile, these strategies can be naturally utilised in various heuristic or evolutionary algorithms to support their operators, such as the crossover, mutation, shifting and swapping of operations, *etc*. Due to the feature, new heuristic or evolutionary algorithms can be conveniently developed in future based on the algorithm and strategies adopted in the chapter.
- The comparisons made for this approach with the other modern optimisation algorithms for various case studies and working conditions can indicate the characteristics of the algorithms clearly. The comparisons and analysis can be regarded as a good reference for users to choose a suitable algorithm to meet their specific requirements.

References

[12.1] Tonshoff, H.K., Beckendorff, U. and Andres, N., 1989, "FLEXPLAN: A concept for intelligent process planning and scheduling," In *Proceedings of the CIRP International Workshop*, Hannover, Germany, pp. 319–322.

[12.2] Zhang, H. and Mallur, S., 1994, "An integrated model of process planning and production scheduling," *International Journal of Computer Integrated Manufacturing*, 7(6), pp. 356–364.

[12.3] Zijm, W.H.M., 1995, "The integration of process planning and shop floor scheduling in small batch part manufacturing," *Annals of CIRP*, **44**, pp. 429–432.

[12.4] Sormaz, D. and Khoshnevis, B., 2003, "Generation of alternative process plans in integrated manufacturing systems," *Journal of Intelligent Manufacturing*, **14**, pp. 509–526.

[12.5] Kumar, M. and Rajotia, S., 2003, "Integration of scheduling with computer aided process planning," *Journal of Materials Processing Technology*, **138**, pp. 297–300.

[12.6] Kumar, M. and Rajotia, S., 2006, "Integration of process planning and scheduling in a job shop environment," *International Journal of Advanced Manufacturing Technology*, **28**(1–2), pp. 109–116.

[12.7] Zhang, Y.F., Saravanan, A.N. and Fuh, J.Y.H., 2003, "Integration of process planning and scheduling by exploring the flexibility of process planning," *International Journal of Production Research*, **41**(3), pp. 611–628.

[12.8]   Brandimarte, P. and Calderini, M., 1995, "A hierarchical bicriterion approach to integrated process plan selection and job shop scheduling," *International Journal of Production Research*, **33**(1), pp. 161–181.

[12.9]   Morad, N. and Zalzala, A., 1999, "Genetic algorithms in integrated process planning and scheduling," *Journal of Intelligent Manufacturing*, **10**, pp. 169–179.

[12.10]   Kim, Y.K., Park, K. and Ko, J., 2003, "A symbiotic evolutionary algorithm for the integration of process planning and job shop scheduling," *Computers & Operations Research*, **30**, pp. 1151–1171.

[12.11]   Yan, H.S., Xia, Q.F., Zhu, M.R. Liu, X.L. and Guo, Z.M., 2003, "Integrated production planning and scheduling on automobile assembly lines," *IIE Transactions*, **35**, pp. 711–725.

[12.12]   Zhang, X.D. and Yan, H.S., 2005, "Integrated optimisation of production planning and scheduling for a kind of job-shop," *International Journal of Advanced Manufacturing Technology*, **26**(7–8), pp. 876–886.

[12.13]   Shen, W., Wang, L. and Hao Q., 2006, "Agent-based distributed manufacturing process planning and scheduling: a state-of-the-art survey," *IEEE Transactions on Systems, Man, and Cybernetics*, **36**(4), pp. 563–577.

[12.14]   Gu, P., Balasubramanina, S. and Norrie, D.H., 1997, "Bidding-based process planning and scheduling in a multi-agent system," *Computers & Industrial Engineering*, **32**(2), pp. 477–496.

[12.15]   Shen, W., "Distributed manufacturing scheduling using intelligent agents," *IEEE Expert/Intelligent Systems*, **17**(1), pp. 88–94.

[12.16]   Li, W.D., Ong, S.K. and Nee, A.Y.C., 2002, "A hybrid genetic algorithm and simulated annealing approach for the optimisation of process plan for prismatic parts," *International Journal of Production Research*, **40**(8), pp. 1899–1922.

[12.17]   Li, W.D., Ong, S.K. and Nee, A.Y.C., 2004, "Optimisation of process planning using a constraint-based tabu search method," *International Journal of Production Research*, **42**(10), pp. 1955–1985.

[12.18]   Kirkpatrick, S., Gelatt, Jr. C.D. and Vecchi, M.P., 1983, "Optimisation by simulated annealing," *Science*, **220**, pp. 671–680.

[12.19]   Kennedy, J. and Eberhart, R.C., 1995, "Particle swarm optimisation," *IEEE Proceedings of the International Conference on Neural Networks IV*, Perth, Australia (Piscataway: IEEE Service Centre), pp. 1942–1948.

13

A Study on Integrated Process Planning and Scheduling System for Holonic Manufacturing

Nobuhiro Sugimura, Rajesh Shrestha, Yoshitaka Tanimizu and Koji Iwamura

Osaka Prefecture University, Sakai, Osaka 599-8531, Japan
Email: sugimura@me.osakafu-u.ac.jp

Abstract

New architectures of manufacturing systems have been proposed aiming at realising more flexible control structures of manufacturing systems, which can cope with dynamic changes in volume and variety of products, and also with unscheduled disruptions. The objective of the research is to develop an integrated process planning and scheduling system, which is applicable to the holonic manufacturing systems (HMS). A basic architecture for the HMS is proposed to determine both suitable sequences of the machining equipment needed to manufacture the products and suitable production schedules for the machining equipment. In particular, procedures are developed to generate suitable production schedules and to modify the process plans based on the scheduling results.

13.1 Introduction

Recently, automation of manufacturing systems in batch productions has been much developed aimed at realising flexible small-volume batch productions. The control structures of the developed manufacturing systems, such as FMS (flexible manufacturing system) and FMC (flexible manufacturing cell), are generally hierarchical. The hierarchical control structure is suitable for economical and efficient batch productions in the steady state, but not adaptable to very small batch productions with dynamic changes in the volumes and the varieties of the products.

Computer systems and manufacturing cell controllers have recently made much progress, and individual computers and controllers are now able to share the decision-making capabilities in the manufacturing systems. The network architectures are widely utilised for the information exchange in design and manufacturing, and some standardised models, such as STEP [13.1] and the CNC data model [13.2], have been developed for the information exchange through the information networks for design and manufacturing.

New distributed architectures of manufacturing systems are therefore proposed, aiming at realising more flexible control structures of the manufacturing systems, in order to cope with the dynamic changes in the volume and the variety of the products and also with the unforeseen disruptions, such as failures of manufacturing equipment and interruption by high-priority jobs. They are so-called autonomous distributed manufacturing systems, biological manufacturing systems, random manufacturing systems and holonic manufacturing systems [13.3]–[13.11].

The objective of the research is to develop an integrated process planning and scheduling system, which is applicable to the holonic manufacturing systems (HMS). The HMS was proposed through an international cooperative research project by the HMS consortium. A basic architecture and systematic methods are proposed to determine suitable sequences of the machining equipment needed to manufacture the products and also suitable production schedules for the machining equipment in the HMS.

The following issues are discussed in this chapter:

1. literature review;
2. basic architecture of integrated process planning and scheduling system;
3. procedures to generate suitable process plans, which include machining sequences of machining features of a product and sequences of machining equipment needed to manufacture the machining features;
4. procedures to select a suitable combination of process plans of individual products and production schedules of machining equipment by applying the GA (genetic algorithm) and the dispatching rules; and
5. procedures to modify the process plans based on the scheduling results.

13.2 Literature Review

New distributed architectures of manufacturing systems have been proposed and many research works have been carried out, aimed at realising more flexible control structures of the manufacturing systems. An autonomous distributed architecture of the manufacturing systems was discussed and applied to the real-time scheduling of manufacturing systems [13.3]. A distributed method was proposed in the paper to carry out the scheduling process based on the decision making of the constituent equipment of the manufacturing system and the coordination among the equipment. New architectures of the manufacturing systems termed "biological manufacturing systems" were also proposed [13.4][13.5], in which the planning and execution of the manufacturing processes are carried out in a distributed manner. The biological manufacturing systems are controlled by the decision-making criteria and the information, which are obtained through the investigation of the living things like animals and their organs. Distributed planning and execution architectures have also been proposed and applied to the whole enterprises [13.6], assembly systems [13.7], and flexible machining systems [13.8].

In the early 1990s, an international cooperative research consortium named "Holonic Manufacturing System" or HMS was established and it has carried out active research works dealing with the modelling, planning and execution processes in HMS. Typical examples of the HMS researches are presented in [13.9]–[13.11].

As regards the process planning and scheduling methods, multi-agent systems (MAS) have been widely applied because of their flexibility, reconfigurability, and scalability [13.12]–[13.14]. A multi-agent system was developed to generate the process routes and the schedules through the contract net bids [13.15]. The system addresses some practical issues for merging the process planning with the shop-floor scheduling such as the part's feature representation and the operation specification. A cascading auction protocol was proposed as a framework for integrating process planning and shop-floor control [13.16]. The integration of the real-time online process planning and the shop-floor control is accomplished progressively through a recursive auction process. IDCPPS [13.17] is an integrated, distributed and cooperative process planning system. The process planning tasks are separated into three levels, namely, initial planning, decision making, and detail planning. The responsibilities of each level are very similar to what has been presented in [13.18]. The integration with the scheduling is considered at each stage with the process planning.

A new distributed process planning methodology was proposed by integrating machining feature-based planning, function block-based control, and agent-based distributed decision making [13.19]. It proposes to use two-level decision making, which are supervisory planning and operation planning. The function block-based integration architecture of the system with manufacturing scheduling and execution control was discussed in a separate paper [13.20]. Another multi-agent architecture was presented to determine operation routes and schedules in [13.21]. This approach separates the rough process-planning task as a centralised shop-floor planner from the detailed process planning conducted through agent negotiations. An agent-based integrated dynamic process planning and scheduling system was developed to increase the responsiveness of the manufacturing systems [13.22]. An agent-based approach for dynamic integration of process planning and scheduling functions was also proposed [13.23]. A review paper [13.24] describes the complexity of the manufacturing process planning and scheduling problems, and reviews the research literature on manufacturing process planning, scheduling as well as their integration, particularly on agent-based approaches to these difficult problems. Major issues in these research areas are discussed, and research opportunities and challenges are identified.

The authors of this chapter have also proposed systematic methods for an integrated system of process planning and production scheduling in the holonic manufacturing environment [13.25]–[13.28].

13.3 Process Planning for Holonic Manufacturing Systems

13.3.1 Holonic Manufacturing Systems

The HMS consortium has developed the following definitions to help the common understanding of the HMS [13.25].

1. Holon: An autonomous and cooperative building block of a manufacturing system for transforming, transporting storing and/or validating information and physical objects. A holon consists of an information processing part and often a physical processing part. A holon can be a part of another holon.
2. Autonomy: The capability of an entity to create and control the execution of its own plans and strategies.

3. Cooperation: A process whereby a set of entities develops mutually acceptable plans and execute these plans.
4. Holarchy: A system of holons that can cooperate to achieve a goal or objective. The holarchy defines the basic rules for cooperation of the holons and thereby limits their autonomy.

Figure 13.1 shows a basic architecture of the holonic manufacturing system (HMS) for the machine products. The system is basically divided into the physical processing part and the information processing part, and both parts consist of a set of holonic components. The physical processing part transforms the blank materials to the final products through the autonomous and cooperative activities of the holonic components. The data required in the physical processes are planned and generated in the information processing part, which also consists of a set of holonic components.

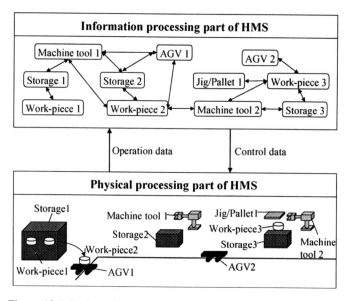

Figure 13.1. Basic architecture of HMS (holonic manufacturing system)

The holonic components in the HMS are basically divided into two classes based on their roles in the manufacturing processes. They are,

1. Resource holons that transform, transport, store and/or validate the products, and
2. Job holons that are transformed by the manufacturing resources from the blank materials to the final products.

The resource holons include all the manufacturing equipment, such as machine tools, assembly stations, tools and jigs, and also the decision-making units, such as process planning, scheduling and equipment control units.

The job holons include all the objects transformed, transported, stored and validated by the resource holons, in order to manufacture the final products.

13.3.2 Integrated Process Planning and Scheduling

The tasks of the process planning and scheduling systems are defined as follows;

1. *Process planning system*: The process planning system generates suitable process plans for the individual products to be manufactured. The process plans give suitable sequences of manufacturing equipment needed to manufacture the machining features of the products, and the machining time of the machining features. The process planning is carried out for the individual products, based on the geometric and technological information represented by the CAD model and the objective functions of the individual products, such as the shop time and the machining cost.

2. *Scheduling system*: The scheduling system determines suitable production schedules of manufacturing equipment in the HMS for manufacturing a set of products. The production schedules give the loading sequences of the products to the manufacturing equipment and the starting times of the individual manufacturing processes of the products. The production schedules are verified based on such objective functions as total makespan and tardiness against due date.

As regards the scheduling problems in HMS, a distributed real-time scheduling system has been proposed in [13.3][13.10] aimed at generating a suitable production schedule when executing the manufacturing processes in the HMS. However, it is assumed that the process plans of the products are predetermined and fixed in the planning phase independently from the production schedules. So, integration of process planning and scheduling in the planning phase is discussed here, which generates both suitable process plans and production schedules for a set of products to be manufactured, before executing the manufacturing processes in the HMS.

Figure 13.2 shows a basic architecture of the integrated process planning and scheduling system proposed here. In what follows, the products to be manufactured are called "jobs". Two types of holons are considered in the research to realise the integrated system shown in Figure 13.2. They are, job holons and scheduling holons. The job holons represent the individual products to be manufactured in the HMS, and carry out the process planning task to determine suitable machining sequences of the machining features and suitable sequences of the manufacturing equipment needed to manufacture the jobs.

The feasible process plans generated by the individual job holons are transmitted to the scheduling holon for production scheduling of the HMS. Following this, the scheduling holon determines suitable production schedules of all manufacturing equipment in the HMS, based on the feasible process plans of all the jobs to be manufactured.

13.3.3 Target System Configuration

The basic configuration of the target HMS considered in the research is shown in Figure 13.3. The HMS consists of a set of machine tools, preparation stations, input

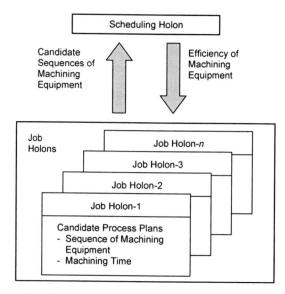

Figure 13.2. Integrated system for process planning and scheduling

and output buffers, AGVs (automated guided vehicles), fixtures and tools. The inputted jobs are transformed to the finished jobs through the manufacturing processes shown in Figure 13.3.

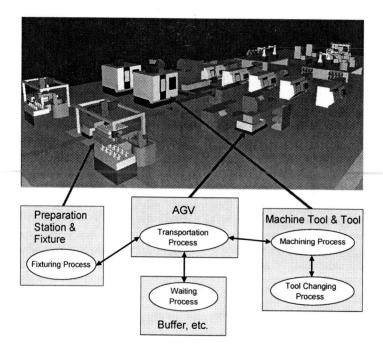

Figure 13.3. Target manufacturing system and manufacturing processes

The inputted jobs are firstly fixed on the fixtures at the preparation stations and transmitted to the input buffer by the AGVs. After the fixturing process, both the jobs and the fixtures are processed as single objects in the HMS. The jobs are then transported to the machine tools in order to carry out the machining processes required. If the refixturing is required to change the positions and the orientations of the jobs against the fixtures, the jobs are transported to the preparation stations and the refixturing process is carried out there. When all the required machining processes of the jobs are finished, the finished jobs are transported to the preparation stations to separate the jobs from the fixtures, and the jobs are transmitted to the output buffers.

13.4 Process Planning by Job Holons

13.4.1 Input Information

The task of the process planning system for the machine products is to select suitable machining sequences of the machining features and sequences of the machining equipment for a target product to be machined in the HMS. The process planning task is carried out by the individual job holons, and the following procedures have been proposed to complete the process planning task [13.25]:

1. extraction of machining features to be machined from the product model;
2. selection of feasible machining equipment for individual extracted machining features; and
3. selection of suitable machining sequences of the machining features and suitable sequences of the machining equipment.

The third step is discussed here to realise a systematic method to select a suitable process plan based on the shop time and the machining costs of the products. The input of the third step is the information about the machining features of the job, which is summarised as follows.

(a) *Feasible machining equipment for machining features*
The individual machining features MF_i have feasible machining equipment ME_{ik}, which are represented by the combinations of the machine tools MT_{ik}, the fixtures FI_{ik}, and the cutting tools CT_{ik}, as shown in Figure 13.4(a).

(b) *Machining time of machining features*
The machining time MAT_{ik} of the machining features MF_i are specified for all the feasible machining equipment ME_{ik}.

(c) *Preference relation among machining processes of machining features*
The preference relations specify the constraints on the machining sequences, as shown in Figure 13.4(b). For example, a plane face MF_1 should be machined before the machining processes of all the holes MF_2 and MF_3 on the plane face. The preference relations are given in the binary relation form as shown in the following:

$$MF_i < MF_j \ (i \diamond j) \tag{13.1}$$

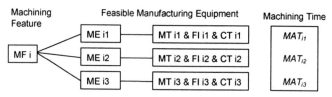

(a) Manufacturing equipment for machining feature

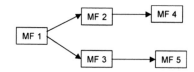

(b) Preference relations among machining features

Figure 13.4. Information about machining features

where, MF_i and MF_j are the machining features, and the MF_i should be machined prior to the MF_j.

(d) *Arrival time TS of the job to the input buffer*
    The workpiece of the target product is then inputted to the HMS, and the manufacturing process may be started at the time *TS*.

The following constraints are considered from the viewpoint of the manufacturing resources and their schedules.

1. *Machine tools and tools*: Individual machine tools have fixed sets of tools in their ATCs (automatic tool changers), and the tool changing time is constant for all the machine tools.
2. *Preparation stations and fixtures*: All the preparation stations have the same functions, and the operation time for fixturing and refixturing processes is constant. Enough fixtures needed for all the jobs are stored at the preparation stations.
3. *AGVs and buffers*: Enough AGVs and buffers are equipped in the HMS. The transportation time of the AGVs among the input and output buffers, the preparation stations and the machine tools are constant.
4. *Production schedules of resources*: All the resources mentioned above have predetermined production schedules to carry out the manufacturing process of the products, except the target jobs. Therefore, the manufacturing processes of the target product shall be allocated to the equipment, when it is free.

13.4.2 Objective Functions

Two types of objective functions are considered here for the process planning tasks. They are the shop time and the machining cost of the target job. The manufacturing processes of the jobs are classified into five types, shown in Figure 13.3, in order to evaluate the objective functions.

1. *Machining process*: Shape generation processes of the jobs with use of the machine tools and the tools.
2. *Fixturing process*: Fixturing and refixturing processes of the jobs with use of the preparation stations and the fixtures.
3. *Tool changing process*: Tool changing processes with use of the ATCs of the machine tools.
4. *Transportation process*: Transportation processes among the machine tools, the preparation stations and the buffers.
5. *Waiting process*: The other processes mentioned above.

The processing time of the manufacturing processes are given by the following:

MAT_{ik}:   Machining time of machining feature MF_i by machining equipment ME_{ik}.

$STT(= constant)$:   Transportation time among the machine tools, the preparation stations and the buffers.

$SFT(= constant)$:   Processing time for fixturing and refixturing at the preparation stations.

$SCT(= constant)$:   Tool changing time at the machine tools.

WT: Waiting time.

The shop time ST of the target job is given by,

$$ST = TF - TS = \sum MAT_{ik} + \sum STT + \sum SFT + \sum SCT + \sum WT \qquad (13.2)$$

where,

ST: Shop time of the individual job.
TF: Finishing time of all the manufacturing processes of the job.
TS: Arrival time of the job.

The finishing time TF and all the processing time in Equation (13.2) are obtained after determining both the machining equipment for the individual machining features and the sequence of the machining equipment.

The machining cost of the target job is evaluated based on both the manufacturing process time and the operation costs per unit time of the manufacturing resources. Table 13.1 shows the manufacturing processes and the required resources. In the table, the circles indicate the manufacturing resources required for the individual manufacturing processes. The total machining cost CO of the target job is given by,

$$CO = \sum TT_p \left(\sum CO_{pq} \right) \qquad (13.3)$$

where,

CO: Machining cost of the target job.
TT_p: Manufacturing process time. The index p specifies the types of manufacturing processes shown in Table 13.1.
CO_{pq}:   Operation costs per unit time of the manufacturing resources. The index q gives the resources required for the operation p shown in Table 13.1.

Table 13.1. Manufacturing processes and required resources

Resources \ Operations	Machining process (p=1)	Fixturing process (p=2)	Trans. process (p=3)	Tool changing process (p=4)	Waiting process (p=5)
Machine tools (q=1)	O			O	
Prep. stations (q=2)		O			
Buffers (q=3)					O
AGVs (q=4)			O		
Fixtures (q=5)	O	O	O	O	O
Tools (q=6)	O			O	

The following equation is proposed as the objective function OF for the process planning system,

$$OF = W_s ST_n + W_c CO_n \qquad (13.4)$$

where, W_s and W_c are the weighting factors for the normalised shop time ST_n and the normalised machining cost CO_n, respectively. The normalised shop time and machining cost are defined in the following equations,

$$ST_n = ST / \sum_i Ave_k(MAT_{ik}) \qquad (13.5)$$

$$CO_n = CO / \sum_i Ave_k(MAT_{ik} \cdot CO_{ik}) \qquad (13.6)$$

where,

$\sum_i Ave_k(MAT_{ik})$: Sum of the average machining times of the individual machining features.

$\sum_i Ave_k(MAT_{ik} CO_{ik})$: Sum of the average machining costs of the individual machining features.

13.4.3 Procedures Based on GA and DP

The problem to be solved in the process planning system is to determine suitable machining sequences of the machining features and suitable sequences of the machining equipment for all the machining features of the target job. The constraints on the problem to be considered are the preference relation among the machining features and the schedules of the machine tools and the preparation stations. A procedure shown in Figure 13.5 is proposed to select suitable machining sequences of the machining features and suitable sequences of the machining equipment, simultaneously. The procedure includes the following two steps.

STEP-1: *Generation of machining sequences by applying GA (genetic algorithm)*
    Sets of feasible machining sequences of the machining features are generated by applying the GA, which satisfy the constraints on the preference relation among the machining features. An optimum sequence of the machining equipment and the objective function value are then determined for the individual machining sequence

in the next step, in order to select suitable machining sequences for generating the next generation of the population in GA.

STEP-2: *Selection of optimum sequence of machining equipment by applying DP (dynamic programming)*
An optimum sequence of the machining equipment is generated by selecting suitable machining equipment for each machining feature. The DP method is applied to minimise the objective function given in Equation (13.4). Suitable machining sequences and suitable sequences of the machining equipment can be obtained by repeating the STEP-1 and STEP-2 as shown in Figure 13.5.

In STEP-1, the gene L_Z of the individuals Z in GA are described in the following form, which represents a candidate of the machining sequence of the machining features.

$$L_Z = \{l_{z1}, l_{z2}, ..., l_{zm}\} \tag{13.7}$$

where, l_{zi} is the ID of the machining feature to be machined in the ith process in the machining sequence. The individuals Z represent a feasible machining sequence of the machining features.

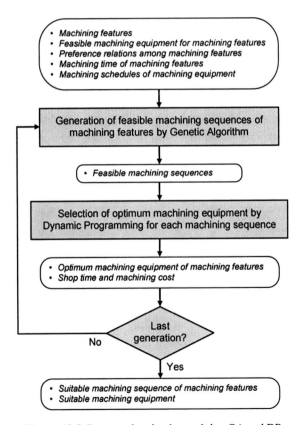

Figure 13.5. Process planning by applying GA and DP

In the first step of GA, the individuals in the initial population are generated randomly as the first parent generation. If the generated individuals do not satisfy the preference relations among the machining features, they are removed. In the second step, a set of offspring is generated from the parent population by applying the generation operators of the crossover and the mutation. The offspring that does not satisfy the preference relations is also removed.

The optimum sequence of the machining equipment is selected for the individual machining sequence in STEP-2, and all the individuals of the parent and the offspring are evaluated from the viewpoint of the objective function, by applying the DP. Suitable individuals are selected to survive to the next generation.

The parameters needed for GA have been evaluated based on various numerical experiments. The parameters shown in Table 13.2 are applied to the GA processes at present [13.25].

Table 13.2. Parameters for GA processing

Population size (= No. of initial generations)	20
No. of offspring generated by crossover	18
Crossover probability	0.5
No. of offspring generated by mutation	4
Probability of mutation	0.1
Total generation	2000

The problem to be solved in STEP-2 is to select a suitable sequence of the machining equipment from all the feasible machining feature sequences, as shown in Figure 13.6. The nodes ME_{ij} in Figure 13.6 show all feasible machining equipment, being the combinations of machine tools, fixtures, and tools. The arcs between the nodes ME_{ij} and ME_{i+1k} specify the increase of the objective function value $\Delta OF_{ij \cdot i+1k}$, which depends on both the ME_{ij} and ME_{i+1k}. $\Delta OF_{ij \cdot i+1k}$ is evaluated as follows. The increase of the shop time $\Delta ST_{ij \cdot i+1k}$ is given as

$$\Delta ST_{ij \cdot i+1k} = SUT_{ij \cdot i+1k} + WT_{i+1k} + MAT_{i+1k} \tag{13.8}$$

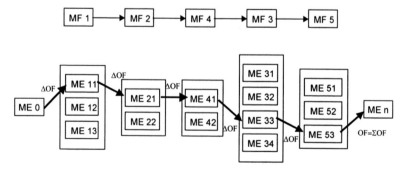

Figure 13.6. Machining sequence and sequence of machining equipment

where, MAT_{i+1k} is the machining time of the $(i+1)$th machining feature MF_{i+1} by the kth feasible machining equipment ME_{i+1k}. $SUT_{ij\cdot i+1k}$ and WT_{ij} are setup time and waiting time between the machining processes of MF_i and MF_{i+1}.

The setup time $SUT_{ij\cdot i+1k}$ is given by the following equations based on the assumptions on the operation time of the fixturing, tool changing and transportation processes.

(1) If refixturing is required,

$$SUT_{ij\cdot i+1k} = 2\ STT + SFT \tag{13.9}$$

(2) If machine tool is changed,

$$SUT_{ij\cdot i+1k} = STT \tag{13.10}$$

(3) If tool is changed,

$$SUT_{ij\cdot i+1k} = SCT \tag{13.11}$$

The weighting time WT_{i+1k} is evaluated based on both the finishing time of the machining process of MF_i and the production schedules of the kth feasible machining equipment ME_{i+1k} of machining feature MF_{i+1}.

As regards the machining cost, the increase of machining cost ΔCO is calculated based on $SUT_{ij\cdot i+1k}$, WT_{i+1k}, and MAT_{i+1k} by applying Equation (13.3).

The dynamic programming (DP) method is adopted for selection of an optimum sequence of the machining equipment that minimises the objective function, including both the shop time and the machining cost of the target jobs.

13.5 Scheduling by Scheduling Holon

A procedure shown in Figure 13.7 is proposed to generate suitable production schedules for all jobs. The job holons firstly generate suitable process plans based on their objective functions by applying the method presented in Section 13.4, and send N candidate process plans to the scheduling holon. Following this, the scheduling holon selects a combination of the process plans of all the job holons and generates a production schedules for the combination. The scheduling holon, then, sends the feedback information to the individual job holons, and the job holons modify their process plans based on the feedback information. The procedures shown in Figure 13.7 are iterated until a suitable combination of the production schedule and the process plans are obtained. This section presents the scheduling process by the scheduling holon and the process plan modification by job holons.

13.5.1 Objective Functions

The following objective functions are considered for the scheduling holon of the HMS.

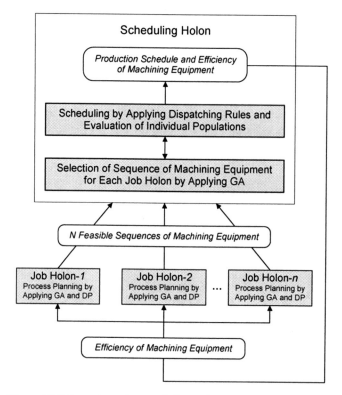

Figure 13.7. Procedures for scheduling and process plan modification

(1) *Makespan: MS*

The make span is the total time between the start time of the first job and the finish time of the last job.

$$MS = TFL - TSF \tag{13.12}$$

where,

    TSF: Starting time of the first machining process of the first job.
    TFL: Finishing time of the last machining process of the last job.

(2) *Total machining cost: TMC*

The total machining cost is the sum of the machining costs of all the jobs, as given in the following. The machining cost of the individual job is calculated, by applying Equation (13.3), as the sum of the cost of all the manufacturing processes of the job.

$$TMC = \sum_i CO_i \tag{13.13}$$

where,

    CO_i: Machining cost of the job i given by Equation (13.3).

(3) *Weighted tardiness cost*: WTC

The weighted tardiness cost is the total penalty cost that would be required due to the delay in the due date of all the jobs.

$$WTC = \sum_i W_i \times \max(0, C_i - d_i)$$
(13.14)

where,

W_i: Weight or delay penalty per unit time charged, if the ith job is not completed before its due date d_i. W_i is set to be 1 in the present research.

C_i: Completion time of the ith job.

d_i: Due date of the ith job.

13.5.2 Scheduling Method Based on GA and Dispatching Rules

A genetic algorithm (GA)-based method is adopted for selecting a combination of process plans, as the first step of the scheduling process. The chromosomes L_s of the individuals S in GA are described in the following form in order to represent a combination of the candidate process plans of all the jobs.

$$L_s = (l_{s1}, l_{s2}, ..., l_{si}, ..., l_{sM})$$
(13.15)

where,

$i = 1, 2, ..., M$: ID of the jobs.

$l_{si} = j$ ($j = 1, 2, ..., N$): ID of one of the candidate process plans of the ith job.

The individual job holons send N candidate process plans to the scheduling holon. The individuals in the initial population are generated by combining the candidate process plans randomly, as the first parent generation of the GA.

After generating the chromosomes L_s of the individuals S representing a combination of the process plans, the scheduling holon shall generate a suitable production schedule of the HMS, *i.e.* the loading sequences of all the machining processes of the jobs to the machining equipment.

At this stage, the scheduling holon shall consider a large number of machining equipments and jobs to be manufactured in the HMS. Therefore, the problem considered here is a large job-shop scheduling problem, and requires a long computational time to arrive at an optimal solution.

A set of dispatching rules is adopted, in the research, for solving the scheduling problems in a short computational time. The dispatching rules give the priority to one job against all the candidate jobs that are waiting for the machining process of the machining equipment [13.29]. The dispatching rules are utilised at the time when the machining equipment is free and some jobs are waiting for the machining equipment.

Let the kth process of the ith waiting job be denoted by MA_{ik} ($i = 1, 2, ..., m$) and its processing time of the machining process be MAT_{ik} ($k = 1, 2, ..., n_i$). Three different dispatching rules are applied to the waiting jobs. These rules have been widely used for the large-scale job-shop scheduling problems. The following gives the dispatching rules considered in this research.

(1) SPT (shortest processing time)

This rule selects a job that has the shortest processing time of the next machining processes. The ith job with the minimum value of the processing time of the next process MAT_{ik} is chosen as per SPT.

(2) SPT/TWKR (shortest processing time / total work remaining)

This rule selects a job that has the lowest ratio of the processing time of the next machining process and the total processing time of the remaining machining processes.

If the processing time of the next process is MAT_{ik} then the ith job with the minimum value A_i in Equation (13.16) is chosen as per SPT/TWKR.

$$A_i = MAT_{ik} / \sum_{\alpha=k}^{ni} MAT_{i\alpha} \qquad (13.16)$$

where, n_i is the total number of machining process of the ith job.

(3) Apparent tardiness cost (ATC)

The ATC rule is a composite dispatching rule that takes into account the due dates, the weights and the processing times of the jobs. It gives a priority index called the ATC index for all the jobs. The ATC index is by

$$ATC_i = \frac{WAT_i}{MAT_{ik}} exp(-\frac{max((d_i - R_i - t),0)}{\delta \times p}) \qquad (13.17)$$

where,

ATC_i: ATC index for the ith job.

WAT_i: Delay penalty per unit time charged, if the ith job is completed after its due date d_i. WAT_i is set to be 1 in the present research.

MAT_{ik}: processing time of the next machining process of the ith job.

$R_i = \sum_{\alpha=k}^{ni} MAT_{i\alpha}$: Total remaining processing time of the ith job.

d_i: Due date for the ith job.

t: Present time.

δ: Parameter that determines the impact of weights versus the slack, and δ shall be between 0.1 and 5.

p: Average of the total remaining processing time of all waiting jobs.

13.5.3 Process Plan Modification

After generating the production schedule, the individual job holons modify their process plans, based on the scheduling results. The modification process of the process plans is described below.

The problem to be solved here is to redistribute the workloads of the busy machining equipment, which is the bottleneck from the viewpoints of makespan and tardiness. Therefore, the following procedures are proposed for the job holons to modify their process plans.

STEP-1: *Evaluation of efficiency of machining equipment*
    The efficiency of the individual machining equipment is verified based on the scheduling results. The efficiency *EF* of individual machining equipment is defined by the following equation,

$$EF = \sum MAT \, / \, MS \qquad\qquad (13.18)$$

where,

$\sum MAT$: Sum of the processing time of the equipment.

MS : Makespan.

    The efficiency *EF* of the individual machining equipment describes the levels of the workloads.

STEP-2: *Modification of processing time of machining equipment*
    The processing time of the individual machining equipment is modified based on their efficiency, which corresponds to the workloads. The modified processing time represent only the virtual processing time, which is referred to only in the modification of the process plans. The modified processing time is given by the following:

$$MAT'_{ik} = MAT_{ik} \, /(1 - EFA^{\lambda})$$

$$EFA^{\lambda} = \sum_{T=1}^{\lambda} EF^{T} \, / \, \lambda \qquad\qquad (13.19)$$

where,

λ : Number of iterations of the scheduling and the process plan modification shown in Figure 13.7.

MAT_{ik} : Initial value of the processing time of the individual machining equipment.

MAT'_{ik} : Modified value of the processing time of the individual machining equipment.

EF^{T} : Efficiency of the machining equipment obtained by the *T*th iteration of the scheduling process.

EFA^{λ} : Averaged efficiency of the machining equipment obtained after λ iterations of the scheduling process.

    In this research, the modified processing time of the individual machining equipment is estimated based on the averaged efficiency obtained after λ iterations, in order not to make the efficiency EF^{T} so sensitive to the modification process.

STEP-3: *Modification of process plans by individual job holons*
    The individual job holons modify their own process plans based on the modified processing time given in Equation (13.19). The modification is carried out by applying the methods introduced in Section 13.4.3. By iterating the steps mentioned above, the production schedules and the process plans are modified.

13.6 Case Studies

13.6.1 Process Planning

A prototype of the process planning system has been implemented on a PC by using C++ language. Some case studies have been carried out to verify the effectiveness of the procedures proposed here.

Figure 13.8 shows an example of the product model of machine products for the case study. This figure also shows the machining features extracted from the product model by applying the extraction procedures proposed in our previous paper [13.25]. Thirty-two machining features were extracted in this case. The preference relations among the machining features are summarised in Table 13.3.

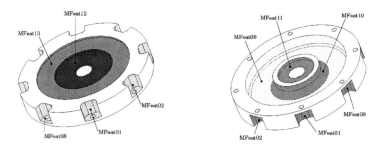

Figure 13.8. Target product

Table 13.3. Preference relations among machining features

Preceding	Succeeding
MF11	MF30
MF30	MF25
MF23	MF32
MF32	MF17
MF20	MF16
MF16	MF21

Suitable machining sequences and sequences of machining equipment were generated as shown in Figure 13.9, based on the information about the machining features. As shown in the figure, the machining processes of the machining features are allocated to the machine tools when they are free, and the schedules of the refixturing and transportation processes are also considered.

Table 13.4 summarised the effect of the weighting factors W_s and W_c in Equation (13.4). As shown in the table, when the W_s is increased, the shop time of the job is decreased. This means that the proposed method selects a suitable machining sequence of machining features and a suitable sequence of machining equipment based on the weighting factors of the shop time and the machining cost of the target jobs.

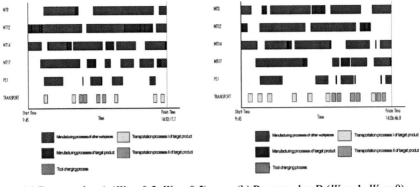

(a) Process plan A (*Ws* = 0.5, *Wc* = 0.5)   (b) Process plan B (*Ws* = 1, *Wc* = 0)

Figure 13.9. Results of process planning

Table 13.4. Effect of weighting factors

Weighting factor	$W_s = 1$ $W_c = 0$	$W_s = 0.5$ $W_c = 0.5$	$W_s = 0$ $W_c = 1$
Shop time	16,906.8	18,497.7	25,013.8
Machining cost	58,306.4	36,014.2	33,354.1

13.6.2 Verification of Dispatching Rules

A prototype system of the integrated process planning and scheduling system has been implemented using C++ language. Some case studies have been carried out to verify the effectiveness of the procedures proposed here.

The conditions of the case studies for process planning and scheduling are summarised below.

1. No. of job holons: 10;
2. No. of candidate process plans of each job holon: 5;
3. No. of machining equipments: 7.

The scheduling system is first verified through the numerical experiments, in order to find suitable combinations of the dispatching rules and the objective functions. In these case studies, the production schedules are generated for the individual combinations of the objective functions and the dispatching rules mentioned in Sections 13.4.1 and 13.4.2, by applying the GA and dispatching rule-based scheduling method. Four cases of the combinations of the process plans are considered as the input for the scheduling process, and the modification of the process plans is not carried out.

Table 13.5 summarises the scheduling results for four cases of experiments. In the table, the numbers show the number of the cases where the dispatching rules in the top row generate the most suitable schedules from the viewpoints of the objective functions given in the left column. As shown in the table, the most suitable combinations of the dispatching rules and the objective functions are,

(a) SPT/TWKR – Makespan,
(b) SPT – Total machining cost, and
(c) ATC – Weighted tardiness cost.

Table 13.5. Scheduling results for various dispatching rules

Objective functions	Dispatching rules		
	SPT/TWKR	SPT	ATC
Makespan	3	0	1
Total machining cost	0	4	0
Weighted tardiness cost	0	1	3

The results are almost the same as those presented in the literature [13.29], which deals with the dispatching rules for the job-shop scheduling problems. Therefore, these combinations are applied in the following case studies to further verify the effectiveness of the process plan modifications.

13.6.3 Verification of Process Plan Modification

Some case studies have been carried out to verify the effectiveness of the process plan modifications.

In the case studies, the individual job holons first generate five candidate process plans by applying the procedures presented in Section 13.3.3, and send the process plans to the scheduling holon. The scheduling holon secondly determines the most suitable combinations of the process plans and the schedules, by applying the procedures based on the genetic algorithm and the dispatching rules given in Section 134.2. Following this, the individual holons thirdly modify their process plans referring to the effectiveness of the individual machining equipment, by applying the procedure presented in Section 13.4.3.

The second and the third steps are iterated, and a most suitable combination of the process plans and the schedule are generated. The number of iterations of the second and third steps is set to be 30, based on the numerical experiments previously carried out.

Figure 13.10 shows a typical example of the schedules, which are obtained before the process plan modifications and after the process plan modifications. As shown in the figure, the loads concentrated on the bottleneck machining equipment are distributed by the process plan modifications, and the makespan of the HMS is improved.

The effects of the process plan modifications are summarised in Table 13.6 for 15 cases. The table shows the comparison between the objective function values obtained in the first iteration and those obtained after 30 iterations. It is shown, through the case studies, that the objective function values are improved by iterating the scheduling process and the process plan modifications. This means that the process plan modification is effective in improving the production schedules of the HMS.

The total computation time of the above-mentioned examples was about 10 min for generating the final process plans and production schedules through 30 iterations

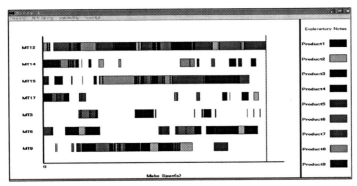

(a) Before modification of process plans

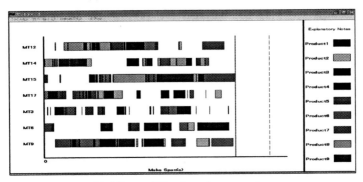

(b) After modification of process plans

Figure 13.10. Example of improvement of schedules

Table 13.6. Summary of improvement of schedules

Objective functions (dispatching rules)	Case	Before modification of process plans	After modification of process plans
Makespan (SPT/TWKR)	(1)	4.38×10^4	3.10×10^4
	(2)	2.57×10^4	2.17×10^4
	(3)	3.15×10^4	1.94×10^4
	(4)	2.26×10^4	1.82×10^4
	(5)	2.42×10^4	1.89×10^4
Total machining cost (SPT)	(1)	6.72×10^5	6.23×10^5
	(2)	4.36×10^5	4.50×10^5
	(3)	3.94×10^5	3.79×10^5
	(4)	4.63×10^5	4.69×10^5
	(5)	4.17×10^5	4.02×10^5
Weighted tardiness cost (ATC)	(1)	5.36×10^4	1.68×10^4
	(2)	1.19×10^4	0.17×10^4
	(3)	4.05×10^4	0.12×10^4
	(4)	0.16×10^4	0.0085×10^4
	(5)	0.88×10^4	0.0043×10^4

of the process planning and the scheduling. The computation time is too long for the real-time or dynamic process planning and scheduling tasks, which are carried out concurrently with the manufacturing processes. However, it is not so long from the viewpoint of the offline process planning and scheduling tasks, which are carried out before executing the manufacturing processes.

13.7 Conclusions

This chapter presents an integrated process planning and scheduling system for the HMS. The following remarks are concluded:

1. The basic architecture for integrated process planning and scheduling system was discussed from the viewpoint of the distributed decision making of the job holons and the scheduling holon.
2. A systematic method was proposed to select suitable machining sequences of the machining features and suitable sequences of the machining equipment based on the genetic algorithm (GA) and dynamic programming (DP). The proposed approach provides a method to select solutions that can minimise the objective function including both the shop time and the machining cost.
3. A systematic method was also proposed to select suitable combinations of the process plans generated by the individual job holons and to generate suitable production schedules for all the manufacturing equipment by applying the genetic algorithm and the dispatching rules.
4. An iterative method was developed to modify the process plans based on the scheduling results. This method enables the individual job holons to modify their process plans in a distributed manner.
5. A prototype of the integrated process planning and scheduling system has been implemented and applied to the case studies. The results of the case studies show that the proposed method is able to select a suitable combination of the process plans and to generate a suitable production schedule. In particular, the process plan modification method is effective to obtain the most suitable combinations of process plans of the individual job holons and the production schedule of the HMS.

References

[13.1] ISO, 1994, "ISO 10303 Industrial automation systems and integration – Product data representation and exchange," *ISO*.
[13.2] ISO, 2001, "ISO 14649 Industrial automation systems and integration – Physical device control - Data model for computerized numerical control," *ISO*.
[13.3] Moriwaki, T. and Sugimura, N., 1992, "Object oriented modelling of autonomous distributed manufacturing system and its application to real-time scheduling," In *Proceedings of the ICOOMS '92*, pp. 207–212.
[13.4] Okino, N., 1992, "A prototyping of bionic manufacturing system," In *Proceedings of the ICOOMS '92*, pp. 297–302.

[13.5] Ueda, K., 1992, "An approach to bionic manufacturing systems based on DNA-type information," In *Proceedings of the ICOOMS '92*, pp. 303–308.

[13.6] Warnecke, H.J., 1993, *The Fractal Enterprise*, Springer-Verlag, New York.

[13.7] Wiendahl, E.P. and Garlichs, R., 1994, "Decentral production scheduling of assembly systems with genetic algorithm," *Annals of the CIRP*, **42**(1), pp. 389–396.

[13.8] Iwata, K. and Onosato, M., 1994, "Random manufacturing system: a new concept of manufacturing systems for production to order," *Annals of the CIRP*, **43**(1), pp. 379–384.

[13.9] Wyns, J., 1996, "Workstation architecture in holonic manufacturing systems," In *Proceedings of the 28th CIRP International Seminar on Manufacturing Systems*, pp. 220–231.

[13.10] Sugimura, N., Moriwaki, T., Hozumi, K. and Shinohara, Y., 1996, "Modelling of holonic manufacturing system and its application to real-time scheduling," *Manufacturing Systems*, **25**(4), pp. 1–8.

[13.11] Sugimura, N., Tanimizu, Y. and Yoshioka, T., 1999, "A Study on object oriented modelling of holonic manufacturing system," *Manufacturing Systems*, **29**(2), pp. 155–159.

[13.12] Baker, A.D., 1996, "Metaphor or reality: a case study where agents bid with actual costs to schedule a factory," *Market-based Control: A Paradigm for Distributed Resource Allocation*, Clearwater, S.H., (ed.), pp. 185–223.

[13.13] Parunak, V.D., 1987, "Manufacturing experience with the contract net," *Distributed Artificial Intelligence*, Huhns, M.N., (ed.), pp. 285–310.

[13.14] Shaw, M.J., 1987, "A distributed scheduling method for computer integrated manufacturing: the use of local area networks in cellular systems," *International Journal of Production Research*, **25**(9), pp. 1285–1303.

[13.15] Gu, P., Balasubramanian, S. and Norrie, D.H., 1997, "Bidding-based process planning and scheduling in a multi-agent system," *Computers and Industrial Engineering*, **32**(2), pp. 477–496.

[13.16] McDonnell, P., Smith, G., Joshi, S. and Kumara, S.R.T., 1999, "A cascading auction protocol as a framework for integrating process planning and heterarchical shop floor control," *International Journal of Flexible Manufacturing Systems*, **11**(1), pp. 37–62.

[13.17] Chan, F.T.S., Zhang, J. and Li, P., 2001, "Modelling of integrated, distributed and cooperative process planning system using an agent-based approach," *Proc. of Inst. Mech. Eng., Part B: Journal of Engineering Manufacture*, **215**(B10), pp. 1437–1451.

[13.18] Huang, S.H., Zhang, H.C. and Smith, M.L., 1995, "A progressive approach for the integration of process planning and scheduling," *IEE Transactions*, **27**(4), pp. 456–464.

[13.19] Wang, L. and Shen, W., 2003, "DPP: an agent-based approach for distributed process planning," *Journal of Intelligent Manufacturing*, **14**(5), pp. 429–440.

[13.20] Wang, L., Hao, Q. and Shen, W., 2003, "Function block based integration of process planning, scheduling and execution for RMS," In *Proceedings of the CIRP 2nd International Conference on Reconfigurable Manufacturing*. (CD-ROM)

[13.21] Denkena, B., Tonshoff, H.K., Zwick, M. and Woelk, P.O., 2002, "Process planning and scheduling with multiagent systems," *Knowledge and Technology Integration in Production and Services – Balancing Knowledge and Technology in Product and Service Life Cycle*, pp. 339–348.

[13.22] Lim, M.K. and Zhang, D.Z., 2004, "An integrated agent based approach for responsive control of manufacturing system," *Computers and Industrial Engineering*, **46**, pp. 221–232.

[13.23] Wong, T.N., Leung, C.W., Mak, K.L. and Fung, R.Y.K., 2006, "Dynamic shopfloor scheduling in multi-agent manufacturing systems," *Expert Systems with Applications*, **31**(3), pp. 486–494.

[13.24] Shen, W., Wang, L. and Hao, Q., 2006, "Agent-based distributed manufacturing process planning and scheduling: a state-of-the-art survey," *IEEE Transactions on Systems, Man, and Cybernetics, Part C: Applications and Reviews*, **36**(4), pp. 563–577.

[13.25] Sugimura, N., Hino, R. and Moriwaki, T., 2001, "Integrated process planning and scheduling in holonic manufacturing systems," In *Proceedings of the ISATP 2001*, pp. 250–255.

[13.26] Sugimura, N., Shrestha, R. and Inoue, J., 2003, "Integrated process planning and scheduling in holonic manufacturing systems – optimisation based on shop time and machining cost," In *Proceedings of the ISATP 2003*, pp. 36–41.

[13.27] Shrestha, R., Sugimura, N. and Inoue, J., 2003, "A study on process planning system for holonic manufacturing – process planning considering both machining time and machining cost," In *Proceedings of ICLEM21*, pp. 753–758.

[13.28] Sugimura, N., Shrestha, R. and Inoue, J. 2004, "A study on integrated process planning and scheduling system for holonic manufacturing system," In *Proceedings of the 37th CIRP International Seminar on Manufacturing Systems*, pp. 323–329.

[13.29] Vepsalainen, A.P.J. and Morton, T.E., 1987, "Priority rules for job shops with weighted tardiness costs," *Management Science*, **33**(8), pp. 1035–1047.

14

Managing Dynamic Demand Events in Semiconductor Manufacturing Chains by Optimal Control Modelling

Yon-Chun Chou

National Taiwan University, Taipei, Taiwan
Email: ychou@ntu.edu.tw

Abstract
This chapter describes a new planning method for proactively responding to dynamic events that pose a threat to degenerate the supply services of manufacturing chains. A dynamic system model for manufacturing chains under exogenous demand shocks is first described. As the impact of dynamic events is most severe when manufacturing chains are in the full-load state, a procedure is next presented for constructing full-load production functions of flexible capacity. The third part of this chapter develops an optimal control model for integrating the production functions of multiple production units. Given a nonordinary event of demand shock, this method can be used to determine if the shock can be absorbed by the manufacturing chains, without degrading its fulfilment services. The background of this chapter is the semiconductor manufacturing industry. Demand shock in the distribution channel of microchip products is described and characteristics of semiconductor manufacturing are incorporated in modelling and method development. Numerical examples are given to demonstrate the working of the method.

14.1 Introduction

A manufacturing chain is a large production system that provides manufacturing as a service. Manufacturing services have broader performance requirements than pure manufacturing. The focus of manufacturing is usually on improving factory efficiency. In contrast, the challenges of manufacturing services lie in effective collaboration between engineering, production and distribution units, in service monitoring and control, and in achieving reliable codevelopment and fulfilment performance. The performance requirements of manufacturing services are more stringent, but their operating environment is less favourable. Many manufacturing chains are faced with a high level of uncertainty in product demand, manufacturing process, and production operation, and uncertainty fosters complex dynamic behaviour of the system. In manufacturing, a common solution to cope with uncertainty is to reserve safety resources, such as protective capacity [14.1] and safety work-in-process (WIP) inventory [14.2], as safeguards. However, if a manufacturing chain is lean in capacity or inventory, a significant event that arises from the inherent sources of uncertainty will have an impact not only on the origin production unit but also on other parts of the chain. Manufacturing chains are systems that are not only larger than factories but also highly dynamic. The predictability of performance emerges as grand challenges of system research. In this chapter, a supply chain model and a new method of proactive production control are described for managing dynamic events that pose a risk to degenerate the supply services of manufacturing chains. Semiconductor manufacturing is the problem background for the development and application of the model and method. Characteristics of semiconductor manufacturing are briefly described next.

Semiconductor manufacturing chains are composed of four stages of manufacturing plants: wafer fabrication, wafer sorting, packaging and product testing. Wafer fabrication plants are complex job shops that contain several hundred machines and the manufacture of a silicon wafer requires several hundred process steps. Scores of products can be found to exist in a factory and thousands of work-in-process lots compete for the same set of machines, creating complex capacity allocation and job scheduling problems [14.3][14.4]. Because of lengthy process routing and prominent queuing characteristics [14.5], the average flow time normally exceeds one month, making flow time control an important performance requirement of wafer fabrication. Wafer sorting, also called circuit probing (C/P), is an inspection and screening process. Circuits on the wafers are tested on functionality to distinguish between good and bad dies (*i.e.* microchips on silicon wafers). The manufacturing systems at wafer sorting plants have the characteristics of non-identical (*i.e.* unrelated) parallel machine stations. Job orders are assigned to testing machines in batches. There may be significant setup times between job types. In the packaging stage, dies are sawed off and packaged using bumping, wiring, and/or bonding processes through a sequence of operations. The last stage is the test and inspection process of final products. The factory setting is similar to that of wafer sorting, except that work-in-process materials are packaged microchips instead of wafers. Material handling and tooling is more complex and varied.

The semiconductor industry has a vertically disintegrated structure that is made up of several stages of engineering firms and manufacturing plants. At each of the four manufacturing stages, there are many plants that are owned either by integrated design and manufacturing (IDM) companies or by foundry manufacturers. There are many possible relationships between partners of the chains. Figure 14.1 depicts some examples of supply chains. Manufacturing services can be provided by a wafer fabrication plant (fab), a circuit probe plant, a packaging plant, or some combination of the plants. A variety of service control schemes are possible. For instance, the last row of the figure shows that the control is delegated to the fab and final-test plant, with the fab controlling C/P operation and the final-test plant controlling packaging operation.

The organisation of the semiconductor industry can be roughly divided into two segments: supply chains and demand chains. On the supply side, a manufacturing chain can be regarded as a system that offers manufacturing services to its customers. For foundry manufacturers, their customers include pure design companies and some IDM companies. For IDM companies, the customers of the manufacturing services are their product divisions. On the demand side, microchip products are usually sold through several tiers and channels of a distribution network. The changes in inventory level in the channels are very dynamic. It should be noted that the channels are a marketplace, and they are usually not subjected to the central planning of any one firm. The volatility and unpredictability of demand in the semiconductor industry is noted for frequent occurrences of overstock and understock in the channels, even with good demand forecast tools. As it happens, overstock or understock can be severe and firms in the demand chains will react by cancelling orders or placing rush orders, which have a direct impact on the nominal production of the supply chains. Production planning and control processes at the supply chain level are usually managed by IC design companies or manufacturing companies. However, because of fragmented ownerships, integration of planning and control has been a persistent problem and challenge.

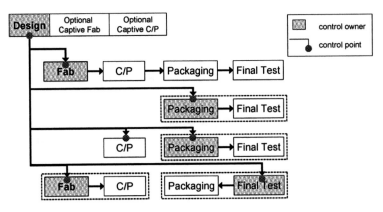

Figure 14.1. Semiconductor manufacturing chains

Semiconductor manufacturing employs continuously advancing process technologies to fabricate complex microchips by using delicate machines in very large plants. It is a complicated operation plagued by all sorts of uncertainty in process yield, equipment reliability, microchip demands and product mixes. At the same time, the cost of equipment is very high and has a high risk of obsolescence. Many processing or testing machines have a unit cost of several to ten million US dollars. Manufacturing resources must be kept very lean. As a result, resource reserves are not a preferred solution to cope with uncertainty. Other systematic solutions must be provided, but planning tools and methods are still in short supply.

Depending on the level of uncertainty, three classes of dynamic events can be distinguished: deviation, disruption, and disaster. Deviation events refer to shifts in system parameters such as machine availability and rework probability. With proper system design, manufacturing operation usually can absorb or quickly recover from deviation events without missing performance targets. Disruption events are those events that are so significant and far reaching that normal operation is disrupted. For example, if product mixes or demand have changed significantly, it will take a significant time for the system to recover from deteriorating performance. During the recovery time, performance becomes unpredictable and remedial actions must be taken in order to bring the system back to a more stable state. The last class of uncertain events is called disasters. Events such as large-scale industrial accidents will make catastrophic and long-lasting impacts on plant operation and functioning of manufacturing chains.

Uncertainty is a prominent theme in the literature of both supply chain management and factory management. Variability in demand, lead time, and order quantity has long been recognised as the major contributing factor to the bullwhip effect phenomenon. A distinction can be made between time-varying models and dynamic models of the vast literature of models on inventory control, scheduling, production rate, work release control, ordering policy, *etc*. If the variables of a decision model vary with time, then the variables are time-varying and the model is a time-varying model. If, for example, demand is modelled as a random variable that follows a certain probability distribution and multiple time periods are under consideration, then demand is time-varying. If a model has several time-varying variables and the variables are interrelated, the model is dynamic. It is clear that not all time-varying models are dynamic models by this definition; but dynamic models are necessarily time-varying models.

Production models of supply chains can be analysed by their two components: production functions of the constituent production units, and a mechanism for linking up the production functions of multiple units. In economics, a production function is a mapping from capital and labour inputs to output. Similarly, a production function in the supply chain domain is a mapping from production states and decisions to performance measures of production. Because there are a multitude of performance measures and many production states and decisions have an effect on performance, the form of production functions can be complicated and varied. There are two approaches to modelling production functions: the time delay approach and the clearing function approach. In time delay models, the emphasis is on modelling the passage time that a job will spend in passing through a factory or production unit. The second modelling approach is to view production as an endeavour of clearing the WIP stock (by processing them and sending them out of the factory). The emphasis of modelling is placed on identifying the nonlinear, state-dependent relationship between output (such as throughput) and input (such as system state variables). Traditional production functions are mappings from the WIP level to throughput [14.6]. Throughput is not completely determined by or is not proportional to the work-release rate. As the work released to a production unit approaches its capacity ceiling, there might be efficiency loss attributable to congestion or process interference effects. This type of models is called clearing functions or output functions [14.6]–[14.8]. These functions are normally increasing in input factors and are concave in shape.

Fluid models and mathematical programs are two of the most popular mechanisms for linking up production functions. Ordinary differential equations have been used to describe the relationship between work-in-process levels, flow rates and production decisions. Analytic or approximate solutions can usually be solved for the steady state of fluid networks [14.9][14.10]. For instance, Nagatani and Helbing [14.9] used differential equations to characterise the relationship between inventory level, order generation, and job arrival of multiple nodes, and then analysed the stability of linear supply chains. Lefeber *et al.* [14.11] pointed out that steady state fluid models are throughput oriented and that the time delay aspect had been overlooked. To remedy this deficiency, they developed a model of partial differential equations to keep track of the progression of production in the time dimension. Mathematical programs make use of difference equations of state variables and control variables. The resultant models can easily include detailed aspects of production such as a materials-acquisition plan, production schedule, and ordering policies, at the expense of higher computation complexity.

In the literature of supply chain management, the type of uncertainty addressed by the time-varying models and dynamic system models is mostly deviations. Process yield and rework were modelled as random variables and stochastic dynamic programs were used by Gong and Matsuo [14.12] to determine the optimal release policy in multiple time periods for multi-product, multi-machine, re-entry-type production systems. Demands or job arrivals are commonly modelled by random variables, stochastic processes, or smoothed forecast process [14.13]. In addition to the time-varying property, models of dynamic systems capture the dynamic interactions between variables. Ni *et al.* [14.14] describe a linear dynamic system for spare part management in a repair network. Disruption events are harder to model using either time-varying models or dynamic system models. Simulation has been used to investigate the effect of some disruption events. Rose [14.15] constructed a simplified plant model of simulation and investigated the behaviour of wafer fabrication when a large quantity of work-in-process suddenly emerges. His plant model is made up of bottleneck workstations as well as a time delay unit that represents all other non-bottleneck stations. Dümmler [14.16] did detailed simulation of wafer fabrication. His paper describes the transient behaviour of state variables and performance measures after the work-release rate is changed and reports on comparing several scheduling policies in such a situation.

How to respond to the occurrence of significant events that pose a risk to degenerate the supply services is a major problem in supply chain management. Such events are very frequent but unpredictable. In this chapter, a dynamic control model will be described for managing disruption events. It should be noted that the model is not intended to be used to "control" the supply chains. Supply chains, being a distributed system and marketplace, are probably not susceptible to being automatically controlled. The intended use of the model is to enhance the capability to manage disruption events so that they become deviation events. If the effect of a class of disruption events can be mitigated, they will no longer be disruption events. Instead, their remnant uncertainty will be at the level of deviations. If that happened, then we could say that supply chain control has been improved. The remainder of this chapter is organised as follows. In Section 14.2, the problem description is provided along with an analysis of the volatility of channel inventory in the semiconductor industry. Sections 14.3 and 14.4 are devoted to the description of production functions and the linking mechanism of the proposed supply chain model. In Section 14.5, numerical examples are provided to demonstrate how the model could be applied and the type of what-if questions that could be answered.

14.2 Problem Description

Figure 14.2 is a representation of a 2-node manufacturing chain model for semiconductor manufacturing. A node can be considered as a production unit of a manufacturing shop or plant and the numerical subscript k is used to index the nodes. The input to the chain is a demand process $d(t)$. Release rate decision $r_1(t)$ is made based on $d(t)$. Normally, $r_1(t)$ will follow $d(t)$, but with some smoothing or time lag. Each node ($k = 1$ or 2) has a capacity C_k, a state variable of work-in-process W_k, and an output rate $\Omega_k(t)$. The release rate decision at node 2 also closely follows $\Omega_1(t)$. The output of the last shop is sent to the distribution channel that comprises microchip product companies, electronic manufacturing service companies, distributors and retailers. The channel inventory is represented as $I(t)$. Market demand is usually different from the forecast $d(t)$. Their difference is represented by noise $\phi(t)$.

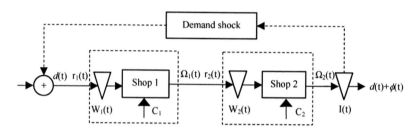

Figure 14.2. A dynamic system representation of semiconductor manufacturing chains

Being at the starting point of the supply chains for the computer, consumer electronics, communication equipment and many other industries, the demand facing semiconductor manufacturing is very dynamic. It is a repeated phenomenon that channel inventory swings from excessive to insufficient and back to excessive. Such swings are quite random and might be very significant. Chou et al. [14.17] analysed the worldwide demand of microchips and the annual throughput of some major semiconductor manufacturing firms, and found that the demand process could be suitably modelled as geometric Brownian motion processes and, with a drift of approximately 0.2, the variation parameter reaches 0.3. Channel inventory information is so valuable that it is monitored, marketed and sold as market intelligence data. Since replenishing channel inventory is a source of demand generation, the inventory level will affect the demand $d(t)$ as indicated by the outer loop. This feedback, however, is not continuous, but is in the form of disruption events. If the ϕ is a simple noise or if the manufacturing chain has ample built-in reserves, the system will be stable. If the ϕ is not a white noise and the chain is lean, the system is prone to be unstable.

Channel inventory information might come as a shock input to the manufacturing chain. When channel inventory is discovered to be lower than what has been anticipated or planned for, it can be inferred that true market demand of the end products is stronger than expected. Urgent orders thus will follow to replenish the inventory. The predictability of manufacturing services will be dependent on the characteristics of $d(t)$. When $d(t)$ has a large fluctuation, the controllability of fulfilment services will be reduced. The impact of disrupting events is the most severe when the chain is fully loaded. We assume that the chain is highly loaded and demonstrate how to assess and mitigate the impact of shock input by proactively coordinating the production decisions at the node levels. Figure 14.3 illustrates a general framework for managing dynamic events in the full-load state of the manufacturing chain.

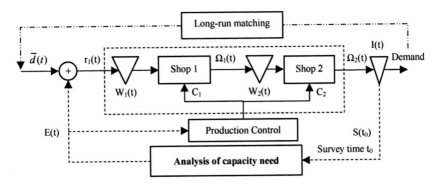

Figure 14.3. Proactive production control for managing dynamic events

Suppose that channel inventory is surveyed at discrete time t_0. The discovered excessive inventory is represented as $S(t_0)$ or simply S, which can be regarded as a scalar (for one product type) or a vector (for multiple product types). It is converted to some anticipatory demand forecasts $E(t)$ over a short run horizon $[t_0, t_0+T]$ after its nature of state is analysed. This analysis process is likely to involve management judgments. The anticipatory demand is a second source of demand stream. It is distinct from the nominal demand input $\overline{d}(t)$, but is merged with $\overline{d}(t)$. (Note that in Figure 14.2, the $d(t)$ includes the demand shock, whereas in Figure 14.3 the demand shock is separated from the nominal stream of demand.) The production control function takes $E(t)$ as input information and coordinates the production decisions at the shops. The hypothesis of this chapter is that system stability can be enhanced and performance improved by putting in place a shorter feedback loop and by coordinating $E(t)$ with control variables of the production shops. The problem setting can be summarised as follows.

(1) The occurrence of dynamic events can not be predicted. But when a dynamic event occurs, it becomes useful leading information to the manufacturing chain in planning and tuning the supply capability. Let the time of occurrence be t_0.

(2) In the long run, the demand input $\overline{d}(t)$ is matched with market demand. But in the medium run, the difference might be significant. This implies that in the short run ϕ is a random variable, in the medium run $S = \int_0^{t_0} \phi(t)dt \neq 0$, but in the long run $\int_0^{\infty} \phi(t)dt = 0$.

(3) The demand input to the system has two parts. One is nominal demand $\overline{d}(t)$ and the other is anticipatory demand $E(t)$, $t \in [t_0 \;\; t_0 + T]$, the latter arising from the dynamic event.

A supply chain model will be described in Sections 14.3 and 14.4. For the remainder of this section, the phenomenon of channel inventory is described. In the semiconductor supply chains, channel inventory includes those stockpiles of microchips in foundry manufacturers, IDMs, electronic manufacturing service companies and distributors. All companies pay attention to the level of the channel inventory. The needs have been so great that there are market intelligence companies who actively monitor the channel inventory and sell the compiled information as information services. For instance, Figure 14.4 shows the excessive inventory data published by iSuppli® over a duration of two years. A closer examination of the data will reveal the impact of channel inventory information on the supply chain production. The inventory under discussion is excess inventory, rather than the absolute inventory. That is, it refers to the inventory in excess of the normal seasonal inventory level. The average excess inventory was 1.6 billions (B) in Q3 of 2004 and it was predicted at that time that the inventory will decrease to 1.5B in the following quarter. However, in Q4 of 2004 the excess inventory turned out to be 1.0B. Given that the total revenue was approximately 200B a year, if the inventory adjustment of 1.5B were to be completed in one month, the impact on the supply chain would amount to a capacity impact of 9% (12/200×1.5) at the aggregate level. Considering that aggregate demand is less volatile than individual demands, Figure 14.4 provides evidences on the disrupting effect on the supply chain. It also shows that inventory is very dynamic; what is realized could be quite different from what has been planned for or predicted.

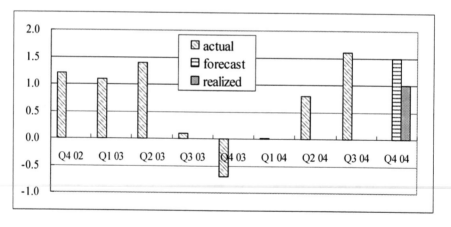

Figure 14.4. Worldwide channel inventories (in $B USD, compiled in this study)

In practice, it can take one month, one quarter or longer to adjust the inventory level back to its "norm". The channels are a large and complex dynamic system. Adjusting channel inventory is never a coordinated effort; each company will analyse the inventory situation from its own perspective and takes an action that best suits its own business objective. For foundry manufacturers, inventory adjustment in the downstream channels has two implications. If the excessive inventory is low or negative, urgent orders are likely to arrive from customers. Since capacity tends to be lean, the existing capacity might be further strained. If the excessive inventory is high, customers are likely to cancel or delay orders. In this case, foundry manufacturers should slow down the pace of deploying new capacity. For IDM companies, a major concern will be on the robustness of supply performance that is essential to product and business planning. When the supply chain is overloaded, its supply performance will suffer and is manifested either in delayed delivery or variability of delivery. Therefore, both customers of and manufacturers in a supply chain will be concerned with the overall workload profile over the whole chain when the channel inventory undergoes adjustment.

As there are many dynamic demand events, either indigenous or exogenous, a management concern would be on their impact on service performance of the manufacturing chains. It would be desirable to plan for some scenarios for such disrupting events and this requires a new supply chain model and planning tool.

14.3 Full-load Production Functions

A production function is a mathematical relationship between system output (such as throughput rate and flow time) and system input and state (such as input rate, capacity, and variety). A wafer fabrication plant is a system that exhibits a significant queuing delay phenomenon. As in any queuing system, the cycle time is dependent on system parameters. In practice, factories use scheduling [14.18] and work release policies (such as the CONWIP policy) to control the escalation and variation in cycle time. Therefore, except for the very short term, cycle time could be treated as fixed. Achieving a fixed cycle time, however, requires endeavour in managing bottleneck machines, improvising alternative machines, and reducing variations in capacity availability and job arrival.

Because the investment cost of equipment is very high, wafer fabrication plants are lean in manufacturing resources. Plants can often be found to be at the full-load states, in which demand exceeds capacity or the throughput is severely constrained by bottlenecks. Conventional production functions normally are concerned with situations in which workload is less than capacity. There is a need to develop new production functions that are suitable for full-load states. Figure 14.5 illustrates our concept of a full-load production function and how such a function is to be constructed. The production function is made up of two segments. There is a regular segment for which the workload is less than nominal capacity. This segment starts at the origin and ends at a saturation point. The saturation point can be regarded as an ideal state of operation in which loading is optimal and flow time is under control. (Interested readers can refer to the *Proceedings of International Symposium on Semiconductor Manufacturing* for frequent appearances of papers that describe industry projects on the reduction of flow time and variation of flow time.) Beyond the saturation point, workload will be greater than the nominal capacity and the system is said to be in the full-load state. This state of excessive workload could happen when there is a shift in product mixes or increase in workload due to rework. In order to prevent deterioration of operation performance, additional capacity, in the form of alternative routing, reactive scheduling, or even postponement of maintenance, must be provisioned.

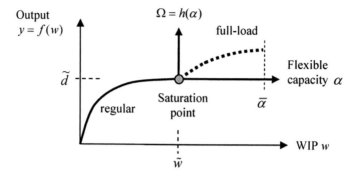

Figure 14.5. Constructing a full-load production function

As it happens, there are usually alternative routings or machines for the manufacture of a product. In the full-load state, products that have been assigned to the bottleneck machines could be rerouted to alternative machines. The load at the bottleneck would be alleviated and system capacity increased. Such an increase in capacity can not be construed as permanent. It is attributable to the flexibility of routing and equipment and it will be called flexible capacity in this chapter. Activating flexible capacity will incur cost, since certain setup operations might be required. In this section, the construction of a full-load production function based on alternative routing will be described. (Alternative routing is also known as machine backup planning or machine grouping in practice [14.3][14.4].) In the full-load state, the mode of manufacturing management should change to that of managing flow time; flow time must be kept under tight control by provisioning additional capacity to the bottleneck. (According to queuing analysis, as the arrival rate approaches the processing rate, flow time will grow exponentially to infinity. But, by proactively activating flexible capacity or dynamically adjusting work input, this has never been allowed to happen in practice.)

Before we describe the full-load production function and its modelling, we must distinguish between average work-in-process and workload. In queuing analysis, WIP level is an indicator of congestion. Both WIP level (N) and flow time (F) are functions of input rate λ and processing rate μ. In addition, N and F are governed by Little's law ($N = \lambda \cdot F$) in steady states. WIP can be interpreted as a resource that is required to maintain nominal production, and λ is the workload for the system. In contrast, clear functions are discrete-time models of production. They describe the efficiency aspect of production. As WIP is increased, variety and complexity increase and efficiency decreases. In a large system, such as wafer fabrication plants, the total WIP in a period (in the discrete model sense) could be approximated by the sum of work released in that period and the base WIP level that is deliberately managed. Because of the subtle difference in the meanings of WIP, the notation w will be used to refer to the WIP in discrete-time models of clearing function (whereas N is the WIP in the steady state of production). That is, as far as clearing functions are concerned $w = N + \lambda$ and the output rate y is a function of w.

In Figure 14.5, there are two coordinate frames. The regular segment of the production function is expressed in the w-y frame and both the WIP level and output rate are expressed in absolute terms. At the saturation point, the output rate is $\tilde{d}$ and the average work-in-process is $\tilde{w}$. In the α-Ω frame, flexible capacity and output rate are expressed in the normalised, marginal terms. Nominal capacity is assumed to equal one (1.0) unit and flexible capacity is represented by the parameter $\alpha \geq 0$. The output rate at the full-load state will be similarly expressed as the marginal output rate Ω. The full-load production function is denoted as $\Omega = h(\alpha)$.

Production functions are increasing and concave. They have the functional form $y = f(w) = aw^2 + bw + c$, with $f'(w) \geq 0$ and $f''(w) < 0$. Therefore, $a < 0$ and $b \geq 0$. In addition, a full-load production function has the following characteristics of utilising flexible capacity.

- The starting point is the saturation point.
- The maximum flexible capacity is finite, which is denoted as $\bar{\alpha}$. Flexible capacity is expressible as a fraction (α) of the nominal capacity of 1.0.
- Flexible capacity has a diminishing effect on increasing the output rate.

These characteristics can be concisely expressed as:

$$f(\tilde{w}) = \tilde{d} \text{ and } \lim_{\alpha \to \alpha} h'(\alpha) = 0$$

These conditions are referred to as geometric properties of production functions.

In the following, the full-load production function is first derived in the w-y coordinates and then transformed to the α-Ω coordinates and instantiated with typical data of a wafer fabrication operation. Given a production function of the form $y = f(w)$, with domain $[w_0, w_u]$ and an initial point (w_0, y_0), the geometric properties require that $2aw_u + b = 0$ and

$$c = y_0 - aw_0^2 + 2aw_u w_0. \text{ Thus,}$$

$$y = aw^2 + bw + c = aw^2 - 2aw_u w + (y_0 - aw_0^2 + 2aw_u w_0)$$
$$= a(w^2 - w_0^2 - 2w_u w + 2w_u w_0) + y_0$$

$$a = \frac{y - y_0}{w^2 - 2w_u w - w_0^2 + 2w_u w_0}$$

The value of a and w_u can be estimated from experiment or simulated data (w_i, y_i), $i = 1, ..., n$.

$$\tilde{a} = \frac{1}{n} \sum_{i=1}^{n} a_i, \text{ where } a_i = \frac{y_i - y_0}{w_i^2 - 2w_u w_i - w_0^2 + 2w_u w_0}$$

$$\breve{w}_u = \max_{i\in(1,...,n)} \{w_i\}$$

Thus, $y = f(w) = \breve{a}w^2 + \breve{b}w + \breve{c}$, where $\breve{b} = -2\breve{a}\,\breve{w}_u$ and $\breve{c} = y_0 - \breve{a}w_0^2 + 2\breve{a}\breve{w}_u w_0$.

The average flow time at wafer fabrication plants is approximately three times that of the total processing time. Let $\tilde{d}$ be the input and output rate, assuming no yield loss. The average WIP equals $3\tilde{d}$. The initial point (w_0, y_0) is the saturation point with $w_0 = 4\tilde{d}$ and $y_0 = \tilde{d}$. Similarly, by substituting $w = 4\tilde{d} + \alpha\tilde{d}$, $\breve{w}_u = 4\tilde{d} + \overline{\alpha}\tilde{d}$ and $y = \tilde{d} + \tilde{d}\Omega$, the full-load production function can be transformed.

$$\tilde{d}\Omega + \tilde{d} = \breve{a}\cdot(\alpha\tilde{d} + 4\tilde{d})^2 - 2\breve{a}\breve{w}_u(\alpha\tilde{d} + 4\tilde{d}) + \tilde{d} - \breve{a}(4\tilde{d})^2 + 2\breve{a}\breve{w}_u \cdot 4\tilde{d}$$

Subtracting by $\tilde{d}$ and dividing by $\tilde{d}$,

$$\Omega = \breve{a}\tilde{d}\alpha^2 + (8\breve{a}\tilde{d} - 2\breve{a}\breve{w}_u)\alpha + (16\breve{a}\tilde{d} - 8\breve{a}\breve{w}_u - 16\breve{a}\tilde{d} + 8\breve{a}\breve{w}_u)$$

Substituting $\hat{w}_u = \overline{\alpha}\tilde{d} + 4\tilde{d}$, and after simplification, it can be obtained that

$$\Omega = (\breve{a}\alpha^2 - 2\breve{a}\,\overline{\alpha}\alpha)\overline{d} \tag{14.1}$$

14.3.1 A Full-load Production Function Based on Alternative Routing

In this section, a full-load production function is constructed for flexible capacity afforded by alternative routing. A data set from the industry is used in the calculation. The data set contains 5 products and 73 machine types. The number of processing steps of the process routings ranges from three to five hundred. Machine data includes such attributes as equipment type, availability and batch size. Using these data, a full-load production function is constructed using the following procedure:

1. For a given total volume of production $\tilde{d}$, generate product-mix scenarios (indexed by s).
2. For each scenario
   2.1 Determine a tool portfolio. The tool portfolio and its associated product mix are regarded as the state of the saturation point.
   2.2 Determine bottleneck machine groups and their alternative machines.
   2.3 Rebalance the workload of the bottleneck machines and their alternative machines to obtain the output rate.
3. Estimate the coefficients of the full-load production function.

If the total volume is divided into 10 parts, and distributed among 5 products, there are 30 possible scenarios of product mixes. Tool portfolio planning is a complex task that involves combinatorial optimisation, economic analysis, and multiple criteria [14.19]. In this work, a static capacity model, as compared to queuing capacity models, is assumed in Step 2.1. Bottleneck machines are identified by ranking machine groups by utilisation and the top 10% are selected. Alternative machines of the bottlenecks are selected from machines of the same types as the bottlenecks but with lower utilisation. In Step 2.3, a mixed-integer program is used to rebalance the workload between the bottlenecks and their alternative machines. This step requires solving two mixed-integer programs in sequence. Subscripts j, k, i and b are used to index product, machine, process step and bottleneck machine, respectively. Let $w_{i,j,k}$ be the standard machine time, $q_{i,j,k}$ be allocated production, C_k be capacity, and N_k be the number of machines. The decision variable $X_{(b,k)}$ refers to the fraction of the available time of machine b that is used to backup machine k. Rebalancing of the workload involves three constraints:

$$\sum_j \sum_i w_{i,j,k} \cdot q_{i,j,k} \le C_k N_k + \sum_{(b,k) \in BR} X_{(b,k)} C_b N_b - \sum_{(k,b) \in BR} X_{(k,b)} C_k N_k \quad \forall k$$

$$X_{(b,k)} \le Y_{(b,k)} \quad \forall (b,k)$$

$$\sum_k q_{i',j,k} \ge (1+\alpha) D_j \quad \forall j, (i' \text{ is a particular step with } w_{i',j,k} \ne 0 \text{ for some } k)$$

$$N_k \in N; \; q_{j,k}, \alpha \in R_+; \; 0 \le X \le 1; \; Y \in \{0,1\}$$

The first constraint is a balance equation for workload and capacity. It can be simplified by evaluating the inner summation over all steps on the left-hand side. The set BR contains pairs of machine (k) and backup machine (b). In the second constraint, the Y variable is a 0–1 indicator variable. The third constraint is used to model the excessive workload α. Using an objective function

$$\text{Min. } Z = \sum_{(b,k) \in BR} Y_{(b,k)}$$

and, for each value of α, if the formulation has a feasible solution Y, then it can be concluded that the excessive workload is manageable through alternative routing. If, on the contrary, there is no feasible solution, then the excessive workload is not workable. In the former case, the maximum output can be obtained by using a second objective function to maximise the output rate y: $\max y = \sum_j \sum_k q_{i',j,k}$.

Using $\tilde{d} = 22000$ and solving each of the 30 scenarios, the output rate y and corresponding α are shown in the following table. For some scenarios, the product mixes are very extreme, with most of the demands concentrating on one or two products; there are no feasible solutions for additional loading of $\alpha = 0.01$. These scenarios are not included in the table.

Table 14.1. Simulated data for the full-load state

Scenario (mix ratio)	Excessive workload α				
	0.0	0.01	0.02	0.03	0.04
4:3:2:1:0	22314	22540	-	-	-
4:3:1:1:1	22605	22803	-	-	-
4:2:2:2:0	24850	25082	25265	25351	-
4:2:2:1:1	22025	22278	-	-	-
3:3:3:1:0	20807	21284	-	-	-
3:3:2:2:0	24071	24395	24559	24681	24753
3:3:2:1:1	22014	22454	22821	23088	23150
3:2:2:2:1	24415	24842	25158	25340	25411
2:2:2:2:2	23441	23849	-	-	-

From the simulated data, full-load production functions can be estimated by using Equation (14.1) and the geometric properties. Three levels of $\tilde{d}$ from 21500 to 22500 have been used in the calculation. The empirical production functions have the following formulas. The data points and functions are displayed in Figure 14.6.

$$\Omega = \begin{cases} -47.964 \cdot \alpha^2 + 5.756 \cdot \alpha & \tilde{d} = 21500 \\ -49.079 \cdot \alpha^2 + 3.926 \cdot \alpha & \tilde{d} = 22000 \\ -50.195 \cdot \alpha^2 + 4.016 \cdot \alpha & \tilde{d} = 22500 \end{cases}$$

It is worth noting that marginal output is greater in magnitude than marginal capacity. This is a testimony to the effect of relieving the bottleneck.

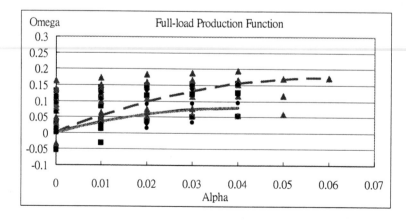

Figure 14.6. Examples of full-load production functions

14.4 A Dynamic System Model

In this section, a basic production model of manufacturing chains is presented. Applications of the basic model will be described in the next section. A dynamic system of Figure 14.2 can be represented by the following equations,

$$\dot{W}_1(t) = \eta(t) - \Omega_1(t) \tag{14.2}$$

$$\dot{W}_2(t) = \Omega_1(t) - \Omega_2(t) \tag{14.3}$$

$$\eta(t) = \overline{d}(t) + E(t) \tag{14.4}$$

$$\dot{I}(t) = \Omega_2(t) - \overline{d}(t) + \phi(t) \tag{14.5}$$

$$\Omega_k(t) = [a_k \alpha_k^2 - 2a_k \overline{\alpha}_k \alpha_k] \tilde{d}_k \qquad k = 1, 2 \tag{14.6}$$

Equations (14.2)–(14.4) are rate balance equations. In Equation (14.5), the rate of change in channel inventory is equal to the difference between $\Omega_2(t)$ and $\overline{d}(t)$, plus a noise $\phi(t)$. In nominal operation $\Omega_2(t)$ is driven by $\overline{d}(t)$; it should revert to $\overline{d}(t)$ in the long run. Equation (14.6) is a full-load production function, in which the parameter $\alpha_k(t)$ is a control variable for activated flexible capacity, $\overline{\alpha}_k$ is the maximum flexible capacity, and $\tilde{d}_k$ is the throughput rate at the saturation point.

Equation (14.4) can be substituted into Equation (14.2) and the production function (Equation (14.6)) can be substituted into Equations (14.2), (14.3) and (14.5). As discussed in Section 14.2, the total output should equal to the total demand in the long run, that is, $\int (\Omega_2(t) - \overline{d}(t)) dt$ should equal 0. In the long run, excessive inventory should also be zero. Thus, $\int \dot{I}(t) dt = 0$ and $\int \phi(t) dt = 0$. In the medium run, however, there will be excessive inventory, accumulated over a certain duration $[0, t_0]$ so that $\int_0^{t_0} \phi(t) dt = s \neq 0$. This quantity s will enter the chain as an anticipatory input over a planning horizon of $[t_0, t_0 + T]$. After substitution, the dynamic system (Equations (14.2)–(14.6)) can be reduced to:

$$\dot{W}_1(t) = \overline{d} + E(t) - [a_1 \alpha_1(t)^2 - 2a_1 \overline{\alpha}_1 \alpha_1(t)] \tilde{d}_1 \tag{14.7}$$

$$\dot{W}_2(t) = [a_1 \alpha_1(t)^2 - 2a_1 \overline{\alpha}_1 \alpha_1(t)] \tilde{d}_1 - [a_2 \alpha_2(t)^2 - 2a_2 \overline{\alpha}_2 \alpha_2(t)] \tilde{d}_2 \tag{14.8}$$

The anticipatory input s is a total amount and let $E(t)$ be the time path of the anticipated demand. Set the event time $t_0 = 0$ without loss of generality. $E(t)$ must satisfy the following equality integral constraint $\int_0^T E(t)dt = s$. Define a state variable $S(t)$ for the remaining amount of the anticipatory input. Therefore, $S(0) = s$, $S(T) = 0$ and

$$\dot{S}(t) = -E(t) \tag{14.9}$$

One more state variable can be defined. However, it is auxiliary and not essential. Let $R_k(t)$ be the cumulative output from shop k.

$$\dot{R}_k(t) = \Omega_k(t)$$

In the dynamic system, the output rate is determined by α_k and the cumulative output can be derived by integrating $\Omega_k(\alpha_k(t))$ over time. Thus, it is sufficient to have state variables W_k and S, without including $R_k(t)$. (In the remainder of this chapter, functional arguments are sometimes dropped for brevity of notation.)

To manage dynamic events, an immediate objective is to solve this dynamic system for $\alpha_k(t)$ at the event time $t_0 = 0$, given $S(0) = s$. If a solution can be found, then it can be said that the impact of the dynamic event is manageable. If no solution can be found, then the dynamic event is disastrous and other means must be provisioned. This objective amounts to finding an admissible control trajectory of $\alpha_k(t)$ subject to some constraints on work-in-process or cumulative output.

The admissible question can be answered by solving an optimal control problem for the dynamic system. Equations (14.7) and (14.8) are equations of motion for W_k . Because activating flexible capacity will incur the cost of machine setup, $\alpha_k(t)$ should not be a free variable.

14.4.1 A Formulation of Optimal Control

A manufacturing chain of two nodes is used to describe a basic model of optimal control in this section. Let the state variable x be $[W_1, W_2, S]^T$ and the control variable u be $[\alpha_1, \alpha_2, E]^T$. It should be noted that variations of the model will define the control variables slightly differently, but this will be described later. We will start with the objective function of the basic model by considering the interest of both customers and manufacturers. Customers would be concerned with feasible $E(t)$ and the output of the chain. Whether an $E(t)$ path is feasible is determined by the existence of solution $\alpha_k(t)$. Since output is determined by the total capacity in the full-load state, it can also be derived from $\alpha_k(t)$. Manufacturers are concerned with utilising flexible capacity and its associated cost. Thus, a general form of the objective function can be expressed as

$$\text{Max}_u \int [\sum_k -g_k(\alpha_k(t)) - h_k(W_k(t))] - m_0 S dt$$

where g_k and h_k are continuous and differentiable cost functions and m_0 is a parameter. The mission of S is to track the depletion of s. Its associated time path $E(t)$ is an input. Therefore, it does not enter into the objective function. It is assumed that the cost function g_k is increasing and convex and that the cost is zero when $\alpha_k(t) = 0$. Although a high level of WIP is not desirable, its associated cost is not as severe as flexible capacity. The cost function h_k is chosen to be linear in excessive work-in-process W_k. There is another justification for this design. As the objective function contains two cost functions, h_k can be considered as a reference base for g_k. In this chapter, g_k and h_k are represented by polynomial functions:

$$g_k(\alpha_k(t)) = b_k \alpha_k^2(t) + c_k \alpha_k(t) \tag{14.10}$$

$$h_k(W_k(t)) = m_k W_k(t) \tag{14.11}$$

where b_k, c_k, and m_k are coefficients, $g_k' \geq 0$ and $g_k'' \geq 0$. This implies that g_k is a convex function with a positive slope. Numerically, $2b_k \alpha_k + c_k \geq 0$ and $b_k \geq 0$. Expressing Equations (14.7)–(14.9) in a concise form, the equations of motion for state variables are:

$$\dot{W}_1(t) = f_1(x,u) \equiv \bar{d}(t) + E(t) - [(a_1 \alpha_1^2(t) - 2a_1 \bar{\alpha}_1 \alpha_1(t)) \tilde{d}_1]$$

$$\dot{W}_2(t) = f_2(x,u) \equiv [(a_1 \alpha_1^2(t) - 2a_1 \bar{\alpha}_1 \alpha_1(t)) \tilde{d}_1] - [(a_2 \alpha_2^2(t) - 2a_2 \bar{\alpha}_2 \alpha_2(t)) \tilde{d}_2]$$

$$\dot{S}(t) = f_3(x,u) \equiv -E(t)$$

The Hamiltonian function is

$$H = \sum_k [-g(\alpha_k(t)) - h(W_k(t))] - m_0 S + \sum_{i=1}^{2} \lambda_i(t) f_i(x,u) + \mu(t) f_3(x,u)$$

where $\lambda_i(t)$ and $\mu(t)$ are co-state variables. The initial state of the system can be expressed as $[\tilde{w}_1, \tilde{w}_2, s]$ for given $\tilde{w}_1$ and $\tilde{w}_2$. The state variable $S(t)$ has a fixed terminal state; its transversality condition is replaced by the condition $S(T) = 0$. The equations of motion for co-state variables are

$$\dot{\lambda}_1(t) = -\nabla_{W_1(t)} H = m_1 \tag{14.12}$$

$$\dot{\lambda}_2(t) = -\nabla_{W_2(t)} H = m_2 \tag{14.13}$$

$$\dot{\mu}(t) = -\nabla_{S(t)} H = m_0 \tag{14.14}$$

Since the terminal conditions for W_k are unconstrained, the transversality condition $\lambda_i(T) = 0$ should hold. Solve for $\lambda_1(t)$, $\lambda_2(t)$, and $\mu(t)$ to obtain:

$$\lambda_1(t) = m_1(t - T)$$

$$\lambda_2(t) = m_2(t - T)$$

$$\mu(t) = m_0 t + k$$

where k is a constant. The transversality condition of $\mu(t)$ is no longer $\mu(T) = 0$ but is replaced by the fixed terminal condition of $S(T) = 0$.

We now turn to the maximum principle condition $Max_u\{H\}$. The function H has four terms. The first term is concave in α_k. The second and fourth terms are not functions of α_k. The functional form of the third term depends on the relative magnitude of λ_1 and λ_2.

$$\frac{\partial H}{\partial \alpha_1} = -2b_1\alpha_1 - c_1 + (\lambda_2(t) - \lambda_1(t))[(2a_1\alpha_1 - 2a_1\bar{\alpha}_1)\tilde{d}_1]$$

$$\frac{\partial^2 H}{\partial \alpha_1^2} = -2b_1 + 2(\lambda_2(t) - \lambda_1(t))a_1\tilde{d}_1 = -2b_1 + 2a_1\tilde{d}_1(m_2 - m_1)(t - T)$$

By definition, $b_1 \geq 0$ and $a_1\tilde{d}_1 \leq 0$. It is desirable to move the work-in-process downstream. Therefore, to impose a higher penalty cost on inventory upstream, $m_1 \geq m_2$ is adopted as a premise. This premise ensures that $\partial^2 H / \partial \alpha_1^2 \leq 0$ and H is concave in α_k. (If $m_1 < m_2$, the convexity of H has to be determined by relative values of the coefficients.) To find the optimum, let $\partial H / \partial \alpha_1 = 0$.

$$-2b_1\alpha_1 + 2a_1\tilde{d}_1(\lambda_2(t) - \lambda_1(t))\alpha_1 = c_1 + (\lambda_2(t) - \lambda_1(t))(2a_1\bar{\alpha}_1\tilde{d}_1)$$

After rearranging terms, $[2a_1\tilde{d}_1(\lambda_1(t) - \lambda_2(t)) + 2b_1]\alpha_1 = 2a_1\tilde{d}_1\bar{\alpha}_1(\lambda_1(t) - \lambda_2(t)) - c_1$

$$\alpha_1^*(t) = \frac{2a_1\tilde{d}_1\bar{\alpha}_1(\lambda_1(t) - \lambda_2(t)) - c_1}{2a_1\tilde{d}_1(\lambda_1(t) - \lambda_2(t)) + 2b_1} = \bar{\alpha}_1 - \frac{2b_1\bar{\alpha}_1 + c_1}{2a_1\tilde{d}_1(m_1 - m_2)(t - T) + 2b_1} \qquad (14.15)$$

Note that $2a_1\tilde{d}_1(\lambda_1(t) - \lambda_2(t)) + 2b_1 \geq 0$. This relation will have implications in the following sections of analysis.

$$\frac{\partial H}{\partial \alpha_2} = -2b_2\alpha_2 - c_2 + \lambda_2(t)[-(2a_2\alpha_2 - 2a_2\bar{\alpha}_2)\tilde{d}_2]$$

$$\frac{\partial^2 H}{\partial \alpha_2^{\,2}} = -2b_2 - 2a_2 \tilde{d}_2 \lambda_2(t) = -2b_2 - 2a_2 \tilde{d}_2 m_2(t-T)$$

As $a_2 \tilde{d}_2 < 0$, $b_2 \geq 0$ and $m_2 \geq 0$, it can be seen that $\partial^2 H / \partial \alpha_2^{\,2} \leq 0$. Let $\partial H / \partial \alpha_2 = 0$.

$$[2b_2 + 2a_2 \tilde{d}_2 \lambda_2(t)]\alpha_2 = 2a_2 \tilde{d}_2 \bar{\alpha}_2 \lambda_2(t) - c_2$$

$$\alpha_2^{\,*}(t) = \frac{2a_2 \tilde{d}_2 \bar{\alpha}_2 \lambda_2 - c_2}{2a_2 \tilde{d}_2 \lambda_2 + 2b_2} = \bar{\alpha}_2 - \frac{2b_2 \bar{\alpha}_2 + c_2}{2a_2 \tilde{d}_2 m_2(t-T) + 2b_2} \qquad (14.16)$$

To find the optimal E path, take the partial derivative of H.

$$\frac{\partial H}{\partial E} = \lambda_1(t) - \mu(t) = m_1(t-T) - (m_0 t + k) = (m_1 - m_0)t - (m_1 T + k) \quad t \in [0 \; T]$$

The sign of $\partial H / \partial E$ is dependent on the relative magnitude of m_0 and m_1. Because m_0 and m_1 are inventory cost and it is desirable to move inventory downstream in manufacturing chains, we will assume $m_0 > m_1$ in the following analysis. (If $m_0 < m_1$ on the contrary, the analysis will be similar but opposite results will be obtained.) The switching function of $\partial H / \partial E$ is determined by $t = (m_1 T + k)/(m_1 - m_0)$. When $t < (m_1 T + k)/(m_1 - m_0)$, $(m_1 - m_0)t - (m_1 T + k) > 0$ and H is linear in E with an upward-sloping curve. When $t > (m_1 T + k)/(m_1 - m_0)$, H is a downward-sloping curve. Since $S(0)=s$, $S(T)=0$ and $\int_0^T E(t)dt = -[S(0) - S(T)]$ $=s$, the value of k can be obtained by solving the equality:

$$\int_0^{\frac{(m_1 T + k)}{(m_1 - m_0)}} \bar{E} dt + \int_{\frac{(m_1 T + k)}{(m_1 - m_0)}}^{T} \underline{E} dt = \bar{E}\frac{(m_1 T + k)}{(m_1 - m_0)} + \underline{E}T - \underline{E}\frac{(m_1 T + k)}{(m_1 - m_0)} = s$$

in which $(m_1 T + k)/(m_1 - m_0) \leq T$. Therefore,

$$k = \frac{(s - \underline{E}T)(m_1 - m_0)}{(\bar{E} - \underline{E})} - m_1 T$$

and by substituting k, $s \leq \bar{E}T$.

Incidentally, it can be seen that k is non-positive ($k \leq 0$, since $s - \underline{E}T \geq 0$).

$$E^*(t) = \begin{cases} \bar{E} & t < (m_1 T + k)/(m_1 - m_0) = \dfrac{s - \underline{E}T}{\bar{E} - \underline{E}} \\[2mm] \underline{E} & t > (m_1 T + k)/(m_1 - m_0) = \dfrac{s - \underline{E}T}{\bar{E} - \underline{E}} \end{cases} \qquad (14.17)$$

where, $\underline{E}$ and $\bar{E}$ denote the lower and upper limits of the domain of $E(t)$, respectively.

A special case:

If $m_0 = 0$, the state variable S will not appear in the objective function. Since $m_1 > m_0$ in this case, the switching function should be revised by a sign change.

$$E^*(t) = \begin{cases} \bar{E} & t > (m_1 T + k)/m_1 = \dfrac{s - ET}{\bar{E} - \underline{E}} \\ \underline{E} & t < (m_1 T + k)/m_1 = \dfrac{s - ET}{\bar{E} - \underline{E}} \end{cases}$$

This implies that $E(t)$ will start at the low level. Because, any input E is added to W_1, a decrease in S will result in an increase in W_1. The work release $E(t)$ is delayed as much as possible, until the time has come to fulfill the constraint of $S(T) = 0$. □

Substituting the control variables (Equations (14.15)–(14.17)) into the Hamiltonian function, the resultant new Hamiltonian function, renamed H^0, contains λ_i and W_k alone, with no control variables. It is straightforward to show that the new function H^0 has the following form (after substituting $\alpha_k *$ and $E*$):

$$H^0 = \tilde{g}(t) - m_1 W_1(t) - m_2 W_2(t) - m_0 S(t) + \lambda_1(t)\tilde{E}(t) + \lambda_2(t)\tilde{f}(t) + \mu(t)\tilde{E}(t)$$

where $\tilde{g}(t), \tilde{E}(t),$ and $\tilde{f}(t)$ are functions of t only.

Since H^0 is concave in state variables for all t for given λ and μ, the *maximum principle* conditions are sufficient for the global maximisation, according to the Arrow sufficiency theorem [14.20].

14.4.2 Closed Control Set

Since the control variable α_k is used to model flexible capacity, there are occasions to impose additional (more stringent) constraints, other than the domain of the production function, on the control variable. In this section, the basic model is extended to include a closed control set. The additional limit on control variables will be treated as inequality constraints:

$$0 \le \alpha_k \le \alpha_k^u$$

Following the modelling approach of [14.20], define a Lagrange function

$$L = H + \gamma_{cl}\alpha + \gamma_{cu}(\alpha_k^u - \alpha)$$

where γ_{cl} and γ_{cu} are row vectors of Lagrange multipliers of dimension k for the lower and upper limits of flexible capacity, and α is α_k in vector form. Let γ_{cl} be $[\gamma_{cl}(1), \gamma_{cl}(2)]$ and γ_{cu} be $[\gamma_{cu}(1), \gamma_{cu}(2)]$ in vector form for $k=2$. It can be seen that the convexity of H is maintained in L. The first-order condition for maximising L calls for, assuming interior solutions,

$$\frac{\partial L}{\partial \alpha_k(t)} = 0 \tag{14.18}$$

$$\frac{\partial L}{\partial \gamma_{cl}(k)} = \alpha_k \ge 0; \quad \frac{\partial L}{\partial \gamma_{cu}(k)} = \alpha_k^u - \alpha_k \ge 0 \tag{14.19}$$

$$\gamma_{cl}(k) \ge 0, \gamma_{cu}(k) \ge 0 \tag{14.20}$$

$$\gamma_{cl}(k)\frac{\partial L}{\partial \gamma_{cl}(k)} = 0, \gamma_{cu}(k)\frac{\partial L}{\partial \gamma_{cu}(k)} = 0 \tag{14.21}$$

The complementary slackness conditions (Equation (14.21)) ensure that the terms in the Lagrange function involving $\gamma_{cl}(k)$ and $\gamma_{cu}(k)$ will vanish in the solution, so that the value of L will be identical with that of H after maximisation. To minimise L, set the first-order condition $\partial L/\partial \alpha_k$ to zero.

$$\frac{\partial L}{\partial \alpha_1(t)} = \frac{\partial H}{\partial \alpha_1(t)} + \gamma_{cl}(1) - \gamma_{cu}(1) = 0$$

$$\frac{\partial L}{\partial \alpha_2(t)} = \frac{\partial H}{\partial \alpha_2(t)} + \gamma_{cl}(2) - \gamma_{cu}(2) = 0$$

Since the constraints $\alpha_k \le \alpha_k^u$ and $\alpha_k \ge 0$ cannot be active simultaneously, there are three possible cases for the values of $\gamma_{cl}(k)$ and $\gamma_{cu}(k)$: (1) $\gamma_{cl}(k) = 0$, $\gamma_{cu}(k) > 0$; (2) $\gamma_{cl}(k) > 0$, $\gamma_{cu}(k) = 0$; (3) $\gamma_{cl}(k) = 0$, $\gamma_{cu}(k) = 0$. Recall $\partial H/\partial \alpha_k$ and denote it as H'.

$$\frac{\partial H}{\partial \alpha_k} \equiv H' = \begin{cases} -[2b_1 + 2a_1\tilde{d}_1(\lambda_1(t) - \lambda_2(t))]\alpha_1 + 2a_1\tilde{d}_1\bar{\alpha}_1(\lambda_1(t) - \lambda_2(t)) - c_1 & k = 1 \\ -(2b_2 + 2a_2\tilde{d}_2\lambda_2(t))\alpha_2 + 2a_2\tilde{d}_2\bar{\alpha}_2\lambda_2(t) - c_2 & k = 2 \end{cases}$$

Let η_k solve for $\partial H/\partial \alpha_k = 0$. For k = 1, the switching function

$$\eta_1 = \frac{2a_1\tilde{d}_1\bar{\alpha}_1(\lambda_1(t) - \lambda_2(t)) - c}{2b_1 + 2a_1\tilde{d}_1(\lambda_1(t) - \lambda_2(t))}.$$

The Hamiltonian H is concave and is maximised at η_1. As noted before,

$$2a_1 \tilde{d}_1 (\lambda_1(t) - \lambda_2(t)) + 2b_1 \geq 0 .$$

That is, the coefficient of α_1 in H' is negative. For any $\alpha_1 < \eta_1$, $H'\big|_{\alpha_1} > H'\big|_{\eta_1} = 0$. Similarly, it can be shown that for any $\alpha_2 < \eta_2$, $H'\big|_{2} > H'\big|_{2} = 0$.

- Consider the case of $\gamma_{cl}(k) = 0$ and $\gamma_{cu}(k) > 0$. It can be solved from Equation (14.21) that $\alpha^* = \bar{\alpha}$. In addition, this is only possible when $H'\big|_{\bar{\alpha}} > H'\big|_{\eta} 0 = 0$, which requires that $\alpha_k^u < \eta_k$. To summarise this case, the optimal control path is $\alpha_k^* = \alpha_k^u$, when $\eta_k \geq \alpha_k^u$.

- Consider the case of $\gamma_{cl}(k) > 0$ and $\gamma_{cu}(k) = 0$. It can be solved that $\alpha_k^* = 0$. Similar to the above case, this is possible only when $H'\big|_{\alpha=0} \leq H'\big|_{\eta} 0 = 0$ and $\eta_k \leq 0$. (Note: The value of 0 is the lower limit of α_k. This case is just the opposite of the previous case.)

- Consider the case of $\gamma_{cl}(k) = \gamma_{cu}(k) = 0$. The Lagrange L reduces to H. The solution is as given by Equations (14.15) and (14.16). This is possible only when η_k is not binding by the control limits. That is, $0 < \eta_k < \alpha_k^u$.

To summarise the three cases, define a switching function η_k that solves for $\partial H / \partial \alpha_k = 0$.

$$\eta_k(t) = \begin{cases} \dfrac{2a_1 \tilde{d}_1 \bar{\alpha}_1 (\lambda_1(t) - \lambda_2(t)) - c_1}{2b_1 + 2a_1 \tilde{d}_1 (\lambda_1(t) - \lambda_2(t))} & k = 1 \\[4mm] \dfrac{2a_2 \bar{\alpha}_2 \tilde{d}_2 \lambda_2(t) - c_2}{2b_2 + 2a_2 \tilde{d}_2 \lambda_2(t)} & k = 2 \end{cases}$$

The optimal control path is:

$$\alpha_k^*(t) = \begin{cases} \alpha_k^u & \text{if } \eta_k(t) \geq \alpha_k^u \\ 0 & \text{if } \eta_k(t) \leq 0 \\ \text{Equations (14.15) and (14.16)} & \text{if } 0 < \eta_k(t) < \alpha_k^u \end{cases} \qquad (14.22)$$

14.5 Numerical Examples and Application

In this section, two numerical examples will be used to demonstrate how the optimal control model can be applied. Assume that the two shops have a capacity to meet nominal demand $\tilde{d}_1 = 22000$ and $\tilde{d}_2 = 22500$, respectively. All demands are normalised and demand $\bar{d} = 0.0$. The full-load production functions are

$$\Omega_1 = -50.562 \cdot \alpha_1^2 + 4.045 \cdot \alpha_1$$

$$\Omega_2 = -51.711 \cdot \alpha_2^2 + 4.137 \cdot \alpha_2$$

The upper limit of flexible capacity is $\bar{\alpha}_k = 0.04$ for $k = 1, 2$. These data yield $a_1 \tilde{d}_1 = -50.562$, $a_2 \tilde{d}_2 = -51.711$, $2a_1 \tilde{d}_1 \bar{\alpha}_1 = -4.045$, and $2a_2 \tilde{d}_2 \bar{\alpha}_2 = -4.137$.

In the first example, the objective function contains no state variable S. In the second example, the state variable S is included. The two examples, as a whole, demonstrate how an S path will make an impact on the production decision α_k.

Example 1: *(The objective functional contains no state variable S)*

This example demonstrates the manufacturers' perspective of using the model. Assume the initial inventory $[\tilde{w}_1, \tilde{w}_2] = [0.13, 0.08]$. Assume that the channel inventory correction is 0.12 over three months. The manufacturers also expect the anticipatory demand stream to vary within a range of 0.035 to 0.045 per time period. Let $s = 0.12$, the planning horizon $T = 3$, and the domain of $E(t)$ is set as $[\underline{E}, \bar{E}] = [0.035, 0.045]$. Since the output rate, cost of flexible capacity and WIP level are the usual concerns of the manufacturers, the following objective function is used, which includes output rate Ω_k, control variable α_k and state variable W_k. The unit cost of inventory is $m_1 = 1.0$ and $m_2 = 0.8$.

$$\text{Max} \quad \int_0^T \sum_{k=1}^{2} [\Omega_k(t) - \alpha_k(t)] - W_1(t) - 0.8 W_2(t) dt$$

Since the output rate Ω_k is expressed as a function of flexible capacity α_k, the objective function can be converted to the standard form of Equations (14.10) and (14.11). That the objective functional contains no terms for variable S is equivalent to setting m_0 to 0,

$$\{\sum_{k=1}^{2} (a_k \alpha_k^2(t) - 2a_k \bar{\alpha}_k \alpha_k(t)) \tilde{d}_k - \alpha_k(t)\} - W_1(t) - 0.8 W_2(t)$$

$$= \sum_{k=1}^{2} (a_k \tilde{d}_k \alpha_k^2(t) - (2a_k \bar{\alpha}_k \tilde{d}_k + 1) \alpha_k(t)) - W_1(t) - 0.8 W_2(t)$$

Therefore, $b_k = -a_k \tilde{d}_k$ and $c_k = 2a_k \tilde{d}_k \bar{\alpha}_k + 1 = -2b_k \bar{\alpha}_k + 1$.

Specifically, $b_1 = 50.562$, $b_2 = 51.711$, $c_1 = -3.045$, $c_2 = -3.137$. Also, $(s - \underline{E}T)/(\bar{E} - \underline{E}) = 1.5$.

The optimal control paths are (Equations (14.15)–(14.17)):

$$\alpha_1^*(t) = \bar{\alpha}_1 - \frac{2b_1\bar{\alpha}_1 + c_1}{2a_1\tilde{d}_1(m_1 - m_2)(t-T) + 2b_1} = .04 - \frac{1.000}{161.798 - 20.225t}$$

$$\alpha_2^*(t) = \bar{\alpha}_2 - \frac{2b_2\bar{\alpha}_2 + c_2}{2a_2\tilde{d}_2 m_2(t-T) + 2b_2} = .04 - \frac{1.000}{351.635 - 82.738t}$$

$$E^*(t) = \begin{cases} 0.035 & t < 1.5 \\ 0.045 & t > 1.5 \end{cases}$$

To obtain performance measures, compute $\Omega_1(t)$ first.

$$\Omega_1(t) = a_1\tilde{d}_1\alpha_1^*(t)^2 - 2a_1\tilde{d}_1\bar{\alpha}_1\alpha_1^*(t)$$

$$= a_1\tilde{d}_1[\bar{\alpha}_1 - \frac{2b_1\bar{\alpha}_1 + c_1}{2a_1\tilde{d}_1(\lambda_1 - \lambda_2) + 2b_1}]^2 - 2a_1\tilde{d}_1\bar{\alpha}_1[\bar{\alpha}_1 - \frac{2b_1\bar{\alpha}_1 + c_1}{2a_1\tilde{d}_1(\lambda_1 - \lambda_2) + 2b_1}]$$

By substituting $b_k = -a_k\tilde{d}_k$ and $c_k = 2a_k\bar{\alpha}_k\tilde{d}_k + 1 = 1 - 2b_k\bar{\alpha}_k$, $\Omega_1(t)$ can be simplified.

$$\Omega_1(t) = -b_1[\bar{\alpha}_1 - \frac{1}{-2b_1(\lambda_1 - \lambda_2) + 2b_1}]^2 + 2b_1\bar{\alpha}_1[\bar{\alpha}_1 - \frac{1}{-2b_1(\lambda_1 - \lambda_2) + 2b_1}]$$

$$= b_1\bar{\alpha}_1^2 - \frac{b_1}{(-2b_1(m_1 - m_2)(t-T) + 2b_1)^2}$$

By substitution of value,

$$\Omega_1(t) = 0.081 - \frac{50.562}{(161.798 - 20.225t)^2}$$

Cumulative output $R_1(t)$ can be computed from $\int_0^t \Omega_1(\tau)d\tau$ with $R(0) = 0$.

$$R_1(t) = \varphi_1 + \int_0^t \Omega_1(\tau)d\tau = \varphi_1 + 0.081t - \frac{1.000}{64.719 - 8.090t}$$

where φ_1 is a constant. Evaluating $R(0)$, $\varphi_1 = \frac{1.000}{64.719} = 0.015$. Thus,

$$R_1(t) = 0.015 + 0.081t - \frac{1.000}{64.719 - 8.090t}$$

Equations (14.7) and (14.8) are used to compute the state variable W_k by integration.

$$W_1(t) = \tilde{w}_1 + \bar{d}t + \int_0^t E(\tau)d\tau - \int_0^t \dot{R}(\tau)d\tau$$
$$= \tilde{w}_1 + \bar{d}t + \underline{E}\min(t, 1.5) + \overline{E}\max(t - 1.5, 0) - R_1(t)$$

$$W_1(t) = 0.115 + 0.035\min(t, 1.5) + 0.045\max(t - 1.5, 0) - 0.081t + \frac{1}{64.719 - 8.09t}$$

Similarly, it can be derived that

$$\Omega_2(t) = a_2\tilde{d}_2\alpha_2 *(t)^2 - 2a_2\tilde{d}_2\bar{a}_2\alpha_2 *(t)$$
$$= b_2\bar{a}_2{}^2 - \frac{b_2}{(-2b_2m_2(t - T) + 2b_2)^2}$$

By substitution of value,

$$\Omega_2(t) = 0.083 - \frac{51.711}{(2351.635 - 82.738t)^2}$$

$$R_2(t) = 0.001 + 0.083t - \frac{1.000}{900.191 - 132.381t}$$

$$W_2(t) = \tilde{w}_2 + R_1(t) - R_2(t) = 0.094 - 0.002t - \frac{1.000}{64.719 - 8.090t} + \frac{1.000}{397.143 - 132.381t}$$

$$S(t) = s - \int_0^t E(\tau)d\tau = s - \underline{E}\min(t, 1.5) - \overline{E}\max(t - 1.5, 0)$$
$$= 0.12 - 0.035\min(t, 1.5) - 0.045\max(t - 1.5, 0)$$

The optimal control paths and the resultant state variables are plotted in Figure 14.7. Because the objective function contains no state variable S, the release of work into shop 1 is delayed as much as possible until it is time to fulfil the requirements of $S(T) = 0$. Since an optimal path exists, it can be concluded that the channel inventory correction will not degrade the quality of supply services.

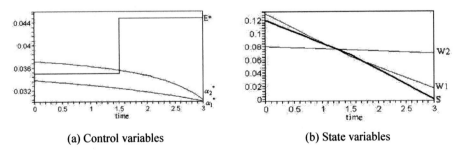

(a) Control variables   (b) State variables

Figure 14.7. Optimal control and state paths (Example 1)

Example 2: *(with state variable S in the objective function and with control limits)*

The $E(t)$ path is a representation of an anticipated demand stream. In this example, the state variable S is included in the objective function to provide a lever for the customer of the supply chain to explore the capability of the chain in coping with given $E(t)$. That is to say, the customer could specify various paths of $E(t)$ in scenario-based proactive planning. In addition to the input data of Example 1, set $\alpha_1^u = 0.034$ and $\alpha_2^u = 0.04$ to impose additional constraints on control variables. Replace the unit inventory costs of state variables by $m_0 = 6$, $m_1 = 4$, and $m_2 = 1$. The objective function is formulated as

$$\text{Max} \int_0^T \sum_{k=1}^2 [\Omega_k(t) - \alpha_k(t)] - 6S(t) - 4W_1(t) - W_2(t)dt$$

The switching function η_1 is

$$\eta_1(t) = \frac{2a_1\tilde{d}_1\bar{\alpha}_1(\lambda_1(t) - \lambda_2(t)) - c}{2b_1 + 2a_1\tilde{d}_1(\lambda_1(t) - \lambda_2(t))} = 0.04 - \frac{1}{1011.240 - 303.372t}$$

For $\eta_1 \geq \alpha_1^u = 0.03$, the threshold value of t can be computed to be 2.784. Therefore, for $t \leq 2.784$, $\alpha_1^* = 0.034$. For $\eta_1 \leq 0$, the threshold value of t is 3.251, which is greater than T. This range of η_1 is not applicable in this example. For $0 < \eta_1 < \alpha_1^u$, Equation (14.15) is applicable. Therefore,

$$\alpha_1^*(t) = \bar{\alpha}_1 - \frac{2b_1\bar{\alpha}_1 + c_1}{2a_1\tilde{d}_1(m_1 - m_2)(t - T) + 2b_1} = 0.04 - \frac{1}{1011.240 - 303.372t}$$

Collecting the three ranges together, the following is obtained:

$$\alpha_1^*(t) = \begin{cases} .034 & \text{if } t \leq 2.784 \\ 0.04 - \dfrac{1}{1011.240 - 303.372t} & \text{if } 2.784 < t \leq 3 \end{cases}$$

The switching function η_2 is

$$\eta_2(t) = \frac{2a_2\bar{\alpha}_2\tilde{d}_2\lambda_2(t) - c_2}{2b_2 + 2a_2\tilde{d}_2\lambda_2(t)} = 0.04 - \frac{1.000}{413.688 - 103.422t}$$

The optimal control path $\alpha_2^*(t)$ can be similarly obtained. Of the three ranges, only one is applicable.

$$\alpha_2^*(t) = \bar{\alpha}_2 - \frac{2b_2\bar{a}_2 + c_2}{2a_2\tilde{d}_2 m_2(t-T) + 2b_2} = .04 - \frac{1.000}{413.688 - 103.422t}, \quad t \in [0, 3]$$

The optimal path of E(t) and the paths of state variables are:

$$E^*(t) = \begin{cases} 0.045 & t < 1.5 \\ 0.035 & t > 1.5 \end{cases}$$

$$R_1(t) = 0.080\min(t, 2.784) + 0.081\max(t - 2.784, 0)$$

$$-\frac{1}{6067.44 - 1820.232\max(t - 2.784, 0)} + 0.0001$$

$$W_1(t) = 0.130 + 0.045\min(t, 1.5) + 0.035\max(t - 1.5, 0) - R_1(t)$$

$$\Omega_2(t) = 0.083 - \frac{51.711}{(413.688 - 103.422t)^2}$$

$$R_2(t) = 0.001 + 0.083t - \frac{1.000}{827.376 - 206.844t}$$

$$W_2(t) = 0.08 + R_1(t) - R_2(t)$$

$$S(t) = 0.12 - 0.045\min(t, 1.5) - 0.035\max(t - 1.5, 0)$$

The optimal paths of control and state variables are displayed in Figure 14.8.

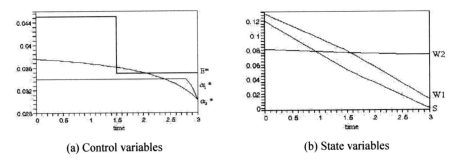

(a) Control variables   (b) State variables

Figure 14.8. Optimal control and state paths (Example 2)

These two examples demonstrate a new planning capability to proactively respond to scenarios of dynamic, interrupting events. The method can be applied to answer the following types of questions in managing manufacturing chains:

- If product demand is 20% higher, does the manufacturing chain have sufficient capability to fulfil it?

- If there is a loss of 15% capacity at a production unit, how would the supply capability be impaired?

14.6 Conclusions

Faced with many uncertainties and disrupting events, production control of manufacturing chains as a large system is a challenging task. In this chapter, a new method based on optimal control modelling is presented for managing dynamic events that pose a risk to degenerate supply services of manufacturing chains. Instead of using a conventional production function, this chapter presents an innovative concept of full-load production function that describes the behaviour of production units in the full-load state. The chapter also presents a dynamic system model for manufacturing chains under exogenous demand shocks and develops an optimal control model for integrating the production functions of multiple production units. We demonstrate the application of this method to evaluating the impact of demand shock events on supply chain production. This model provides a useful tool of proactive planning to respond to dynamic, disrupting events inherent in manufacturing chains.

Acknowledgment

This study is partially supported by funding from the Semiconductor Research Corp. and ISMI, Inc. under contract No. 2004-OJ-1214, and by funding from the National Science Council of Taiwan under grants 93-2213-E-002-010, 94-2213-E-002-013 and -014.

References

[14.1]   Craighead, C.W., Patterson, J.W. and Fredendall, L.D., 2001, "Protective capacity positioning impact on manufacturing cell performance," *European Journal Operational Research*, **134**, pp. 425–439.
[14.2]   Lin, Yu-Hsin, and Lee, Ching-En, 2001, "A total standard WIP estimation method for wafer fabrication," *European Journal of Operational Research*, **131**, pp. 78–94.
[14.3]   Duenyas, Izak, Fowler, John W. and Schruben, Lee W., 1994, "Planning and scheduling in Japanese semiconductor manufacturing," *Journal of Manufacturing Systems*, **13**(5), pp. 323–332.
[14.4]   Johri, Pravin K., 1993, "Practical issues in scheduling and dispatching in semiconductor wafer fabrication," *Journal of Manufacturing Systems*, **12**(6), pp. 474–485.
[14.5]   Connors, Daniel P. and Yao, David, 1996, "A queuing network model for semiconductor manufacturing," *IEEE Transactions on Semiconductor Manufacturing*, **9**(3), pp. 412–427.
[14.6]   Karmarkar, U.S., 1989, "Capacity loading and release planning with work-in-progress (WIP) and leadtimes," *Journal of Manufacturing and Operations Management*, **2**, pp. 105–123.
[14.7]   Missbauer, H., 2002, "Aggregate order release planning for time-varying demand," *International Journal of Production Research*, **40**(3), pp. 699–718.

[14.8]   Zäpfel, G. and Missbauer, H., 1993, "Production planning and control systems including load-oriented order release: problems and research perspectives," *International Journal of Production Economics*, **30–31**, pp. 107–122.

[14.9]   Nagatani, Takashi and Helbing, D., 2004, "Stability analysis and stabilization strategies for linear supply chains," *Physica A: Statistical Mechanics and its Applications*, **335**(3–4), pp. 644–660.

[14.10] Riddalls, C.E. and Bennett, S., 2001, "The optimal control of batched production and its effect on demand amplification," *International Journal of Production Economics*, **72**, pp. 159–168.

[14.11] Lefeber, E., Van Den Berg, R.A. and Rooda, J.E., 2004, "Modelling, validation and control of manufacturing systems," In *Proceedings of the 2004 American Control Conference*, **5**, pp. 4583–4588.

[14.12] Gong, L. and Matsuo, H., 1997, "Control policy for a manufacturing system with random yield and rework," *Journal of Optimization Theory and Application*, **95**, pp. 149–175.

[14.13] Graves, S.C., Kletter, D.B. and Hetzel, W.B., 1998, "A dynamic model for requirement planning with application to supply chain optimization," *Operations Research*, **46**(3), pp. S35–S49.

[14.14] Ni, Ming, Luh, P.B., Xiong, Bo, Chang, Shi-Chung, 2004, "An inventory control policy for maintenance networks," In *Proceedings of International Conference on Intelligent Robots and Systems*, IEEE/RSJ, Sept. 28–Oct. 2, pp. 1238–1244.

[14.15] Rose, O., 1998, "WIP evolution of a semiconductor factory after a bottleneck workcenter breakdown," In *Proceedings of the 1998 Winter Simulation Conference*, pp. 997–1003.

[14.16] Dümmler, M.A., 2000, "Analysis of the instationary behavior of a wafer fab during product mix changes," In *Proceedings of the 2000 Winter Simulation Conference*, pp. 1436–1442.

[14.17] Chou, Yon-Chun, Cheng, C.-T., Yang, Feng-Cheng and Liang, Yi-Yu, 2007, "Evaluating alternative capacity strategies in semiconductor manufacturing under uncertain demand and price scenario," *International Journal of Production Economics*, **105**(2), pp. 591–606.

[14.18] Lu, S.C.H., Ramaswamy, D. and Kumar, P.R., 1994, "Efficient scheduling policies to reduce mean and variance of cycle-time in semiconductor manufacturing plants," *IEEE Transactions on Semiconductor Manufacturing*, **7**, pp. 374–385.

[14.19] Chou, Yon-Chun and Wu, Chuan-Shun, 2002, "Economic analysis and optimization of tool portfolio in semiconductor manufacturing," *IEEE Transactions on Semiconductor Manufacturing*, **15**(4), pp. 447–453.

[14.20] Chiang, Alpha C., 1992, *Dynamic Optimisation*, McGraw-Hill, New York, NY, USA.

15

A Parameter-perturbation Approach to Replanning Operations

Nazrul I. Shaikh[1], Michael Masin[2] and Richard A. Wysk[1,2]

[1] Marcus Department of Industrial and Manufacturing Engineering
The Pennsylvania State University, University Park, PA 16802, USA
Email: rwysk@psu.edu

[2] Department of Industrial Engineering and Management
Technion-Israel Institute of Technology, Haifa 32000, Israel

Abstract
In this chapter, we present operational and planning models for manufacturing systems. These models are developed to show the operational fidelity of the models to various manufacturing processes. The models are initially used to plan production activities for a variety of products that will be semi-automatically manufactured. The same models are then used to show how optimal operational conditions can be developed for a variety of processes (minimum time and cost). The models are then extended to show how the operating conditions can be perturbed so that optimal short-term planning models can be developed.

The approach is illustrated using a family of product family models developed at Penn State University. The Factory for Advanced Manufacturing Engineering (FAME) at Penn State, a sophisticated manufacturing system is used to produce the products in volume, as would be the case for many commercial products.

15.1 Introduction

More often than not, there are discrepancies between what was planned and what can be executed. Unplanned events such as machine breakdowns, delays in receiving and dispatch of shipments, and order changes are more or less ubiquitous. If service level requirements (SLR) are associated with the successful implementation of these plans, then the discrepancies can be expensive. The planning and execution systems should therefore have some mechanism to deal with the deviations from the expected. Introducing redundancy into the system is the usual approach for buffering the impacts of these unplanned events on the plans. This is done as replanning is considered expensive – both computationally and from implementation perspectives as well as causing system nervousness. In this chapter, we present the *aggregate high fidelity modelling* (AHFM) approach for the production planning problem. As the name suggests, the AHFM approach proposes an aggregation over various planning hierarchies (hierarchies, as defined in hierarchical production planning literature), and the solution is comparable to those obtained through various aggregate planning approaches – the efficacy of the approach, however, lies in the way the AHFM approach facilitates replanning and the systematic handling of deviations from the expected.

The AHFM approach systematically develops a planning model that spans multiple levels of planning hierarchies under a common objective determined by the production planner. It aggregates the flexibility/redundancy that is available at each of these levels of hierarchies into a common pool, and the replanning approach (plan perturbation) then optimally uses this pooled redundancy to buffer the impact of the unplanned events. In this way, the AHFM formulation accentuates the available redundancy in a manner that allows replanning and rescheduling on the fly, and plan perturbation – the associated solution approach, seeks neighbourhood optimality that minimises system nervousness when a transition is made from one production plan to the other. The proposed approach is found to generate superior production plans and shows linear to low-order polynomial growth in complexity as the problem size grows.

Following the introduction section, the rest of the chapter is organised into 4 sections. In Section 15.2, we describe the AHFM approach for production planning. Section 15.2 comprises of 3 subsections describing the model formulation, implementation, and scalability details. Section 15.3 links the impact that the unplanned events can potentially have on the planning problem, and puts the replanning problem into perspective. The specific case of the addition of a new order to an existing production plan is taken up for a detailed analysis and discussion in this section. Section 15.4 discusses the extension of the AHFM approach to other planning domains; with emphasis on flight scheduling and rescheduling. The discussion and conclusions are presented in Section 15.5.

15.2 AHFM Approach

The traditional approach to production planning applies functional decomposition to planning tasks: first, process engineers locally optimise all process parameters, including the selected process plan and cutting speeds, and, only then do production engineers optimise the production quantities and schedules in order to obtain the maximum profit when all process variables are considered as fixed parameters. It creates a master and slave hierarchy where the solutions at one level form the starting point for decisions to be made at the next level.

The AHFM approach is an extension of the *high fidelity modelling* (HFM) approach taken at Penn State [15.1]–[15.4]. HFM contains detailed simulations of work-floors, and, given the global system objectives, integrates and optimises work-floor operations with lower-level process control of resources and higher-level production and supply chain planning and control. In the AHFM approach, only "hard" parameters, such as tools and material characteristics are fixed, while most of the other parameters such as cutting speeds, feed rates, and depth of cut are considered as being soft parameters, and are modelled as global variables in the aggregate model.

In order to fully appreciate the AHFM approach, there is a need to understand the entire process and identify the operational parameters that are in fact variables.

In this chapter, the AHFM methodology is illustrated in a manufacturing environment starting from an existing high-level plan, through low-level components, to the integrative high fidelity model. The model will consist of an aggregative production planner for manufacturing based on a real implementation in the FAME lab at the Pennsylvania State University. The layout of the shop floor is shown in Figure 15.1. There are several kinds of Haas CNC machines, such as vertical milling centres (VF-0 and VF-3), and turning centres (SL-20 and SL-30), all of which have the capability to manufacture a variety of prismatic and rotational parts with reliability and accuracy. An Arena RT simulation-based control system is utilised for computer-integrated control of the work cells.

The main products are Penn State-based souvenirs. This work will be focused on one of the souvenirs – three parts from a Penn State chess set, shown in Figure 15.2.

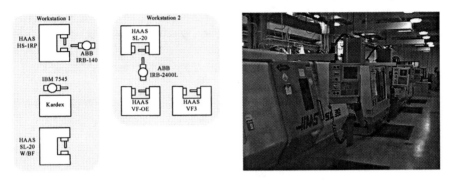

Figure 15.1. Example layout of Penn State CIM Lab

Part *King*   Part *Bishop*   Part *Pawn*

Figure 15.2. Example parts (PSU chess set)

15.2.1 AHFM for Production Planning

Parameters that directly impact process speed, such as spindle speed or feed in machining, are an obvious choice. However, there are less-obvious parameters that impact operational planning models. For example, different sample plans for quality assurance can provide the same level of quality; in the semiconductor industry these sample plans have a direct impact on the production flow of wafers and, consequently, on the production control models. The cost versus throughput rate tradeoff of raw material selection in discrete manufacturing, concentration selection in process industries, or universal versus specialised operator selection in services, can be optimally solved only in an operational integrative model and not during the design stage as is usually done. The following steps are proposed for the implementation of the AHFM approach.

1. *Traditional model*: Formulate the problem using existing modelling procedures for problem optimisation.
2. *Visible and hidden parameters*: List all of the parameters used in the model, including hidden ones (*e.g.* chosen process plan/path for manufacturing each part in the scheduling problem).
3. *Potential for parameter variation*: Identify the "soft" parameters that were fixed in previous design steps. Reduce the set of variable parameters using local optimisation models.
4. *Integrative model*: Incorporate all/most of the remaining "soft" parameters in the original model.
5. *Implementation*: Develop optimisation method/heuristic solutions to solve the extended problem.

15.2.1.1 Traditional Model

Traditionally, during the *process* planning stage, several process alternatives are checked and then one alternative, considered as the best, is chosen. Consequently, the *production* planning assumes a single process plan and known (fixed) processing times for all the parts that are to be manufactured. Let G be the set of part types; J_g be the set of process plans for part type g, $g \in G$; I_j be the set of process steps of process plan j; M be the set of available machines. We define t_{ijm} as the effective processing time of part type g, $g \in G$, using process plan j, $j \in J_g$, on machine m, $m \in M$, in the ith processing step, $i \in I_j$; x_j as the number of parts made using process plan j; k_g as the synchronising coefficient between part types that defines the required production volume ratios, *e.g.* for part "*King*" it is 1.0, for part "*Bishop*" it is 0.5, and for part "*Pawn*" it is 0.125; t_{ijm} as the machining time for part using process plan j on machine m. In traditional production planning, $|J_g| = 1$ for all part types $g \in G$, *i.e.* there is only a single process plan for each part. The P1 model maximises the throughput of chess sets, constrained by the capacity of the individual resources, K_m.

P1 Model:

$$\text{Max } k_1 \sum_{j \in J_1} x_j \tag{15.1}$$

Subject to

$$k_1 \sum_{j \in J_1} x_j = k_g \sum_{j \in J_g} x_j \qquad \forall g \in G \mid g > 1 \tag{15.2}$$

$$\sum_{i \in I_j} t_{ijm} = t_{jm} \qquad \forall g \in G, j \in J_g, m \in M \tag{15.3}$$

$$\sum_{g, j \in J_g} x_j t_{jm} \leq K_m \qquad \forall m \in M \tag{15.4}$$

$$x_j \geq 0 \quad \forall g \in G, j \in J_g \tag{15.5}$$

Equation (15.1) defines the objective function of maximum productivity – production quantities of any part can be used since all parts are synchronised through Equation (15.2). Equation (15.2) ensures that while making a chess set, for every "*King*" produced during a given production schedule, two "*Bishops*", and eight "*Pawns*" are also produced. Equation (15.3) finds the total processing time of each process plan j on each machine m. Equation (15.4) ensures that the production capacities of the individual machines are not exceeded. A schedule to manufacture x_j parts is then generated.

15.2.1.2 Visible and Hidden Parameters

Machining time t_{ijm}, synchronising coefficient k_g and resource capacity K_m are straightforward parameters used in the aggregative production-planning model P1. However, one of the hidden assumptions of the model is a given process plan for each part. This process plan should be an additional parameter taken into account.

15.2.1.3 Potential for Parameter Variation

The synchronising coefficient k_g is a part of the product (chess set) specification and, therefore, cannot be varied. Other parameters should be considered for parameter variation.

(1) Machining time

Selection of appropriate machining parameters is critical as they determine the effective processing times of the product. These parameters are selected through rule-based, data-driven or model-based approach. A model-based approach helps to determine the operating parameters that correspond to parameters that minimise the production cost per component or minimise the effective production time given in Equations (15.6) and (15.7).

$$C_{ijm} = \frac{C_b}{N_b} + C_m t_{ijm} + C_r \left(\frac{t_{ijm}^{(m)}}{T_r} \right) \tag{15.6}$$

$$t_{ijm} = t_{ijm}^{(m)} + t_{ijm}^{(h)} + t_r^{(t)} \left(\frac{t_{ijm}^{(m)}}{T_r} \right) \tag{15.7}$$

where C_b is a batch setup cost, N_b refers to the number of parts in the batch, C_m is a cost of machining operation on machine m, C_r is a cost of tool r, $t_{ijm}^{(m)}$ is the machining time for the ith processing step of the jth process plan on machine m,

$t_{ijm}^{(h)}$ is the material handling time (usually a constant) for the ith processing step of the jth process plan on machine m, $t_r^{(t)}$ is the tool change time (usually a constant) for the rth tool, T_r is the tool life of the rth tool. In our production environment, the same setup time is required for each part in the batch, therefore, C_b is proportional to the batch size, and N_b should be set to one. For the general case, batch sizes should be part of the decision variables as widely discussed in lot-splitting/lot-streaming literature [15.5][15.6]. Minimisation of the operating costs, C_{ijm}, or effective processing time, t_{ijm}, is taken as the objective and is solved as a constrained minimisation problem, where the typical constraints are of the form presented in Equations (15.8)–(15.12) [15.7]:

Constraints on the Spindle Speed (depends on the machine):

$$v_m^{min} < v_{ijm} \tag{15.8a}$$

$$v_{ijm} < v_m^{max} \tag{15.8b}$$

Feed constraint (depends on the machine):

$$f_m^{min} < f_{ijm} \tag{15.9a}$$

$$f_{ijm} < f_m^{max} \tag{15.9b}$$

Cutting force constraint:

$$K_F f_{ijm}^q d_{ijm}^w < F_{ijm}^{max} \tag{15.10}$$

Power constraint:

$$P_{ijm} = F_{ijm} v_{ijm} = K_F f_{ijm}^q d_{ijm}^w v_{ijm} < P_{ijm}^{max} \tag{15.11}$$

Surface finish constraint:

$$K_S f_{ijm}^h D_{ijm}^{-1} < R_{ijm}^{max} \tag{15.12}$$

where, K_s and h are specific coefficient and exponents of surface roughness constraint; K_F, q, and w are specific coefficient and exponents of cutting force constraint; and D_{ijm} is the tool diameter used for the ith operation of the jth process plan on machine m. The tool diameter is usually given and set to the largest feasible size for the given operation, since, as follows from Equation (15.12), tools with higher values of D_{ijm} can have higher feed rates, f_{ijm}. The feed rate, f_{ijm}, cutting speed, v_{ijm} and the depth of cut, d_{ijm}, determine the machining time, $t_{ijm}^{(m)}$, and the tool life, T_r. The general relationship between the machining time and tool life with the machining conditions is presented in Equations (15.13) and (15.14).

$$t_{ijm}^{(m)} = \frac{K_{ij} L_{ijm} D_{ijm}}{nf_{ijm} v_{ijm}} \tag{15.13}$$

$$T_r = \frac{K_r}{v_{ijm}^{\alpha'} f_{ijm}^{\beta'} d_{ijm}^{\gamma'}} \tag{15.14}$$

where, L_{ijm} is the tool path length for the ith operation of the jth process plan on machine m, K_r is the Taylor's tool life constant for tool r; α', β', and γ' are speed, feed and depth of cut exponents for tool r for operation i, and n is the number of teeth. Solving a mathematical model using Equation (15.6) or (15.7) as the objective and Equations (15.8a)–(15.14) as the domain constraints gives the values of the machining parameters v_{ijm}, f_{ijm}, and d_{ijm} for the ith operation of the jth process plan on machine m. In order to reduce the number of cuts, the depth of cut d_{ijm} should be fixed at the maximum allowable limit dictated by the tool properties and material properties. Between the two remaining decision variables, the spindle speed v_{ijm} and the feed parameter f_{ijm}, the following property essentially reduces the decision space to the spindle speed only.

Property: If the objective is to minimise a function of production time and/or production cost described by Equations (15.6) and (15.7), then in the optimal solution at least one of the Constraints (15.8a), (15.9b), (15.10), and (15.12) is bounding, i.e. either the feed parameter f_{ijm} is set to its maximum value or the spindle speed v_{ijm} is set to its minimum value or both.

Proof: Let the optimal solution spindle speed and feed parameters be v_{ijm} and f_{ijm}, respectively. If at least one of Constraints (15.8a), (15.9b), (15.10), and (15.12) is bounding, the property holds. If none of these constrains is bounding, we get a contradiction as follows. Let us simultaneously decrease v_{ijm} by $(1+\varepsilon)$ to $v_{ijm}/(1+\varepsilon)$ and increase f_{ijm} by $(1+\varepsilon)$ to $f_{ijm}/(1+\varepsilon)$, where ε is a small number. If Constraints (15.8a), (15.9b), (15.10), and (15.12) are not bounding they should continue to be satisfied with the new spindle speed and feed parameters. Constraint (15.11) holds, too, since $q \leq 1$ [15.7]. The exponents in the tool life Equation (15.14) have the following relation: $\beta' > \alpha'$, therefore the new tool life will be longer while, according to Equation (15.12), the machining time is unchanged. Consequently, the new production time and production cost are better than the original ones – in contradiction to the assumption that we have an optimal solution with the spindle speed and feed parameter, equal to v_{ijm} and f_{ijm}, respectively. ☐

Using typical values for the tool life coefficients and other parameters, f_{ijm} usually is set to the upper bound value based on the surface finish constraint before v_{ijm} reaches its lower bound. Consequently, we can substitute Equations (15.13) and (15.14) into Equations (15.6) and (15.7) and obtain the following generic form:

$$C_{ijm} = \frac{a_{ijm}^{(C)}}{v_{ijm}} + b_{ijm}^{(C)} v_{ijm}^{\alpha_{ijm}} + c_{ijm}^{(C)} \tag{15.15}$$

$$t_{ijm} = \frac{a_{ijm}^{(t)}}{v_{ijm}} + b_{ijm}^{(t)} v_{ijm}^{\alpha_{ijm}} + c_{ijm}^{(t)} \tag{15.16}$$

where, $a^{(C)}$, $b^{(C)}$, $c^{(C)}$, $a^{(t)}$, $b^{(t)}$, $c^{(t)}$ are coefficients defined by the part and tool materials, manufacturing process, and machining and tool change costs; their values are found using manufacturing data handbooks and equipment costs. Constraints (15.8a)–(15.12) can be reduced to Constraint (15.17):

$$v_{ijm}^{min} \leq v_{ijm} \leq v_{ijm}^{max} \qquad \forall i,j,m \tag{15.17}$$

where, v_{ijm}^{min} and v_{ijm}^{max} are the lower and the upper feasible bounds of v_{ijm} based on the above constraints.

Lemma 1. Cost is a convex function of the spindle speed.
Proof:

$$C_{ijm}^0 = -\frac{a_{ijm}^{(C)}}{v_{ijm}} + b_{ijm}^{(C)} v_{ijm}^{\alpha_{ijm}} + c_{ijm}^{(C)} \tag{15.18}$$

$$\frac{dC_{ijm}^0}{dv_{ijmt}} = -\frac{a_{ijm}^{(C)}}{v_{ijm}^2} + b_{ijm}^{(C)} \alpha_{ijm} v_{ijm}^{(\alpha_{ijm}-1)} \tag{15.19}$$

$$\frac{d^2 C_{ijm}^0}{dv_{ijm}^2} = \frac{2a_{ijm}^{(C)}}{v_{ijm}^3} + b_{ijm}^{(C)} \alpha_{ijm} (\alpha_{ijm}-1) v_{ijm}^{(\alpha_{ijm}-2)} \underset{\substack{\alpha_{ijm}>1 \\ a,b,v>0}}{\geq} 0 \tag{15.20}$$

Therefore, C_{ijm}^0 is a convex function of positive spindle speed $v_{ijm}^{(C)}$. □
We can find the optimal spindle speed, $v_{ijm}^{(C)}$, as follows.

$$\frac{dC_{ijm}^0}{dv_{ijm}^{(C)}} = -\frac{a_{ijm}^{(C)}}{v_{ijm}^{(C)2}} + b_{ijm}^{(C)} \alpha_{ijm} v_{ijm}^{(C)(\alpha_{ijm}-1)} = 0 \tag{15.21}$$

$$v_{ijm}^{(C)} = \sqrt[{(\alpha_{ijm}+1)}]{\frac{a_{ijm}^{(C)}}{b_{ijm}^{(C)}}} \tag{15.22}$$

Lemma 2. Processing time is a convex function of the spindle speed.
Proof:

$$t_{ijm}^0 = -\frac{a_{ijm}^{(t)}}{v_{ijm}} + b_{ijm}^{(t)} v_{ijm}^{\alpha_{ijm}} + c_{ijm}^{(t)} \tag{15.23}$$

$$\frac{dt_{ijm}^0}{dv_{ijm}} = -\frac{a_{ijm}^{(t)}}{v_{ijm}^2} + b_{ijm}^{(t)} \alpha_{ijm} v_{ijm}^{(\alpha_{ijm}-1)} \tag{15.24}$$

$$\frac{d^2 t_{ijm}^0}{dv_{ijm}^2} = \frac{2a_{ijm}^{(t)}}{v_{ijm}^3} + b_{ijm}^{(t)} \alpha_{ijm} (\alpha_{ijm}-1) v_{ijm}^{(\alpha_{ijm}-2)} \underset{\substack{\alpha_{ijm}>1 \\ a,b,v>0}}{\geq} 0 \tag{15.25}$$

Therefore, t_{ijm}^0 is a convex function of positive spindle speed $v_{ijm}^{(t)}$. □

We can find the optimal spindle speed, $v_{ijm}^{(t)}$, as follows.

$$\frac{dt_{ijm}^0}{dv_{ijm}^{(t)}} = -\frac{a_{ijm}^{(t)}}{v_{ijm}^{(t)2}} + b_{ijm}^{(t)}\alpha_{ijm}v_{ijm}^{(t)(\alpha_{ijm}-1)} = 0 \qquad (15.26)$$

$$v_{ijm}^{(t)} = \sqrt[1/(\alpha_{ijm}+1)]{\frac{a_{ijm}^{(t)}}{b_{ijm}^{(t)}}} \qquad (15.27)$$

(2) Alternative process paths/plans
In the traditional approach, as in selecting the machine parameters, a process path/ plan that minimises either processing time or production cost per part is usually chosen. The operations routing summaries (ORS) are specific to the shop floor and take into consideration the machines and the tooling that is available. Multiple ORS are possible for the same part. However, in practice, usually one ORS (typically the one that is most economic) is selected and used to generate the process plan. If the machining time is made a decision variable, multiple ORS can be selected. An alternative ORS that can be used for machining the pieces in the chess set is presented in [15.3], Figures B1–B4 and Tables B1–B3.

(3) Machine capacity
In traditional aggregative planning, designers use conservative estimates of the nominal machine capacity based on historic performance. Detailed scheduling tries to make the actual plan as close (*i.e.* as feasible) as possible to the original aggregative plan. If there is not enough capacity, *e.g.* due to machine and tool breakdowns, finite buffers and/or an inefficient schedule, the production plan is scaled down to meet the available capacity. Preventive maintenance, buffer design and production scheduling may directly affect the effective machine capacity. However, in the relatively simple manufacturing facility used in this example, the effective variability of processing times is the main reason why the nominal capacity may be overestimated.

15.2.1.4 Building Integrative Models

The main theme of the integrative models is putting together the profit objective and decision variables identified in the previous step.

Integration of machining time and alternative process paths/plans
    The traditional P1 model finds the production quantities x_j when the processing times t_{ijm} and process plan j for part type g are fixed. Model P1 can be modified into Model P2 according to the AHFM approach by combining Equations (15.15) and (15.16) with the P1 model:

P2 Model:

$$\text{Max } pk_1\sum_{j\in J_1}x_j - \sum_{g,j\in J_g,i\in I_j,m}x_j\left(\frac{a_{ijm}^{(C)}}{v_{ijm}} + b_{ijm}^{(C)}v_{ijm}^{\alpha_{ijm}} + c_{ijm}^{(C)}\right) \qquad (15.28)$$

Subject to

$$k_1 \sum_{j \in J_1} x_j = k_g \sum_{j \in J_g} x_j \qquad \forall g \in G \mid g > 1 \tag{15.29}$$

$$\sum_{g, j \in J_g, i \in I_j} x_j \left(\frac{a_{ijm}^{(t)}}{v_{ijm}} + b_{ijm}^{(t)} v_{ijm}^{a_{ijm}} + c_{ijm}^{(t)} \right) \leq K_m \qquad \forall m \in M \tag{15.30}$$

Constraint (15.17) defines the feasible range of spindle speed for each operation

$$x_j \geq 0 \quad \forall g \in G, j \in J_g \tag{15.31}$$

where p is the selling price of the chess set. In AHFM, the cutting parameter selection is done during production planning, providing an opportunity to improve the system's global performance.

15.2.2 Solution Approach to AHFM

We suggest an iterative procedure for capacity adjustment. In the iterative procedure, detailed simulation-based scheduling checks the feasibility of the solution of model P2 using traditional heuristics. If there is no feasible solution, the capacity is reduced and model P2 is solved again. In order to be effective, the resulting solution should differ from simply scaling down the number of units in the original production plan. In order to be efficient, the procedure should converge rapidly. It is found that the iteration procedure is efficient – just one iteration is sufficient, but not effective – the resulting production plan is completely equivalent to scaling down of the original one.

Once the processing steps in the alternative routes in P2 have been determined, the a, b and c coefficients for Equations (15.15) and (15.16) are determined. The parameters that are used for determining a, b and c depend on the tool material, workpiece material, manufacturing process and machining and tool change costs, and have been obtained from the data handbooks. Four cases were considered for analysis.

1. *Case 1 – Traditional*: When the production plan is determined based on machining parameters that are considered fixed and a single process plan per part is used. In this case, decisions made during the process planning stage are propagated to the production planning stage as fixed parameters. Given the processing times and paths, the production planning module finds the maximum production quantities using the P1 model. Given fixed sales prices and unconstrained demand, P1 model is equivalent to profit maximisation.

2. *Case 2 – Variable Machining Time*: When the production plan allows variable machining times, but only one process plan per part is used. In this case, only process paths are fixed during the process planning stage, while the cutting speeds (and resulting processing times and costs) are determined by the P2 model during the production planning stage.

3. *Case 3 – Variable Process Plan*: When the production plan allows multiple process plans per part, but machining parameters are considered as fixed. In this case, during the process planning stage, a set of possible process plans is developed with fixed cutting speeds in each. How much each process plan is used is determined by the P2 model during the production planning stage.

4. *Case 4 – Variable Process Plan and Machining Time*: When the production plan allows multiple process plans per part and allows variable machining times as well. In this case, during the process planning stage, a set of possible process plans is developed. All other process decisions, *i.e.* how much each process plan is used and what the cutting speeds are used, are done by the P2 model during the production planning stage.

For determining the actual throughput and profitability, detailed schedules have to be generated and the actual throughput needs to be determined. Popular heuristics are incorporated into the simulation model to generate feasible schedules wherever possible. The original production quantities were scaled down if they did not meet the available machines' capacity. The results from the simulation were used to generate the operating schedule.

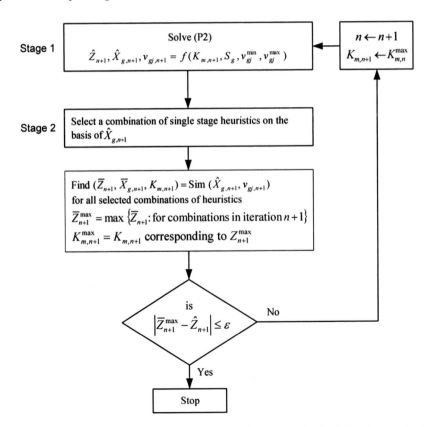

Figure 15.3. Spatiotemporal partition algorithm for composite high-fidelity production planning

A spatiotemporal partition heuristic (see Figure 15.3) is proposed wherein mathematical modelling is used for taking global decisions, selecting process plans and process parameters, and generating strict bounds on the solution. Simulation is used for developing a detailed model of the shop floor and its control, taking time-dependent decisions, determining actual machine capacities, and generating schedules. The hybrid model calls the mathematical model and the simulation model alternately, the output of one taken as the input for the other until there is a convergence, or the desired level of productivity has been reached.

- *Stage 1*: The solution to P2 is independent of sequencing constraints, and the solution therefore is the upper bound on the profitability of the shop floor for the machining capacity. Mathematical modelling is used in this case as the constraints and the objective function are quasiconvex, and nonlinear programming tools can converge to a good solution quickly.
- *Stage 2*: Simulation is used to generate a dynamic model that takes as input the process plans, the processing times, and the lot sizes from the static model and uses the information to determine the actual processing capability of the shop floor. A detailed shop-floor model is created to represent the actual shop floor, and the constraints therein. The parameters generated by the P2 model are taken as inputs and the simulation is executed to determine the minimum shop-floor capacity that is required to complete the lots request. If the required shop-floor capacity for satisfying the demands set by P2 is higher than the capacity that is available (due to interactions, starvation at a bottleneck, constraints due to material handler, *etc.*), the machining capacity in the P2 model is reduced, and a new request is generated by the P2 model. The process is repeated until the actual shop-floor capacity is greater than or equal to (within an epsilon) the shop-floor capacity in the P2 model and all the lots can be completed within the required time span.

15.2.2.1 Implementation Details

Some of the implementation issues encountered are as follows:

Stage 1: The P2 model is developed and solved using GAMS. The model, though nonlinear, is convex or quasiconvex at least for some ranges of cutting speeds. Standard packages such as CONOPT, MINOS, and SNOPT can be applied for nonlinear optimisation. It takes as input the sequences that are possible for a given part, machine availability and capacity, and the parameters for determining the process parameters (a, b, and c). The outputs of the model are the lot sizes that need to be made by the various process plans and the processing times for the parts in the machines that maximise the profits at the enterprise level.

Stage 2: A detailed simulation model was developed for a cluster of three machines: SL20, VF3, and VFOE. It was assumed that the shop floor is a job shop and any sequence involving the three machines is allowable. Some of the other assumptions that are made in the model are as follows:

- The setup times are included in the processing times.
- A part recaptures a machine immediately after a processing step if it requires the same machine. This assumption allows grouping of processing times at a

single machine for operations that are in sequence. Because of this assumption, the processing times of individual steps are not required for scheduling and the sum of all processing times in a group is directly taken as the output from the GAMS model.

- Transportation time between machines was assumed to be 0.
- The system was modeled as a deterministic system and different heuristics were introduced as queue-selection rules.

Initially, to determine the machining capacity, it is required that the throughput be maximised. The SPT queue-selection rule was used at each machine so as to ensure this. Different queue-selection rules were also tried, however, the completion time with other rules was much higher. Once the parts are created, they are sent to the machine according to the sequence they are supposed to follow (each part has a predefined sequence).

After the required number of parts have been created and sent to the required sequence, the parts either get processed, or wait in the system. At this stage, there are interactions between parts, and it is this interaction that causes the actual machining capacity to be below the theoretical values. Machines may be blocked or starved at particular time instants due to the processing sequences.

The machine capacity in this model is in terms of time for which it is available. If the simulations run length exceeds the time span, the iterative procedure mentioned in Figure 15.3 is then employed to decrease the machining capacity in the P2 model. The P2 model is executed again and a new set of optimal processing parameters, lot sizes and sequences are derived for the decreased machining parameters. A *"Results.txt"* file is updated and the simulation model is rerun. This process is repeated until the simulation run time is smaller than or equal to the time span of the planning horizon.

At the end of the iterative procedure, a file *"Start.txt"* will contain a feasible schedule. Another file *"Out.txt"* has data of the profitability that can be achieved using the above procedure. At this point, it can be inferred that the actual machine availability of the machines is only for approximately 82% of the time. For the rest of the time, they are starving.

It may be noted that the bottlenecking machine, M1 is now utilised for approximately 83% of the time. The parts are sent to machines M2 or M3 after processing at M1. For the rest of the time span, it is starved. This is necessary to ensure that all the parts are completed within the specified time span.

CONOPT with a GAMS interface has been used for optimisation [15.2], Arena 7 for simulation [15.6] and MINITAB 14 for statistical analysis [15.3], respectively. The following results are obtained for each case.

1. *Case 1*: The maximum profits that can be generated when different handbook values of spindle speeds are selected after a single process plan has been selected *a priori* are presented in Figure 15.4(a). As can be seen, the profitability can considerably deteriorate if bad machining parameters are chosen. During the process planning stage, usually two spindle velocities are considered – the slow spindle speed with lowest cost per part and the fastest allowed spindle speed. Both reduce the potential profitability by tens of per cent!

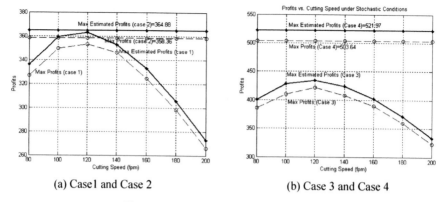

(a) Case1 and Case 2   (b) Case 3 and Case 4

Figure 15.4. Profits versus cutting speed

2. *Case 2*: When machining parameters are varied, the maximum achievable profit is $\hat{Z} = 364.85$ and $\hat{x} = 94$. Simulation was then used to determine the operating schedule and the actual machining capacity under shop-floor conditions. The actual number of parts that can be completed is $x = 92$, thereby generating actual profits of $Z = 358.36$. These are presented in Figure 15.4(a) as a horizontal line. Even though the 1.5% difference between profit in this case and the best profit in the previous case is relatively small, moving the cutting speed decision into the production planning stage eliminates the possibility of choosing a bad speed that can cost tens of per cent.

3. *Case 3*: The maximum profits that can be generated when different handbook values of spindle speeds are selected (multiple process plans) are presented in Figure 15.4(b). The cutting speeds were fixed at 8 levels in the operating region to determine the profitability that can potentially be achieved for those fixed operating parameters. These estimates are made under conditions of zero machine starvation or blocking. Actual profits, however, are slightly lower than these estimated values and are determined by simulating the shop condition (see Figure 15.4(b)) shown in Table 15.1; the profits in Case 3 dominate the profits in Case 1 by about 15% since a more balanced utilisation of capacity is possible.

4. *Case 4*: When multiple process plans are allowed and the machining parameters are varied, $\hat{Z} = 521.97$ and $\hat{x} = 133$. The results of the simulation and the comparison of the actual profits with the estimated profits for both fixed and variable parameters are presented in Figure 15.4(b). The profit obtained in Case 4 is better than the highest profit obtained in traditional Case 1 by more than 30%, that is much higher than just the sum of the improvement of Cases 2 and 3. We should note that during the process planning stage, the cutting speed that optimises the profitability in Case 1 is not known, so the actual improvement from Case 1 to Case 4 can be even higher. Variation of cutting speed becomes more important when the production floor is more complex with more execution alternatives. This leads us to a conjecture that a large-scale system would benefit more from

Table 15.1. Combined results

Cutting speed	Case 1			Case 3		
	Math model exp. profits	Actual capacity	Actual profits	Math model exp. profits	Actual capacity	Actual profits
80	335.93	0.9732	326.94	400.275	0.9636	385.72
100	359.31	0.9729	349.58	428.27	0.9574	410.03
120	363.04	0.9731	**353.28**	434.09	0.9719	**421.90**
140	352.89	0.9811	346.25	424.07	0.9608	408.06
160	332.75	0.9752	324.52	402.27	0.9677	389.29
180	305.34	0.9768	298.26	371.41	0.9687	359.80
	Case 2			Case 4		
Var.	364.88	0.9821	**358.36**	521.97	0.9648	**503.64**

the AHFM approach since (a) they have more opportunities for alternative process plans, and (b) local optimisation in large-scale systems would lead to higher imbalance and loss of production capacity.

Overall, adjusting hierarchical decision making regarding process plans and machining parameters has resulted in a 30% increase in the profitability of the shop-floor operation chosen for the example across Case 1 to Case 4. The actual increase in profitability depends on actual parameters, but AHFM production plans *always* dominate the traditional ones and at least in *some* cases the improvement is measured in tens of per cent without any investment required.

15.2.3 Scalability of AHFM

Scalability of the P2 model can be determined in terms of the computational time and effort required to solve the P2 model as the problem size increases. The methodology used here comprises of two steps. The first step comprises of identifying the factors that can influence the size of the P2 solution space, their impact on the computational effort, and their interaction with each other. The second step comprises of building empirical equations for predicting the computational time and effort required for solving the P2 model for a given problem.

Step 1: A designed experiment has been conducted to determine the factors that contribute to the size and complexity of the model, and the interactions of these factors with each other. The number of part types, number of processing steps per part (assuming that all the processing is conducted on unique machines), the number of process plans that are available per part type (assuming that all the process plans for a single scenario are unique and non-overlapping), the range of the machining parameters, and the number of machines, are the factors identified as contributors to the problem size. The factors and their levels are presented in Table 15.2. It may be noted that four of the five factors are independent: the number of machines (E) is basically the product of factors B and C. The response that is being studied is the computational effort that is required to solve the P2 model. The relationship between the computational effort and the computation time is linear and while the computational time is dependent upon the system configuration, the computational effort is independent of it. In Figure 15.5(a), we plot the relationship between the numbers of iterations and computing time for a Pentium III, 512 MB RAM, 1.1 GHz computer used for this work.

A 2^{4-1} experiment with two replications was conducted using MINITAB with D = ABC as the design generator for the 4 factors (A, B, C, and D) and the ranges presented in Table 15.2. The half-factorial experiment with two replicates at each of the 8 combinations indicates that the factors D, A, and C, and to a certain extent BC (indicator of the number of machines) have a statistically significant effect on the computational effort. The interactions (other than BC that is statistically significant at 0.1) are statistically insignificant. The effect of process plan complexity is also statistically insignificant as far as computational effort is concerned. The Pareto chart indicating the relative contribution of the factors and their interactions is presented in Figure 15.5(b).

Table 15.2. Process parameters (treatment factors) of interest and their level

Ser.	Factor	Range for DOE (−/+)		Range for regression		No. of levels
		Min	Max	Low	High	
1	Number of part types (A)	3	11	10	100	6
2	Number of operations per part type (B)	3	7	3	20	5
3	Number of process plans per part type (C)	1	3	1	5	5
4	Percentage variability in cutting speed (D)	0	100	0	100	7
5	Number of machines (E = BC)	N/A	N/A	5	40	5

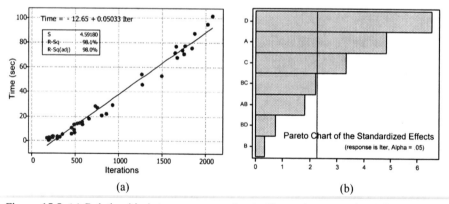

(a)   (b)

Figure 15.5. (a) Relationship between computational effort and computational time, and (b) Pareto chart for effects of the four factors and their interactions on the computational effort to solve P2.

Step 2: To determine the effect that each of these factors, *i.e.* D, A, C, and BC has on the computational effort, regression over a wide range of values that covers most reasonable-size problems was used. The ranges that were considered are provided in Table 15.2. The relationship between the factors D, A, C, and BC and the computational effort required for solving the P2 model (in terms of number of iterations) are presented in Figures 15.6(a)–(d).

Based on the relationships that have been determined in Step 2, it can be concluded that the computational effort is a function of A^2, C, C^2, D and E. The approximate relationship between the computational effort and the statistically significant factors can be determined by fitting a regression equation between the identified factors and their impact and the computation effort. In this integrative regression, C becomes not significant. Using D, A^2, E, and C^2 as the predictors, the regression equation that is obtained is as follows:

$$\text{Iter} = -2408 + 16.5D + 1.03A^2 + 102E - 55.6C^2$$

All the factors D, A^2, E, and C^2 are statistically significant with P values equal to 0.000, the residuals are normally distributed. The adjusted R^2 value, indicative of the quality of the regression estimate is 93.3%. The regression equation indicates that the computational effort scales as a function of $O(A^2+D+E)$ and the number of alternative paths, C, reduces the complexity. The computational complexity of Model P2 enables us to solve industrial-size problems with up to five hundred part types, one hundred machines and a 100% variability range in less than 250,000 iterations that is approximately four hours on a Pentium III, 1.1 GHz PC with 512 MB RAM.

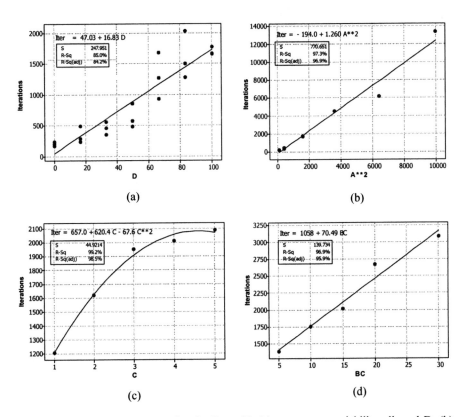

Figure 15.6. Scaling of computational effort with (a) percentage variability allowed D, (b) number of part types A, (c) number of available process plans per part type C, and (d) number of available machines BC = E.

15.3 Plan Perturbation due to New Customers Orders

Most manufacturing firms, be it one of the big three car manufacturers or small shop floors, develop production plans and set production goals that enable them to satisfy the forecasted demand over a given time horizon. The introduction of an additional customer order or the possibility of implementation of unplanned activities within that production plan that may disrupt the original plan is a common occurrence. If additional production capacity is available with the shop floor, the newly introduced task is incorporated in the plan without any difficulty. However, more often than not, the schedule for the shop floor is tight. Shop floors resort to different strategies to deal with the introduction of unplanned activities in the production plan in this scenario. Depending upon the backorder cost of the parts in the original production plan, the planner either drops some part of the previous production plan in order to accommodate the additional demand on resources, or, if the backorder cost is too high, the additional order is not accommodated in that time frame. There is another possibility as well, that though well known, usually goes unexplored – the previous plans can be perturbed, *i.e.* the processing times assigned to the scheduled production can be changed to accommodate the additional load on the machining resources.

Let us consider the following production environment. A company has customers orders D_{nt} for each part n, $n = 1 \ldots N$, in each period t, $t = 1 \ldots T$, and a tentative production plan to produce x_{nt} parts in period t. Production of part n requires I_n process steps; each step i, $i = 1 \ldots I_n$, requires machine y_{ni} for processing time p_{nit}. The machines have a finite capacity K_{mt} for each machine m, $m = 1 \ldots M$, and period t. The holding and backorder costs in period t are h_t and b_t, respectively. A new customer's order for D_n parts arrives. We would like to address the following two questions using the plan-perturbation approach: when could the company supply the order and what will its corresponding operational cost be?

Unlike the common approach of finding an "empty slot" to put the order, we would like to consider a possible change in the process plan when processing time p_{nit} can be changed in order to produce D_n parts in period t. We consider three possibilities:

- Policy P1 – the customer's order can be undivided between periods and the tentative production plan should hold;
- Policy P2 – the customer's order can be divided between periods but the tentative production plan should hold;
- Policy P3 – a completely new production plan could be derived.

15.3.1 Estimation of New Order Cost

Based on the three policies mentioned in the previous section, we have the following algorithms for perturbing the production plans generated by the AHFM approach.

15.3.1.1 Policy P1: Indivisible New Order

(1) *Single period model*
Let us assume that the perturbation cost minimised for time period T when a new indivisible order of product 0 with size x_0 arrives.

Model M1:

$$Z_1^T = \sum_{j=1}^{N} x_{jT} \sum_{i=1}^{o_j} \left(\frac{a_{ijm}^{(C)}}{v_{ijmT}} + b_{ijm}^{(C)} v_{ijmT}^{\alpha_{ijm}} + c_{ijm}^{(C)} \right) +$$

$$\sum_{i,m} x_0 \left(\frac{a_{i0m}^{(C)}}{v_{i0mT}} + b_{i0m}^{(C)} v_{i0mT}^{\alpha_{i0m}} + c_{i0m}^{(C)} \right) \tag{15.32}$$

$$\sum_{j=1}^{N} x_{jT} \sum_{i=1}^{o_j} \left(\frac{a_{ijm}^{(t)}}{v_{ijmT}} + b_{ijm}^{(t)} v_{ijmT}^{\alpha_{ijm}} + c_{ijm}^{(t)} \right) +$$

$$\sum_{i,m} x_0 \left(\frac{a_{i0m}^{(t)}}{v_{i0mT}} + b_{i0m}^{(t)} v_{i0mT}^{\alpha_{i0m}} + c_{i0m}^{(t)} \right) \le K_{mT} \quad \forall m \tag{15.33}$$

Lemma 3. Equations 15.32 and 15.33 are convex with respect to the spindle speeds v_{ijm}.
Proof: Based on Lemmas 1 and 2, cost and processing time are convex functions of v_{ijm}. The sum of convex functions is a convex function. □

Theorem 1. Model M1 can be solved by convex programming.
Proof: Straightforward from Lemma 3. □

Convex programming is scalable for implementing the proposed approach for large-scale real-life problems.

(2) Multiple period cost – due date efficient frontier
The following algorithm provides the decision maker with the cost versus due date efficient frontier when the new order cannot be divided between periods.

Algorithm A1:
    Step 0: Setting the basis: find minimum cost for all operations of the new order, C_{i0m}^0. Find the current cost of existing orders for all time periods $T = 1, ..., TH, C_{ijm}^T$.
    Step 1: For each time period $T = 1, ..., TH$, run Model M1 and find Z_1^T. If infeasible, set $Z_1^T = \infty$.
    Step 2: Set $Z_{P_1}^1 = Z_1^1$. For each time period $T = 2, ..., TH$, find $Z_{P_1}^T$: $Z_{P_1}^T = \min(Z_{P_1}^{T-1} + x_0 h, Z_1^T)$. Going backward, find the production period t_T when the new order is actually processed in order to obtain $Z_{P_1}^T$.
    Step 3: For each time period $T = 1, ..., TH$, find the additional cost for setting the due date of the new order in this period:

$$\Delta C_{P_1}^T = Z_{P_1}^T - \sum_j \sum_i \sum_m C_{ijm}^{t_T} - \sum_i \sum_m C_{i0m}^0 \tag{15.34}$$

Theorem 2. Algorithm A1 finds the optimal solution for Policy P1.
Proof: Algorithm A1 finds the optimal solution when the due date is set to one. By the induction argument, it is easy to see that it finds the optimal solution for all other due dates.   ☐

Unfortunately, straightforward extension of Model M1 to divisible new customer orders is not convex, as shown in the next section.

15.3.1.2 Policy P2: Divisible New Order

For due-date period DD, the model for Policy P2 is as follows:

Model M2:

$$Z_2 = \sum_{T=1}^{TH} Z_2^T = \sum_{T=1}^{TH} \left(\begin{array}{l} \sum_{j=1}^{N} x_{jT} \sum_{i=1}^{o_j} \left(\dfrac{a_{ijm}^{(C)}}{v_{ijmT}} + b_{ijm}^{(C)} v_{ijmT}^{\alpha_{ijm}} + c_{ijm}^{(C)} \right) + \\ \sum_{i,m} x_{0T} \left(\dfrac{a_{i0m}^{(C)}}{v_{i0mT}} + b_{i0m}^{(C)} v_{i0mT}^{\alpha_{i0m}} + c_{i0m}^{(C)} \right) \end{array} \right) \tag{15.35}$$

$$\sum_{j=1}^{N} x_{jT} \sum_{i=1}^{o_j} \left(\dfrac{a_{ijm}^{(t)}}{v_{ijmT}} + b_{ijm}^{(t)} v_{ijmT}^{\alpha_{ijm}} + c_{ijm}^{(t)} \right) +$$

$$\sum_{i,m} x_{0T} \left(\dfrac{a_{i0m}^{(t)}}{v_{i0mT}} + b_{i0m}^{(t)} v_{i0mT}^{\alpha_{i0m}} + c_{i0m}^{(t)} \right) \leq K_{mT} \quad \forall m,T \tag{15.36}$$

$$\sum_{T=1}^{DD} x_{jT} \geq x_0 \tag{15.37}$$

Lemma 4. Equations 15.35 and 15.36 are not convex with respect to the spindle speeds v_{ijm} and x_{0T}.
Proof: Function $x\left(\dfrac{a}{v} + bv^{\alpha} + c\right)$ is not convex with respect to v and x.   ☐

Therefore, Model M2 cannot be solved directly by convex programming. Let us consider discretisation of x_0. This can be done without loss of generality since we produce whole parts only. Let x_{min} be the resolution of the discretization, *i.e.* x_{min} parts are indivisible and should be processed together. When a part of the new order of size x_{prev}^T is already scheduled in period T and the problem is to schedule the new suborder x_{min}, the corresponding model is as follows:

Model M3:

$$Z_3^T = \sum_{j=1}^{N} x_{jT} \sum_{i=1}^{o_j} \left(\dfrac{a_{ijm}^{(C)}}{v_{ijmT}} + b_{ijm}^{(C)} v_{ijmT}^{\alpha_{ijm}} + c_{ijm}^{(C)} \right) +$$

$$\sum_{i,m} \left(x_{prev}^T + x_{min} \right) \left(\dfrac{a_{i0m}^{(C)}}{v_{i0mT}} + b_{i0m}^{(C)} v_{i0mT}^{\alpha_{i0m}} + c_{i0m}^{(C)} \right) \tag{15.38}$$

$$\sum_{j=1}^{N} x_{jT} \sum_{i=1}^{o_j} \left(\frac{a_{ijm}^{(t)}}{v_{ijmT}} + b_{ijm}^{(t)} v_{ijmT}^{\alpha_{ijm}} + c_{ijm}^{(t)} \right) +$$

$$\sum_{i,m} \left(x_{prev}^{T} + x_{min} \right)_0 \left(\frac{a_{i0m}^{(t)}}{v_{i0mT}} + b_{i0m}^{(t)} v_{i0mT}^{\alpha_{i0m}} + c_{i0m}^{(t)} \right) \le K_{mT} \quad \forall m \tag{15.39}$$

Lemma 5. Equations 15.38 and 15.39 are convex with respect to the spindle speeds v_{ijm}.
Proof: Based on Lemmas 1 and 2, cost and processing time are convex functions of v_{ijm}. The sum of convex functions is a convex function. ☐

Theorem 3. Model M3 can be solved by convex programming.
Proof: Straightforward from Lemma 5. ☐

Given the distribution of x_{prev}^{T} and due date DD, the following algorithm finds where to schedule the next x_{min} parts of the new order.

Algorithm A2:
> *Step 0:* Setting the basis: find minimum cost for all operations of the new order, C_{i0m}^{0}. Find the current cost of existing orders for all time periods $T = 1$, ..., TH, C_{ijm}^{T}.
>
> *Step 1:* For each time period $T = 1$, ..., DD, run Model M3 and find Z_3^{T}. If infeasible, set $Z_3^{T} = \infty$.
>
> *Step 2:* Set $Z_4^1 = Z_3^1$. For each time period $T = 2$, ..., DD, find Z_4^{T}:
> $$Z_4^{T} = \min \left(Z_4^{T-1} + x_{min} h, Z_3^{T} \right).$$
>
> *Step 3:* Going backward, find the production period t_T when the new suborder is actually processed in order to obtain Z_4^{DD}.

Lemma 6. Given the distribution of x_{prev}^{T} and due date DD, Algorithm A2 finds the optimal schedule for the next x_{min} parts of the new order.
Proof: Similar to the proof of Theorem 2. ☐

In what follows, Algorithm A3 finds the minimal cost for supplying the complete customer order by due date DD.

Algorithm A3:
> *Step 0:* Setting the basis: find minimum cost for all operations of the new order, C_{i0m}^{0}. Find the current cost of existing orders for all time periods $T = 1$, ..., TH, C_{ijm}^{T}.
>
> *Step 1:* For each $l = 1$, ..., x_0/x_{min}, run Algorithm A2 for scheduling the suborder l. Add the production period to set P.
>
> *Step 2:* When $l = x_0/x_{min}$, set $Z_{P_2}^{DD} = Z_4^{DD}$.

Step 3: Find the additional cost for setting the due date of the new order in this period:

$$\Delta C_{P_2}^{DD} = Z_{P_1}^{DD} - \sum_{T \in P} \sum_j \sum_i \sum_m C_{ijm}^T - \sum_i \sum_m C_{i0m}^0 \tag{15.40}$$

Lemma 7. Given disretisation of x_0 by x_{min} and due date DD, Algorithm A3 finds the optimal solution for Policy P2.

Proof: Let us assume that there is a different solution with better (lower) cost. In this solution, there should be at least one period, say period t_1, with more suborders than in the solution found by Algorithm A3, and should be at least one period, say period t_2, with less suborders than in the solution found by Algorithm A3. If we reschedule the optimal solution taking one suborder from period t_1 to period t_2, the solution is feasible and has the same or lower cost. If the cost is lower, we have a contradiction. If the cost is the same, we repeat this procedure until we either improve the solution or rebuild the solution found by Algorithm A3 – both of them bring us to a contradiction.   □

Algorithm A4 extends A3 to build the cost versus due date efficient frontier for divisible customer orders.

Algorithm A4:
Step 1: For each $DD = 1, ..., TH$, run Algorithm A3.

Theorem 4. Given discretisation of x_0 by x_{min}, Algorithm A4 finds the optimal solution for Policy P2.
Proof: The result directly follows from Lemma 7.   □

15.3.1.3 Policy P3: New Production Plan

This involves solution of the P2 model again (see previous sections).

15.3.2 New Order Insertion Case Study

This subsection focuses on exemplifying the plan perturbation approach for handling a new order insertion. We start with a multiple time horizon planning model that generates a production plan (using AHFM) for the production unit presented in Section 15.2. The specifics of the original plan and the new order are as follows:

1. Planned production for the 5 time periods: with demand as 12, 11, 10, 7, and 9 sets for time period 1–5, respectively.
2. New order size is of 9 units of part type *King*.
3. Holding cost of part type *King* is $0.33/time period for a lot size of 9.

The base case (production plan generated using the AHFM) can be summarised as follows:

Minimum cost for a part type *King*: $7.531

Cost of existing order:

	Period 1	Period 2	Period 3	Period 4	Period 5
Planned sets	12	11	10	7	9
Planned cost	679.778	616.902	558.178	384.657	499.501

Furthermore, the perturbation cost, given that the base case and order size are known, and the state of the system is the same, can be summarised as in Figure 15.7.

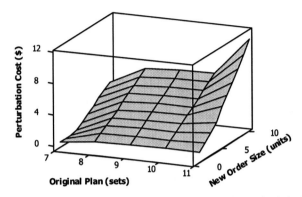

Figure 15.7. Perturbation cost as a function of order size and original plan

We then implement each of the policies 1–2 in cases 1 and 2, respectively.

Case 1: Indivisible new order in a multiple time period model
Implementing Algorithm A1:

Step 1: Identify the perturbation cost for an order of given batch size in each time period for the given planning scenario.

	Period 1	Period 2	Period 3	Period 4	Period 5
Planned sets	12	11	10	7	9
Planned cost	Inf.	694.230	631.522	455.545	572.802

Perturbation Cost:

	Period 1	Period 2	Period 3	Period 4	Period 5
Planned sets	12	11	10	7	9
Perturbation	Inf.	9.549	5.565	3.109	5.522

Step 2: Based on the perturbation costs and the inventory holding costs, identify from the past and the current period, the time period in which the cost for the current time period is the smallest.

	Period 1	Period 2	Period 3	Period 4	Period 5
Planned sets	12	11	10	7	9
Condition	Base	min(Inf., 9.549)	min(12.549, 5.565)	min(8.565, 3.109)	min(6.109, 5.522)
Prod. time	-	2	3	4	5

Step 3: Based on the perturbation cost and the inventory holding cost, determine the cost versus due date frontier.

	Period 1	Period 2	Period 3	Period 4	Period 5
Planned sets	12	11	10	7	9
Perturbation	Inf.	9.549	5.565	3.109	5.522

The results of Case 1 in terms of perturbation cost are shown in Figure 15.8.

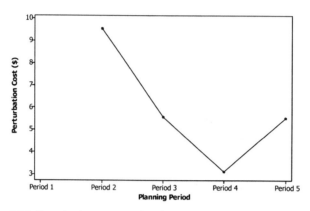

Figure 15.8. Perturbation cost vs. planning period (due-date quote) for Case 1

Case 2: Divisible new order in a multiple time period model

Step 0: Model the base case (same as in the previous case).
Step 1: Determine the perturbation cost matrix.

	Period 1	Period 2	Period 3	Period 4	Period 5
Planned sets	12	11	10	7	9
1 Insertion	Inf.	0.679	0.618	0.000	0.605
2 Insertions	Inf.	1.419	1.236	0.000	1.211
3 Insertions	Inf.	2.225	1.854	0.000	1.876
4 Insertions	Inf.	3.135	2.498	0.225	2.434
5 Insertions	Inf.	4.687	3.116	0.789	3.051
6 Insertions	Inf.	5.706	3.707	1.417	3.669
7 Insertions	Inf.	6.862	4.325	1.981	4.287
8 Insertions	Inf.	8.140	4.942	2.545	4.905
9 Insertions	Inf.	9.549	5.565	3.109	5.522

Step 2: Based on the perturbation cost matrix and the inventory holding costs, identify from the past and the current period, the insertion units in the time periods in which the cost for the current time period is the smallest.

	Period 1	Period 2	Period 3	Period 4	Period 5
Planned sets	12	11	10	7	9
Split	None	None	None	None	4 in 4 and 5 in 5
Cost	Inf.	9.549	5.565	3.109	4.609

Comparisions of perturbation costs in Case 1 and Case 2 at different order sizes, 9 and 25, are illustrated in Figures 15.9 and 15.10, respectively.

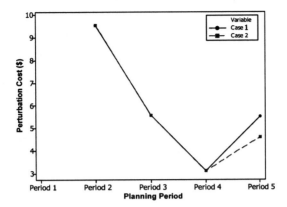

Figure 15.9. Comparison of perturbation costs for Cases 1 & 2 (order size: 9 *Kings*)

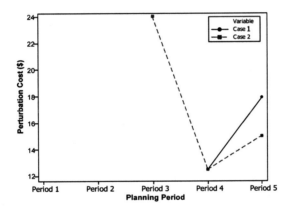

Figure 15.10. Comparison of perturbation costs for Cases 1 & 2 (order size: 25 *Kings*)

15.4 Extending the Applicability of AHFM

The AHFM has been applied to process planning in the food industry as well as for fleet scheduling [15.7]. This section will be focusing on fleet scheduling.

Fleet assignments determine the type of aircraft to operate for flights in a given schedule, subject to a variety of constraints. Airlines usually determine the timings for their flights to respond to time-dependent demand and the requirements of frequency plans, available fleets, and aircraft routings. Nevertheless, delays and rerouting are unavoidable. Adjustments have to be made by altering some flights from their so-called optimal times. Scarce runway time slots represent a resource whose value can be determined from the impact of such rescheduling on the objective function of the original schedule. Cao and Abid [15.8] analysed the relationship between rescheduling of flights and airline profit.

The AHFM principles are applied to a transportation problem in this section. In most aircraft fleet-scheduling models, the aircraft speed is assumed to be a given fixed parameter. This is usually the most economic speed, *i.e.* which reduces fuel consumption per mile. However, most airplanes can fly at higher speeds (at least 5% to 25%) depending on the type of aircraft. Of course, higher speed results in higher costs, sometimes significantly higher, up to 50% from costs of the most economical speed. For the preliminary check of the concept, we extended a basic daily fleet-scheduling problem that can be formulated as follows: given (a) the size of fleet of each aircraft type, (b) daily schedule of flights between airports, (c) costs associated with an assignment of particular aircraft type to particular flight, (d) penalty costs of unscheduled flights, and (e) maintenance time for each aircraft type in each airport, what is the optimal assignment of airplanes to flights that minimises the total cost. When the schedule is feasible, the penalty cost should be high enough to ensure assignment of all flights. However, when the schedule is infeasible for given fleets, the penalty cost can directly affect the optimal solution. The model is solved as a minimum-cost network flow problem where nodes represent airports in each time interval and arcs represent scheduled flights and dwells (where airplanes are on the ground and remain at the same airport until the next time interval) as shown in Figure 15.11.

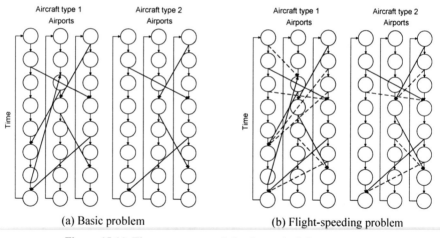

Figure 15.11. Time–space network for fleet-scheduling problems

A small-scale model of three cities and three fleets was taken from the AMPL test problem site [15.9] (Airline fleet-assignment problem, files *fleet2i.mod* and *fleet2.dat*). It was solved using the GAMS interface and OSL solver. In this problem the flight schedule is feasible, *i.e.* all flights can be scheduled. In the parameter variation model, we assume that the flight times can be reduced by 10 minutes with costs increased by 15% or reduced by 20 minutes with costs increased by 50%. As shown in Table 15.3, the total cost was reduced by 25% by allowing speeding of the flights. If we reduce the fleet size, making the flight schedule infeasible, *e.g.* due to technical failures, flight speeding can help in keeping the schedule or, at least, reduce the costs, as shown in Table 15.3.

Table 15.3. Results for the fleet-assignment problem

	Fleet sizes	Traditional approach	Parameter variation
Original problem	72S-6, 73S-6, L10.2	2068.00 (100%)	1576.00 (76%)
Infeasible problem 1	72S-3, 73S-3, L10.1	2353.00 (100%)	1613.65 (69%)
Infeasible problem 2	72S-1, 73S-1, L10.1	3038.00 (100%)	2236.85 (74%)

15.5 Conclusions

In this chapter, the extension of the AHFM approach for integrated planning in a variety of industries was explored. Using a small set of problems, it has been demonstrated that the model formulation, although not generic, can be easily modified to accommodate a variety of system specifics, *i.e.* manufacturing, processing and travel. The problems that have been illustrated reflect the actual characteristic of such systems, but the size of the problems was reduced for illustration. A major part of the focus of this chapter was to uncover the potential savings that may be gleaned from such a technique. Because the opportunity (increased profitability from 25–85% for realistic small-scale problems), there is a strong indication that heuristics for large-scale problems should be developed as the scale exceeds the capacity of today's solvers. The potential savings for integrating the planning domain are too large to ignore this critical aspect of the problem.

The aggregation of the production planning models becomes more and more complex as we increase the number of hierarchies that the aggregate model covers. It is essential to limit the complexity of the problem with an emphasis on the tractability and computational complexity. Analysis of the special properties of the models, such as unimodal regions, is required. At the end of Section 15.2, it is shown that the computational time results for systems with increased size (number of parts, number of alternative plans, variability of the range of parameters and the number of machines). As expected, when the size of the problem increased, the solution time increases. The increase of solution time appears to grow linearly as the problem increases in the number of machines and the number of operations in a part plan. The computational time increased quadratically as the number of parts increased. Solutions for problems of a large size (20 machines and 100 part types) were obtained in less than an hour using a personal computer. It appears that this procedure can easily be scaled to solve real manufacturing problems.

References

[15.1]   Benjaafar, S., 1996, "On production batches, transfer batches and lead times," *IIE Transactions*, **28**(5), pp. 357–362.

[15.2]   Brooke, A., Kendrick, D., Meeraus, A. and Raman, R., 1998, *GAMS: A Users Guide*, GAMS Development Corporation, http://www.gams.com/.

[15.3]   Carver, R.H., 2004, *Doing Data Analysis with MINITAB 14*, Duxbury Press, Belmont, CA, USA; 2nd edn.

[15.4]   Chang, T.C., Wysk, R.A. and Wang, H.P., 1992, *Computer Aided Manufacturing*, Prentice-Hall, Inglewood, NJ, USA.

[15.5]   Johnson, D., 2003, "A framework for reducing manufacturing throughput Time," *Journal of Manufacturing Systems*, **22**(4), pp. 283–298.

[15.6]   Kelton, D.W., Sadowski, R. and Sadowski, D., 2001, *Simulation with ARENA*, McGraw-Hill Education, New York, NY, USA; 2nd edn.

[15.7]   Masin, M., Shaikh, N.I. and Wysk, R.A., 2003, "Aggregative parameter variation in optimisation modelling," *IEEE Transactions on Robotics and Automation*, **19**(4), pp. 529–542.

[15.8]   Cao, J. and Abid, K., 2000, "Value of runway time slots for airlines," *European Journal of Operational Research*, **126**(3), pp. 491–500.

[15.9]   http://www.ampl.com/cm/cs/what/ampl/NEW/ILOG/.

16

STEP into Distributed Manufacturing with STEP-NC

Xun Xu

Department of Mechanical Engineering
School of Engineering, University of Auckland
Private Bag 92019, Auckland, New Zealand
Email: x.xu@auckland.ac.nz

Abstract
As the technological information of enterprises becomes more and more distributed, continuous data acquisition in decentralised computer numerical control (CNC) machine tools and databases is necessary for a distributed manufacturing system. This chapter focuses on an emerging ISO standard, informally known as STEP-NC, and the role it plays in closing the gap between design and manufacturing for a distributed and collaborative manufacturing environment. This new standard defines a new generation of NC programming language and is fully compliant with STEP. STEP-NC brings richer information to the CNC machine tools, hence intelligent machining and control are made possible. Its Web-enabled feature gives itself an additional dimension in that distributed manufacturing can be readily supported. A case study toward the end demonstrates such a system. The system adopts a three-tiered, Internet-based network architecture, so that designing and manufacturing data can be exchanged over the Internet. A STEP-NC adapter has been developed to convert generic manufacturing information (at the task level) into native manufacturing information (at the method level) for a chosen machine tool. A native CNC system database has been developed to capture the conditions and capabilities of a machine tool and cutting tool. This database is the foundation of the whole system since it contains all the native information about a specific CNC machine.

16.1 Introduction

Manufacturing system's configuration has undergone a number of revolutionary changes since the days of craft production in the 1880s. The most recognised was of course the dedicated transfer (machine) line, which enabled mass production at high efficiency and low cost. With the need of the 1970s and 1980s to produce a wider range of differing parts, "flexible" manufacturing was developed to meet the needs of producing smaller batches of different parts. These systems used groups of computer numerically controlled (CNC) machines that could be reprogrammed to make different parts and are often combined with automated transport systems and storage. These CNC machines became the central elements in the systems such as flexible transfer lines, flexible manufacturing systems (FMS) and flexible manufacturing cells (FMC).

However, the new era of manufacturing industry at the turn of the century, is characterised as highly dispersed and changeable. Increasingly, the industry is led by a group of smaller, more agile corporations that springs up for specific purposes, exists while the market sustains the new product and then gracefully disbands as the market changes. Needless to say, manufacturing systems in this new era are required to be robust, distributed and intelligent.

Li *et al.* [16.1] described a distributed manufacturing environment (Figure 16.1), in which factories possessing various machines and cutting tools are at different geographical locations, and different manufacturing capabilities are often selected to achieve the highest production efficiency. When jobs requiring several operations are received, feasible process plans are produced by available factories according to the precedence relationships of those operations. Manufacturing operations can be performed by different machines and cutting tools located at different locations. The final optimal or near-optimal process plan is decided after comparison of all the feasible process plans [16.1]. Different factories may have different optimal ways of producing a part. For example, given the job as shown in Figure 16.1 [16.1] that contains *i* number of features and *j* number of operations, there are three different ways for Factory 1 to produce the part in an optimal fashion with four different machine tools and three different cutting tools. To arrive at a globally optimised solution in such a distributed manufacturing environment, infrastructure, data model and hardware are the three key elements.

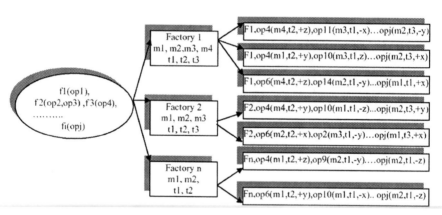

fi – feature ID, opj – operation ID, Factory n – factory ID
mk – machine tool ID, tp – cutting tool ID, –/+x, y, z – tool approach direction (TAD)

Figure 16.1. Distributed manufacturing system

CNC machine tools are the main piece of hardware in any manufacturing system. There are demands and new opportunities to empower the current CNC machines with the much-needed features such as distributability, interoperability and reconfigurability. To this end, the major issues are product data compatibility/ interoperability and machine tool adaptability. Up to now little research has been carried out in this field, but due to the developments of the new CNC data model known as STEP-NC, there has been a surge of research activities in trying to address the above-mentioned issues. This chapter reports on these research activities and tries to address the issues related to distributability, interoperability and reconfigurability for CNC machine tools.

16.2 Impediments of Current CNC Technologies

Today's CNC machine tools are well developed with capabilities such as multi-axis control, error compensation and multi-process manufacture (*e.g.* combining mill/turn/laser and grinding operations). In the meantime, these capabilities have made the programming task increasingly more difficult and machine tools themselves less adaptable. Some effort has been made to alleviate this problem, in particular the trend towards open and distributable architecture control, based on OSACA (Open System Architecture for Controls within Automation systems) [16.2] and OMAC (Open Modular Architecture Controller) [16.3] where third-party software can be used at the controller working within a standard windows operating system. One further recognisable industrial development is the application of software controllers, where PLC logic is captured in software rather than in hardware.

Although these developments have improved software tools and the architecture of CNC systems, vendors and users are still seeking a common data model to support all aspects, in particular the CAM/CNC end, of manufacturing processes with no information loss. Though there are many CAM tools that provide varying shop-floor programming functionalities [16.4]–[16.6], the problem of adaptability and interoperability between systems was and is still seen as one of the key issues in supporting distributed manufacturing.

CNC machine tools complete the product design and manufacturing lifecycle, and more often than not they have to communicate with upstream sub-systems such as CAM, CAPP and even CAD. In the case when neutral data exchange protocols such as SET (Standard Exchange and Transfer), VDA (Verband des Automobilindustrie: German Automobile Industry Association), and IGES (Initial Graphics Exchange Specification) are used, information exchange can happen between heterogeneous CAD and/or CAM systems. This is, however, only partially successful since these protocols are mainly designed to exchange geometrical information and are not totally suitable to all the needs of the CAD/CAPP/CAM industry. Thus, the international community developed the ISO 10303 [16.7] set of standards, well known as STEP (STandard for the Exchange of Product data).

By implementing STEP AP-203 [16.8] and STEP AP-214 [16.9] within CAD systems, the design data exchange barrier is removed. However, data-exchange problems between CAD/CAM and CNC systems remain unsolved. CAD systems are designed to describe the geometry of a part precisely, whereas CAM systems focus on using computer systems to generate plans and control the manufacturing operations according to the geometrical information present in a CAD model and the existing resources on the shop floor. The final result from a CAM system is a set of CNC programs that can be executed on a CNC machine. STEP AP-203 and STEP AP-214 only unify the input data for a CAM system. On the output side of a CAM system, a fifty-year-old international standard ISO 6983 (known as G-Code or RS274D) [16.10] still dominates the control systems of most CNC machines. Outdated yet still widely used, ISO 6983 only supports one-way information flow from design to manufacturing. The CAD data are not utilised at a machine tool. Instead, they are processed by a postprocessor only to obtain a set of low-level, incomplete data that makes modification, verifications and simulation difficult. The changes made at the shop floor cannot be directly fed back to the designer. Hence, invaluable experiences on the shop floor cannot be preserved and reutilised.

The ISO 6983 standard focuses on programming the path of the cutter centre location (CL) with respect to the machine axes, rather than the machining tasks with respect to the part. Thus, it defines the syntax of program statements, but in most cases leaves the semantics ambiguous, together with low-level limited control over program execution. These programs, when processed in a CAM system by a machine-specific postprocessor, become machine dependent. In order to enhance the capability of a CNC machine, CNC controller vendors have also developed their own tailored control command sets to add more features to their CNC controllers to extend ISO 6983. These command sets once again vary from vendor to vendor resulting in further incompatible data among the machine tools.

The current inflexible CNC control regime means that the output from a CAM system has no adaptability, which in turn prevents the CNC machine tools from having any distributability and interoperability. The main reason is that a G-code based part program only contains low-level information that can be described as "how-to-do" information. The CNC machine tools, no matter how capable they are, can do nothing but "faithfully" follow the G-code program. It is impossible to perform online, real-time intelligent control, nor machining optimisation. It is time that a new NC data model was developed that can describe "what-to-do" information, *i.e.* task-level information.

16.3 The STEP-NC Standard

Today, a new standard namely ISO 14649 [16.11]–[16.16] recognised informally as STEP-NC is being developed by vendors, users and academic institutes worldwide to provide a data model for a new breed of intelligent CNCs. The data model represents a common standard specifically aimed at NC programming, making the goal of a standardised CNC controller and NC-code generation facility a reality. Currently, two versions of STEP-NC are being developed by ISO. The first is the Application Reference Model (ARM) (*i.e.* ISO 14649) and the other Application Interpreted Model (AIM) of ISO 14649 (*i.e.* ISO 10303 AP-238 [16.17]). For more information on the use and differences between them readers are referred to [16.18]–[16.20].

Contrary to the current NC programming standard (ISO 6983), ISO 14649 is not a method for part programming and does not normally describe the tool movements for a CNC machine. Instead, it provides an object-oriented data model for CNCs with a detailed and structured data interface that incorporates feature-based programming where a range of information is represented such as the features to be machined, tool types to be used, the operations to perform, and the sequence of operations to follow. Though it is possible to closely define the machine tool trajectory using STEP-NC, the aim of the standard is to allow these decisions to be made at a latter stage by a new breed of intelligent controller – a STEP-NC controller. It is the aim that STEP-NC part programs may be written once and used on many different types of machine tool controllers providing that the machines have the required process capabilities. In doing this, both CNC machine tools and their control programs are made adaptable, interoperable and distributable. Figure 16.2 illustrates that in contrast to a conventional system, both geometric and machining information can now be bidirectionally transferred between a CAD/CAM system and a STEP-NC controller [16.21][16.22]. One critical issue is that the tool-path movement information is optional and ideally should be generated at the machine by the STEP-NC controller.

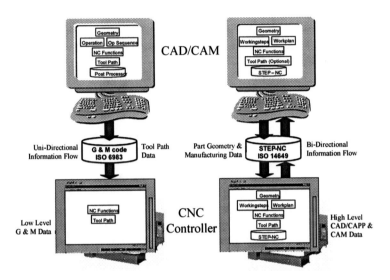

Figure 16.2. Bidirectional information flow with STEP-NC

In STEP-NC, geometric information is defined by machining features (similar to AP-224 [16.23]) with machining operations termed "Workingsteps" performed on one or more features. These *Workingsteps* provide the basis for a "Workplan" to manufacture the component. It is this type of data that is transferred into and out of a STEP-NC intelligent controller. It would be interpreted by the controller, enabling CNC operators to interact at the Workingstep (*i.e.* machining operation) level via an intelligent MDI (Manual Data Interface) or CAD/CAM system at the controller.

To summarise, some of the benefits with using STEP-NC are as follows [16.24]:

- STEP-NC provides a complete and structured data model, linked with geometrical and technological information, so that no information is lost between the different stages of the product development process.
- Its data elements are adequate to describe task-oriented NC data.
- The data model is extendable to further technologies and scalable (with conformance classes) to match the abilities of a specific CAM, SFP (Shop Floor Programming) or NC.
- Machining time for small-to-medium-sized job lots can be reduced because intelligent optimisation can be built into the STEP-NC controllers.

- The postprocessor mechanism will be eliminated, as the interface does not require machine-specific information.
- Machine tools are safer and more adaptable because STEP-NC is independent of machine tool vendors.
- Modification at the shop-floor can be saved and fed back to the design department hence bidirectional information flow from CAD/CAM to CNC machines can be achieved.
- XML files can be used as an information carrier, and hence can enable Web-based distributed manufacturing.

A detailed discussion on value proposition for STEP-NC can be found in a report produced by the OMAC STEP-NC Working Group [16.25] and other publications [16.20][16.21][16.24].

16.4 STEP-NC Implementation Methods

Like other parts of the STEP standard, the STEP-NC data model is constructed based on the EXPRESS language [16.26]. The EXPRESS language is a formal language for the definition of entity-attribute data models. It is a completely generic modeling language and can therefore be used to model data objects of any type. The STEP-NC EXPRESS information model is organised into schemas. These schemas contain model definitions and serve as a scooping mechanism for subdivision of STEP-NC models. EXPRESS also gives STEP-NC an object-oriented flavour. Its inheritance can be illustrated by the definition of manufacturing features defined in STEP-NC [16.12]. For every 2½D manufacturing feature, there has to be a feature placement value. Therefore, it is defined at the top level (in the *Two5D_ manufacturing_feature* entity). This attribute is inherited by all the "child" entities, *i.e.* machining, replicate and compound features. Similarly, each sub-type of machining features will have an elementary surface to define its depth, and it is defined once for all at the *machining_feature* level.

EXPRESS language does not define any implementation methods. Therefore, additional implementation methods are defined to describe STEP-NC instances for building product exchange models, *e.g.* ISO 14649 models and ISO 10303 AP-238 models. There are several implementation technologies available:

1. A product model specific file format called Part 21 physical file [16.27];
2. A variety of programming language bindings that allow an application programmer to open a data set and access values in its entity instances. Bindings have been developed for C, C++ and Java [16.28]–[16.31];
3. Three methods for mapping EXPRESS-defined data into XML, described by Part 28 Edition 1 [16.32]; and
4. The XML Schema-governed representation of EXPRESS described by Part 28 Edition 2 [16.33].

The last two implementation methods are of particular relevance in supporting distributed manufacturing environment, though the first two are more fundamental to STEP-NC implementations.

16.4.1 Part 21 Physical File Implementation Method

STEP Part 21 defines the basic rules of storing EXPRESS/STEP data in a character-based physical file. Its aim is to provide a method so that it is possible to write EXPRESS/STEP entities and transmit those entities using normal networking and communication protocols (*i.e.* FTP (File Transfer Protocol), e-mail and HTTP (Hypertext Transfer Protocol)).

A Part 21 file does not have any EXPRESS schemas included. It only defines the relationships between entities that are defined by the intended external EXPRESS schemas. The Part 21 file format uses the minimalist style that was popular before the advent of XML. In this style, the same information is never written twice so that there is no possibility of any contradictions in the data. Each entity instance in a Part 21 file begins with a unique Entity ID and terminates with a semicolon ";". The Entity ID is a hash symbol "#" followed by an integer and has to be unique within the data exchange file. The Entity ID is followed by an equal symbol ("=") and the name of the entity that defines the instance. The name of the instance is then followed by the values of the attributes listed between parentheses and separated by commas. The following is the excerpt of a STEP-NC ARM file.

```
ISO-10303-21;
HEADER;
FILE_DESCRIPTION(('A STEP-NC milling file'),'1');
FILE_NAME('sample_part1.stp',$,('AUMS'),(''),'Prototype
    Mill','','');
FILE_SCHEMA(('STEP-NC milling schema'));
ENDSEC;
DATA;
// Project and Workplan
#1=PROJECT('Contour',#2,(#3));
#2=WORKPLAN('Work plan',(#4),$,#5);
#3=WORKPIECE('Workpiece',#6,0.01,$,$,#8,());
// Workingsteps
#4=MACHINING_WORKINGSTEP('Rough Contour',#13,#16,#17);
#5=SETUP('main_setup',#44,#48,(#51));
#6=MATERIAL('ST-50','Steel',(#7));
#7=PROPERTY_PARAMETER('E=200000 N/mm^2');
#8=BLOCK('Block',#9,260.000,210.000,110.000);
// Geometric data
#9=AXIS2_PLACEMENT_3D('BLOCK',#10,#11,#12);
............

// Manufacturing features
#16=GENERAL_OUTSIDE_PROFILE('Profile',#3,(#17),#18,#22,$,$,$
    ,$,#23,$,$);
// Operation data
#17=SIDE_ROUGH_MILLING($,$,'Contour
    profile',#38,10.000,#39,#40,#43,$,$,$,20.000,5.000,0.000)
    ;
#18=AXIS2_PLACEMENT_3D('Position of contour',#19,#20,#21);
#19=CARTESIAN_POINT('Position of
    contour',(40.000,90.000,100.000));
```

```
#20=DIRECTION('',(0.0,0.0,1.0));
#21=DIRECTION('',(1.0,0.0,0.0));
#22=TOLERANCED_LENGTH_MEASURE(20.000,$,$,$);
#23=COMPOSITE_CURVE('Contour Profile',(#24,#25,#56),.F.);
............

// Tool data
#40=CUTTING_TOOL('Endmill 10mm',#41,(),(50.000),50.000);
#41=TAPERED_ENDMILL(#42,3,.RIGHT.,.F.,$,$);
#42=TOOL_DIMENSION(10.000,$,$,$,$,$,$);
// Machining technology
#43=MILLING_TECHNOLOGY($,.TCP.,$,3.3333,$,0.10,.T.,.F.,.F.);
#44=AXIS2_PLACEMENT_3D('Reference point to Machine
   zero',#45,#46,#47);
#45=CARTESIAN_POINT('',(20.000,30.000,10.000));
............

#56=COMPOSITE_CURVE_SEGMENT(.CONTINUOUS.,.T.,#57);
#57=POLYLINE('Second cut of the
   contour',(#29,#30,#31,#32,#33,#27));
ENDSEC;
END-ISO-10303-21;
```

16.4.2 Data Access Implementation Methods

The STEP Data Access Interface (SDAI) reduces the costs of managing integrated product data by making complex engineering applications portable across data implementations. Currently, four international standards have been established for SDAI,

- Standard data access interface [16.28];
- C++ language binding to the standard data access interface [16.29];
- C language binding of standard data access interface [16.30]; and
- Java programming language binding to the standard data access interface with Internet/Intranet extensions [16.31].

Each standard defines a specific way of binding the EXPRESS data with a particular computer programming language. Binding is a terminology given to an algorithm for mapping constructs from the source language to the counterparts of another. Generally speaking, the binding defined in SDAI can be classified into early and late binding. The difference between them is whether the EXRESS data dictionary is available to the software applications. There is no data dictionary in an early binding, whereas in a late binding, the EXPRESS schema definition is needed by late binding applications at run time. For example, the SDAI for C++ and Java language binding is a typical early-binding approach; while the SDAI for C language binding is a late-binding approach.

The early-binding approach generates specific data structure according to the EXPRESS schemas and the programming language definitions. The entities defined in EXPRESS schemas are converted to C++ or Java classes. The inheritance properties in the EXPRESS schemas are also preserved in those classes. The advantage of an early binding is that the compiler of the programming language can perform additional type checking. But because of the complexities of EXPRESS schemas, the initial preparation, compiling and linking of an early-binding approach can be time consuming.

The late-binding approach, on the other hand, does not map EXPRESS entities into classes. It uses EXPRESS entity dictionaries for accessing data. Data values are found by querying those EXPRESS entity dictionaries. Only a few simple functions need to be defined in the late-binding approach to get or set values. A late-binding approach is suitable for a programming language that does not have strong type checking such as C language or an environment that may have multiple EXPRESS schemas.

A mixed-binding approach may provide the advantages of an early binding (compile-time type checking and semantics as functions in a class) and late binding (simplicity). For example, a mixed binding takes advantage of the observation that applications rarely use all of the structures defined by an AP AIM (*e.g.* AP-238). The subset of structures that are used, called the working set, can be early-bound, while the rest of the AP is late-bound. All data is still available, but the application development process is simplified. The number of classes and files that are needed are reduced dramatically, resulting in quicker compilations, simpler source control and more rapid development.

16.4.3 XML Implementation Method (*Part 28 Edition 1*)

XML consists of different rules for defining semantic tags that breaks a document into parts and identifies the different parts of the document. Furthermore, it is a meta-markup language that defines a syntax in which other field-specific markup languages can be written [16.34]. Essentially, XML defines a character-based document format. The following is a simple XML document defining a milling cutter,

```
<?xml version="1.0"?>
    <MILLING_TOOL>
        MILL 18MM
    </MILLING_TOOL>
```

XML is flexible because there is no restriction to these tag names. Hence, it is possible to assign more human-understandable tag names in an XML document, while computers just interpret an XML document according to a predefined formula. It is obvious that the use of meaningful tags can make an XML document human understandable as well as computer interpretable.

When representing EXPRESS schemas, Part 28 (Edition 1) [16.32] specifies an XML markup declaration set based on the syntax of the EXPRESS language. EXPRESS text representation of schemas is also supported. The markup declaration sets are intended as formal specifications for the appearance of markup in conforming XML documents. These declarations may appear as part of Document Type Definitions (DTDs) for such documents.

Like the method used in SDAI, STEP Part 28 (Edition 1) [16.32] defined two broad approaches for representation of data corresponding to an EXPRESS schema.

One approach is to specify a single markup declaration set that is independent of the EXPRESS schema and can represent data of any schema. This approach is called XML late binding. The second approach is to specify the results of the generation of a markup declaration set that is dependent on the EXPRESS schema. This approach is called XML early binding. STEP Part 28 (Edition 1) defines one late-binding approach and two early-binding approaches.

16.4.4 XML Implementation Method (*Part 28 Edition 2*)

It has soon become evident that the use of DTD syntax to specify mappings of EXPRESS to XML as prescribed in Part 28 (Edition 1), results in a sub-optimal solution. Recognising the limitations of the first edition, such as those discussed in the previous section, ISO has begun to work on the second edition of Part 28 employing W3C XML Schema. The Part 28 (Edition 2) EXPRESS-to-XML-schema mapping and configuration language [16.33] is still under development. The main theme of the new implementation method is its two-level method. At the lower level CAD authoring systems can continue to read and write STEP datasets. The only difference on this level is that these datasets can now have an XML format to make them more compatible with the higher level. At the upper level the datasets are modularised by inserting information from the mapping tables into the XML data to explain the meaning of each entity sequence. The new method can open up the definition of an Application Protocol into a series of interconnected XML schemas.

This method is implemented using two languages, a configuration language for describing how to map EXPRESS information into an XML-defined form, and the existing STEP mapping table language converted into an XML form.

16.4.5 Recap – Issues Concerning STEP-NC in XML Format

The introduction of XML technology aids the development of the new type of machine tool system based on STEP-NC. XML has been extensively researched in various fields. Therefore, there are many software packages and guides that are readily available as well as reliable. Because of this, development of a system using XML becomes much easier and faster.

However, STEP-NC defines and stores CAD/CAM and CNC data as a graph structure. This structure allows an item to be referenced by one or more items. XML has a tree structure that does not take multiple references. The generation of an individual element in every duplicated reference of an item will be the simplest mapping method. But in this case, the size of a generated XML file will increase and an operator may be easily confused by the duplicates. This means that unnecessary items should not be duplicated in the conversion process. These may be duplicated, for the convenience of the intelligent controller, and may include,

- Any item that is connected to several different Workingsteps. For example, a machining feature referenced by a rough and finish machining Workingstep; and
- Any geometrical items.

16.4.6 Recent Research Publications

There has been a trend of using XML [16.32] (or rather ISO 10303 Part 28) instead of EXPRESS language (or ISO 10303 Part 21 [16.27]) to represent the STEP-NC information. The reason for this is obvious. The XML processing ability can easily support the e-manufacturing scenario and in particular distributed process planning activities. CNC machine tools can share information with other departments in and outside the company over the Internet/Intranet.

Wosnik *et al.* presented a generic approach to pre-process and feedback process data from servo drives to CNCs and CAPP systems in order to enable the higher-level functionality of these systems [16.35]. This approach rests on the application of open digital servo drives in machine tools and the design of application-dependent algorithms to process and exchange drive signals for both online and offline optimisation of machining processes. Also discussed are the effects of the proposed methodology on design characteristics of open servo drives and CNC kernels as well as on the data interfaces. The methodologies in their paper are exemplified by their application in drive-spanning process force reconstruction.

Fichtner *et al.* [16.36] used agent and neural network technologies to acquire and prepare distributed NC information for STEP-NC based process planning. The paper suggests that the problem of current technological databases can be solved with the help of agent technology and machine learning for distributed information on shop-floor and planning departments. As a result, a combination of agent-based organisation and self-learning of feature-based technological information was introduced.

A 5-axis STEP-NC milling machine that is run by STEP-NC in XML format has been developed at the ERC-ACI (Engineering Research Centre for Advance Control and Instrumentation) in Seoul National University [16.37]–[16.40]. The controller of this 5-axis STEP-NC milling machine can execute machining operations automatically. The interpreter for STEP-NC in XML format has a user-friendly interface. The system was tested using two XML files, one 2.5D and the other 3D. These two files were written using an XML editor from the developed interpreter. Both test files were machined automatically by the STEP-NC controller, free of G-code.

The above research outcomes have made initial inroads into supporting distributed CAPP based on STEP-compliant data models. There is, however, a lack of effort to address the problems at the system level whereby a "bigger picture" of STEP-compliant distributed CAPP is envisaged and experimented. The following sections present such a piece of research.

16.5 A STEP-compliant CAPP System for Distributed Manufacturing

The STEP-compliant CAPP system developed at the University of Auckland [16.41] [16.42] aims to support a distributed manufacturing environment. The system adopts a three-tiered, Web-based network architecture (Figure 16.3). The client tier consists of a set of applications and a Web browser, enabling interactions between users and the system. The process plans that are used as an input to a CNC system are described in accordance with the STEP-NC AIM standard. Instead of low-level information as stipulated by ISO 6983, higher-level information, such as machining features, Workingsteps and Workplans, is used to constitute a process plan. A database structure has been developed for both generic and native manufacturing information and XML is used to represent the STEP-NC information in these databases.

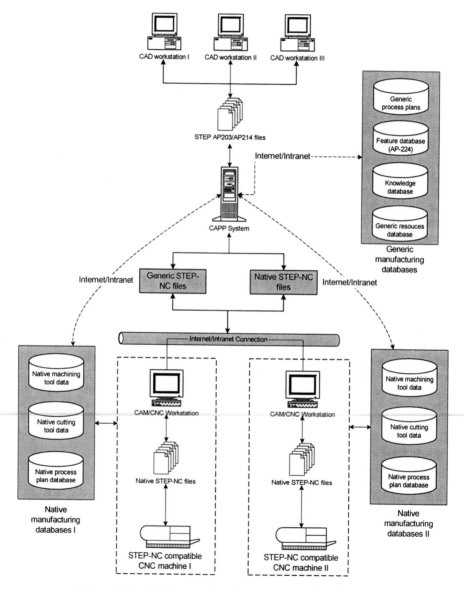

Figure 16.3. A STEP-compliant distributed manufacturing model

In the design phase, STEP AP-203 or AP-214 is used as the neutral data format to exchange design data between different CAD systems or between CAD and CAPP systems. Two different manufacturing information databases (generic and native) co-exist to support CAPP data exchange. The generic manufacturing databases contain abstract information about machine tools and cutting tools of any kind. Hence, process plans generated using generic manufacturing resources cannot be executed directly at the shop floor. This is because a STEP-NC based process plan at this stage has only information about "what-to-do", *i.e.* the tasks. Examples of what-to-do information include machining features and the description of requirements of machine tool(s) and cutting tool(s). At this stage, no information about selection and determination of specific machine tool(s) and cutting tool(s) is present in the STEP-NC program. Generic process plans are therefore machine independent. The native manufacturing databases reflect the actual conditions and capabilities of a shop floor including existing machine tools and cutting tools that can be used for "populating" and optimising a generic process plan so as to generate native process plans for final execution. To this end, a native manufacturing database can be considered as a "DNA" bank for all the available manufacturing facilities.

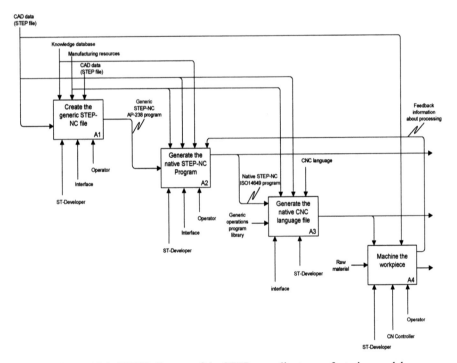

Figure 16.4. IDEF0 diagram of the STEP-compliant manufacturing model

As the CAPP system is designed based on the three-tiered architecture, it has the ability to switch between different manufacturing databases to generate generic or native STEP-NC files to provide a maximal flexibility to support distributed and collaborative manufacturing. As shown in dashed lines in Figure 16.3, when the CAPP system is connected to a generic database, the output STEP-NC files will be universal and machine tool independent. Under this condition, the CAM system can later populate and optimise a generic STEP-NC file based on the native manufacturing database on the shop floor to obtain a suitable STEP-NC file for a specific CNC machine. When the CAPP system is directly connected to a native manufacturing database, it will be able to optimise the machining sequence, select machine tools and cutting tools at the process planning stage and generate a STEP-NC file that can be directly used by a targeted CNC machine. Figure 16.4 shows the detailed information flow in the system.

In this scenario, CAM systems are more likely to be integrated with STEP-NC enabled CNC machines or rather their controllers. The main functions of a CAM system are therefore to optimise the generic STEP-NC information and offer data connections to a CAPP system instead of calculating tool trajectories and generating CNC programs, which will be handled by the built-in functions of the STEP-NC controller.

16.5.1 System Model

The abstract model of the proposed STEP compliant manufacturing system is illustrated in Figure 16.5. In order to support a distributed manufacturing environment, the system is of a three-tiered network hierarchy.

The client tier is effectively a GUI, consisting of a set of applications and a Web browser to enable interactions between users and the system. The main functions of the client tier are to analyse the necessary interactions between users and the entire system as well as to provide an effective way to realise the interactions using existing technologies.

The business logic tier is the core of the proposed system; it acts as a CAPP server. The main functions in this tier are similar to those of a traditional CAPP system. Common CAPP tasks such as feature recognition, designation of machine tool/ cutting tool and operation optimisations are carried out on the CAPP server.

The data tier supports the CAPP server. It represents generic or local shop-floor information pertaining to process planning. By switching between different data sources at the data tier, the CAPP system can better support collaborative manufacturing.

(1) Client Tier – User Interface
The client tier directly interacts with users. As a STEP-compliant system, there are a number of different modules that are needed to provide the required functions in the client tier. They are,

- a user interface that can view process plans through a specific query statement;
- a STEP physical file interpreter that interprets the STEP and STEP-NC structures;
- a GUI that displays 3D models based on the geometrical information in a STEP file;

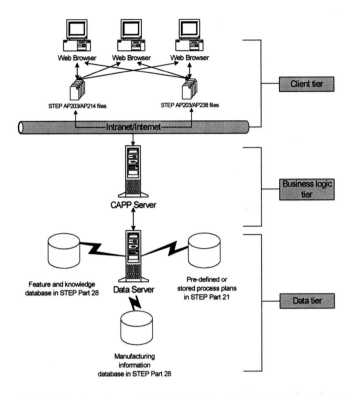

Figure 16.5. Abstract system model of the STEP-compliant manufacturing system

- a module that presents manufacturing plans in STEP-NC terms, *i.e.* Workplans and Workingsteps;
- a module that presents, and allows the user to modify, manufacturing information such as features, machine tools, cutting tools and tolerances in a Workingstep;
- a module that allows users to alter the sequence of Workingsteps and/or Workplans; and
- an XML interpreter that can interpret both the generic manufacturing information from the CAPP server and native manufacturing information from a database in XML format.

The client tier starts with listing the existing process plans in a process plan database through a query statement. When a specific process plan is chosen, Workplans and the solid models in the plan can be presented to the user via GUI. At this stage, the XML interpreter provides the interpreted XML manufacturing information to the clients. XML DTD, XSLT and XML Schema and/or the controls for keeping the XML data are retrieved from the manufacturing databases in a desired manner. In doing so, the manufacturing information within the XML data can be easily used to modify the current process plan. In response to the requirements from the client tier, the most suitable framework in which these modules can be implemented is a Web browser.

(2) *Business Logic Tier – CAPP Server*

Two different types of business logic tiers are represented as shown in Figures 16.6 and 16.7. The difference between them is the way in which the Workingstep optimiser works with the process planner. In Model I, the CAPP server is able to access different native manufacturing resources to generate different native process plans, hence an "integrated" scenario. Model II on the other hand, generates the generic and native process plans in tandem. The latter is likely to be generated at the shop floor. Therefore, it supports an "interfacing" scenario.

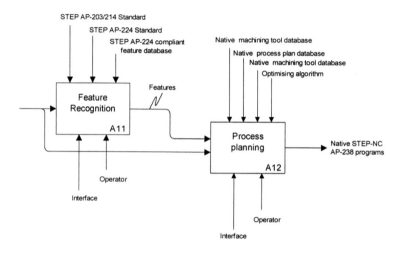

Figure 16.6. CAPP server Model I (Integrated model)

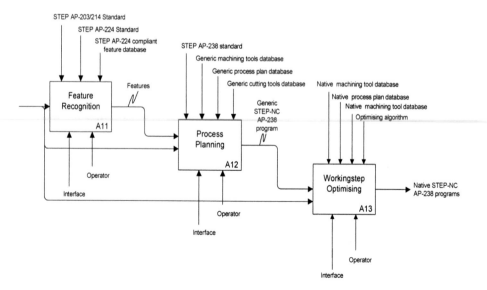

Figure 16.7. CAPP server Model II (Interfacing model)

In both models, feature recognition is the first task. The inputs to this module are data files conforming to ISO 10303 AP-203 or AP-214. The controls include ISO 10303 AP-203, AP-214, AP-224 and a feature database compatible with AP-224. AP-203 and AP-214 are used to describe the pure geometrical information of a part, whereas AP-224 is used to describe machining features in a process plan. The goal of using AP-224 as a control here is to provide a universal and STEP-compliant machining feature library in place of different proprietary feature libraries from different system vendors. AP-224 is also used to define STEP-NC machining features. The feature recognition module may have two different modes, automatic and manual.

The main function of the process planning module in Model I is to assign manufacturing resources to the features generated by the feature recogniser. The controls of the process-planning module include ISO 10303 AP-238 (STEP-NC AIM), Workingstep optimising algorithms and native manufacturing resources databases. The native manufacturing resources databases conform to ISO 14649 Part 111 (tools for milling) and Part 121 (tools for turning). As the manufacturing information is stored in STEP Part 28 (Edition 1) XML format, the connection between the CAPP server and the native resources databases is via the Internet. If the native shop-floor manufacturing resources are connected, the process planning module can directly assign specific manufacturing resources such as machine tools and cutting tools to each feature for creation of Workingsteps. Workingsteps are optimised and properly sequenced to generate a process plan that can be executed immediately at the shop floor.

In Model II, the Workingstep optimisation mechanism is separated from process planning and forms a new module. This may be due to the fact that the native manufacturing resources are still pending. In this case, the outputs of the process-planning module are generic process plans. They will be "populated" by a "native STEP-NC adapter" with the information from an identified native manufacturing resource to give a native process plan. Essentially, generic process planning is a process of "enriching" the machining features, represented as the AP-224 format in this case, with the necessary syntax information to form entities defined by STEP-NC, *e.g.* Workplans and Workingsteps. Some preliminary decisions such as Workingstep ordering and setup planning will be mainly based on the feature information and the information from a generic manufacturing database. Note that whatever decision is reached at this stage, changes can be easily made once the native manufacturing information becomes available. Many information slots in an STEP-NC file will remain empty or carry default values at this stage. This is intended by the standard, that is STEP-NC has the ability to just model what-to-do information. A detailed description about the above-mentioned native STEP-NC adapter is given in the next section.

To recap, both Models I and II seem to perform similar functions. There is, however, a fundamental difference between them. In Model I, the output from the CAPP server is a specific, "how-to-do" process plan, which can be directly used by a specific manufacturing facility. This "how-to-do" information may not be used by other manufacturing facilities. In Model II, there is an intermediate result from the process-planning module, *i.e.* a generic process plan. This generic process plan contains "what-to-do" instead of "how-to-do" information. It is therefore machine tool independent. The "what-to-do" information maintains its generic nature until the last moment when the CAM system of the chosen machine tool populates it with the native manufacturing information so as to generate a specific (how-to-do) process plan. Therefore, Model II possesses the required flexibility and portability to support collaborative and distributed manufacturing.

(3) *Data Tier – Data Model*

The databases in the data tier are constructed by applying the Part 28 (Edition 1) rules to the EXPRESS schemas. For example, the feature database is constructed by applying the Part 28 rules to ISO 10303 AP-224 schemas and the cutting tool database is constructed by applying the Part 28 rules to ISO 14649 Part 111 and Part 121 schemas. The following XML codes from the cutting tool database define a centre drill, which can be displayed in a Web browser (Figure 16.8).

```
<STEP-XML xmlns:ceb="urn:iso10303-28:ceb">
<cutting_tool ceb:id="66" ceb:copies="4">
  <id>CENTER_DRILL_5MM</id>
  <its_tool_body>
     <center_drill ceb:id="65" ceb:copies="4">
        <dimension>
           <tool_dimension ceb:id="59" ceb:copies="4">
              <diameter>5.000000</diameter>
              <tool_top_angle>0.000000</tool_top_angle>
              <tipcutting_edge_length>0.000000
                 </tipcutting_edge_length>
              <edge_radius>0.000000</edge_radius>
              <edge_center_vertical>0.000000
                 </edge_center_vertical>
              <edge_center_horizontal>0.000000
                 </edge_center_horizontal>
           </tool_dimension>
        </dimension>
        <number_of_teeth>2</number_of_teeth>
        <hand_of_cut>
           <hand>right</hand>
        </hand_of_cut>
     </center_drill>
  </its_tool_body>
  <overall_assembly_length>50.000000
     </overall_assembly_length>
  <angle_for_spindle_orientation>0.000000
     </angle_for_spindle_orientation>
  <tool_holder_diameter_for_spindle_orientation>0.000000
     </tool_holder_diameter_for_spindle_orientation>
</cutting_tool>
</STEP-XML>
```

The RDBMS (Relational Database Management System) is used in the data tier. In order to keep the original structure within an XML document, XML documents are stored as a whole in the RDBMS or as an external file outside the RDBMS. Once such a database is constructed, the information required by the CAPP server can be carried by the XML documents and transferred via the Internet. The XML documents are readily viewable in Web browsers and/or interpreted by a STEP-XML interpreter in the CAPP server to obtain specific manufacturing information.

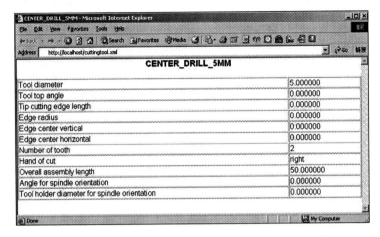

Figure 16.8. Cutting tool information in XML format shown in a Web browser

16.5.2 Native STEP-NC Adaptor and Native CNC Databases

An important part of the CAPP server Model II is the STEP-NC Adapter. The STEP-NC Adapter works with a native CNC system database and a Human–Machine (HM) Interface (Figure 16.9). The CNC database stores the detailed information about machine tools, cutting tools, setup, *etc*, that the adapter uses to "populate" the generic information (stored in a generic STEP-NC program) so as to generate a new NC program specific to, or customised for, the targeted NC machine tool. The HM interface can display the details of "what-to-do" information stored in the program.

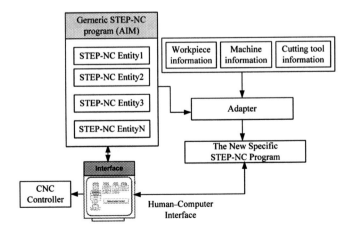

Figure 16.9. Logical structure of the native STEP-NC adaptor

CNC machining centres are vendor specific and vary in their hardware configuration and control software. Numerous data are preloaded onto a machine tool during production to cater for different machine functions. In order for the adapter to work with different machine tools, it is essential to have a database structure that is capable of modelling the native information of different machine tools. Based on this database, the adapter can be developed with ease. When a new machine tool arrives, the only task would be to store the information of the new machine tool into the native database. The adapter, which has been developed to work with the database of a fixed structure, does not need to be altered. The remaining part of this section describes the information that is necessary to construct the database.

There are two categories of information about a machining centre that need to be modelled in the database, (1) mechanical components of a machining centre, *i.e.* the physical configuration of the machining centre, and (2) the activities of the machining centre that may be controlled, and the data that is needed to control it, *i.e.* control and data components of a machining centre.

(1) *Physical Configuration of a Machining Centre*

Five different types of information have been considered at the present time, these are: Workspace, Travel, Linear axes, Tool magazines, Tool handling systems and Accessories.

The machine Workspace is defined as a bounded plane or volume in which the tool and workpiece can be positioned and through which controlled motion can be invoked. It is one of the basic pieces of information for defining the maximum size of a workpiece. Travel designates a maximum distance that the cutting tool(s) can reach in the X, Y and Z directions, respectively. Together with the workpiece, it is one of the basic pieces of information for defining the maximum machining size of a workpiece. Linear axes, which are often mutually orthogonal, denote the relative linear motion of the tool (and/or workpiece) provided by the machine's independent mechanisms. Information about all cutting tools is stored in the Tool magazines and Tool handling system. Accessories of a machining centre contain all the hardware that is necessary for a machine to be robust, efficient, reliable and accurate. Some examples are, the pallet loading system, tailstock in a lathe or fixtures for a milling centre.

(2) *Control and Data Components of a Machining Centre*

The control and data components of a machining centre contain information about the controller, feed rate, maximum spindle speed, coordinate system and tool change parameters. The coordinate system defines the base of a machining centre. When an NC instruction is generated, the geometry of the workpiece must be transformed into this coordinate system. The tool change parameters include the tool-changing times and tool-changing sequences.

16.5.3 System Development

In the interest of space, only two types of development work are discussed. The objectives are to (a) enable a process planner to view and manually edit the existing process plans/STEP-NC programs in STEP AP-238 Part 21 file format using the manufacturing resources provided in STEP Part 28 XML format, and (b) enable access to, and modification of, manufacturing databases across the Internet.

(1) *Client Tier Implementation*
The prototype of the client tier has been developed and implemented under the Microsoft Windows® environment. All client applications are unified within Microsoft Internet Explorer® 6.0. A set of development tools and technologies are used,

- Microsoft Visual C++® 6.0 and Microsoft Foundation Classes (MFC®);
- ST-Developer® [16.43] and STIX® [16.43];
- OpenGL® (Open Graphics Library);
- ActiveX® technology.

ST-Developer® is a software development package for developing and working with STEP applications. It offers libraries for reading, writing, processing and checking STEP data of Part 21 formats. It also provides EXPRESS early binding with C++ classes and Java classes. These features help to develop additional STEP-compatible applications. Applications written in Visual C++® can offer functions to read and write STEP Part 21 files, as well as create, delete, traverse and change any EXPRESS defined datasets compiled as objects in C++ style. STIX® is a STEP IndeX library for STEP AP-238 from the same company. It contains a C++ library that provides useful functions to process manufacturing data in STEP AP-238 format. Therefore, STIX® simplifies implementation and processing of STEP AP-238 information in programs written in Microsoft Visual C++®.

Figure 16.10 shows the client user interface. A STEP AP-238 file in STEP Part 21 format is represented in an ActiveX control. The left frame is a tree structure listing all the Workplans and Workingsteps in the STEP-NC file, whereas the right frame represents the geometrical model of the finished part. Client users can modify the information such as cutting tools, tolerance and manufacturing technologies.

(2) *Business Logic Tier Implementation*
In the business logic tier, a Web server has been constructed. It utilises the Internet Information Server (IIS®), Visual Basic® IIS application and Active Server Pages (ASP®) to generate dynamic Web pages for different client users. The Web server separates users into two groups: process planners and database administrators.

A process planner can access a list of existing process plans generated by the Web server. The process planner can then choose the desired process plan to modify. Each hyperlink leads to one existing process plan represented in an AP-238 Part 21 physical file and stored in the data tier. Once a process plan is chosen, the process plan file will be downloaded into the ActiveX control and represented in a Web browser for modifications. New process plans can be uploaded back to the data server after modifications.

For database administrators, the tier functions differently. The STEP Part 28 XML documents that contain manufacturing information are parsed by the Web server before transmission. Hence, database administrators can focus on the manufacturing information in the XML file. The generated dynamic Web page for database administration is illustrated in Figure 16.11.

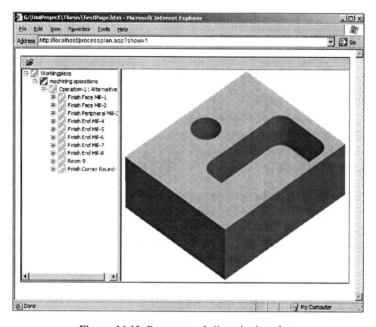

Figure 16.10. Prototype of client tier interface

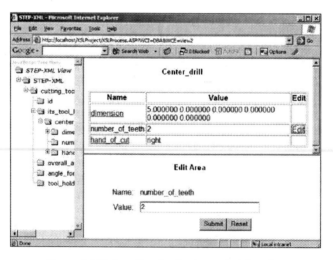

Figure 16.11. Interface for database administration

The left frame in the interface is a tree menu structure presenting the structure of an XML document. The upper-right frame represents the extracted manufacturing information. The lower-right frame is the editing area, in which, the "Name" section represents tags in an XML document, and the "Value" section refers to the detailed manufacturing information in the XML tags. Once the modifications are submitted, the Web server generates a new XML document based on the new values. The corresponding values in the manufacturing databases are also updated to reflect the changes.

(3) *Data Tier Implementation*

The data tier is currently implemented using Microsoft Access 2000®. The manufacturing information is stored as XML files in the operating system on which the data tier is implemented. The complete file paths of those XML files are stored. Some important attributes such as tool diameter, tool length and tool name by which the manufacturing information can be identified are also extracted and stored in the tables to enable more specific query statements. The main benefit of such an implementation method is that the database is easy to construct, and both the original XML structure and the flexibility provided by SQL can still be preserved. For example, a simple query statement, *"Select filepath from drilltools where tooldiameter = 5.0 and overallasslength = 50.0"*, will return the file path for the existing drilling tools with a diameter 5.0 and overall assembly length 50.0. The CAPP server can then extract the detailed manufacturing information from the XML documents according to the file paths.

(4) *STEP-NC Adapter*

The STEP-NC Adapter relates the process plan in the STEP-NC format to the potential manufacturing facilities stored in the database. In other words, the adapter tries to "interpret" the STEP-NC entities in terms of the native machining functions present in the database (Figure 16.12). Whether or not a complete interpretation is possible or the given process plan is achievable, also gives rise to the suitability of a specific machine tool.

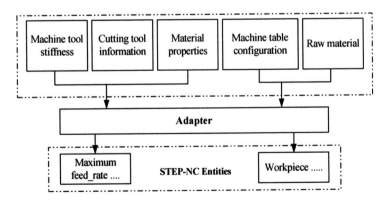

Figure 16.12. Adapter working diagram

In the adapter, the decomposed generic STEP-NC program is mapped to the information stored in the database (Figure 16.13). The outcome is a collection of Workingsteps organised under one or more Workplans for one or more machine tools present in the database. Failure to complete such a mapping process renders a signal indicating an "un-runable" STEP-NC program using the manufacturing facilities selected from the database.

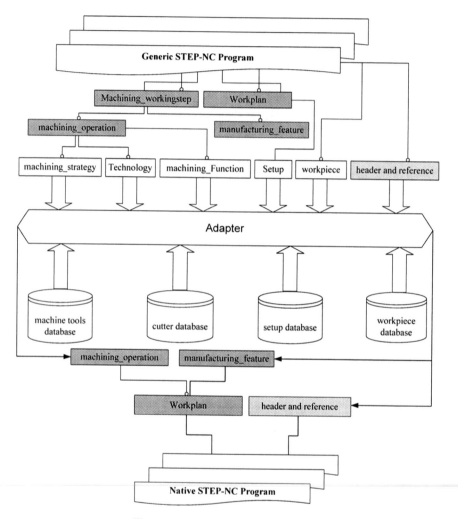

Figure 16.13. Structure of the *Adapter*

(5) *Native CNC Databases*

The Database, built using Microsoft Access®, has an explicit and simple data structure, making it easier for a machine operator to use. To illustrate the structure and use of the database, a CNC lathe has been used.

Machine resource

There are two types of manufacturing information stored in this database; machine tool and machine parameters (Figure 16.14). The manufacturing information in the database has been rearranged into four groups, for which an HM interface has been developed. This interface allows users to retrieve, edit and create database entities.

Cutting tool database
ISO/DIS 14649 Part 121 [16.16] has been followed to define the cutter parameters for the lathe cutting tool database. The tool database provides the information for tool setup in an NC program. Therefore, information such as tool type, tool geometry and tool expected life is also included.

Setup resource database
The setup resource database includes the information related to the machine tool, operator, workpiece and setup coordinate system. Similar to the previous two databases, it is possible to load and modify an existing setup file to suit a new setup.

Material (workpiece) resource
In this database, the workpiece's configuration, its parameters and the material properties are defined. Once a workpiece is defined or loaded, it can be displayed in a GUI.

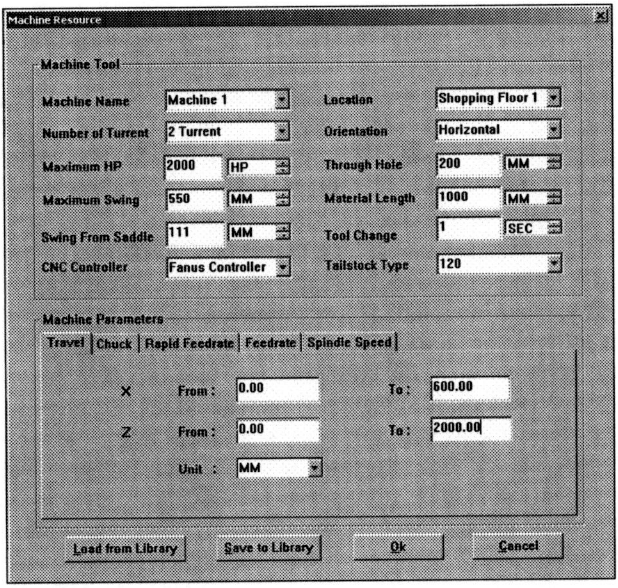

Figure 16.14. Interface for machine resource database

16.6 Conclusions

First published in 1994, the initial scope of STEP covered the geometry, product structure and technical drawings. Since then, the standard has been extensively implemented on major commercial projects in the aerospace, automotive and other industries. Typical implementations are the use of STEP to combine the information on the shape and other characteristics of individual parts with assembly structures to form a single integrated representation of a complex assembly or product. This information is gathered from a range of application systems and consolidated into a STEP file that can be transferred to other companies and unloaded into their corresponding systems. The advantage from combining this data is that it guarantees consistency for information deliveries, and avoids the administrative cost of ensuring that data is consistent between multiple systems.

Now STEP is on the verge of a new success with the release of a specification for defining the data input to CNC controllers – STEP-NC. Currently, these controllers are driven by vector codes developed in the 1960s. STEP-NC provides a unique NC program format for CAPP, CAM and NC, avoids postprocessing and entails a truly exchangeable format. Operators can now be supported at the machine tool level by complete information containing understandable geometry (machining features), task-oriented operations (Workingsteps and Workplans), strategies and tool definitions. CNC machines implementing STEP-NC can have a more open and adaptable architecture, making it easier to integrate with other manufacturing facilities such as workpiece-handling devices. The relatively high level of funding over a number of STEP-NC related projects during a short span of time, sufficiently demonstrated the importance of the STEP-NC related development work. Participation of, and collaboration among, a wide variety of organisations such as end users, academic and research institutions, and manufacturers of CAM systems, controls and machine tools, echoes the significance and relevance of this work, in particular from the industry perspective. STEP-NC comes in two "forms", Application Requirements Model (ISO 14649) and Application Interpreted Model (ISO 1030 AP-238). AP-238 provides an information view of the data, whereas ISO 14649 provides a functional view of the data.

STEP-NC can also support a distributed and collaborative manufacturing scenario. This is demonstrated by the case study. The system in the case study supports a bidirectional information flow throughout the design and manufacturing chain. Design information in its entirety is available in the manufacturing model. Manufacturing information is feature-based and task-oriented. In a collaborative manufacturing environment, designers and manufacturers are often geographically dispersed. Therefore, the framework adopts a three-tiered, Web-based network architecture to provide an open structure for the system. This architecture provides convenient ways in exchanging design and manufacturing data in STEP Part 21 and/or Part 28 file format through the Internet. The client user interface is implemented within a Web browser so that the implementation and maintenance costs can be reduced. Manufacturing information databases implemented in STEP Part 28 XML format enabled the CAPP server to switch between geographically dispersed shop-floor resources through Internet connections to realise collaborative manufacturing.

The STEP-NC adapter plays an essential role in the system as it is responsible for "populating" the generic manufacturing information (what-to-do) to arrive at native manufacturing information (how-to-do). Another key element of the system is the native manufacturing databases. The fact that the information about "what-to-do" and "how-to-do" is separated makes the manufacturing information "mobile".

This feature enables agile, adaptable and distributed manufacturing. The system has a user-friendly interface through which one can easily amend and update a process plan.

It is the author's belief that with the persisting development and implementation of STEP-NC, what is happening may not be simply the reshaping of CNC. It may be the reshaping of manufacturing. In the vision that is now emerging, the CNC machine tool will be playing a more important role than ever in the distributed manufacturing scenario.

References

[16.1] Li, L., Fuh, J.Y.H., Zhang, Y.F. and Nee, A.Y.C., 2005, "Application of genetic algorithm to computer-aided process planning in distributed manufacturing environments," *Robotics and Computer-Integrated Manufacturing*, **21**(6), pp. 568–578.

[16.2] Lutz, P. and Sperling, W., 1997, "OSACA - The vendor neutral control architecture," In *Proceedings of the European Conference on Integration in Manufacturing IiM'97*, Dresden, (Fichtner, D. *et al*. eds.), Selbstverlag der TU Dresden.

[16.3] "Open Modular Architecture Controls: OMAC-HMI, OSACA, JOP-Standard CNC Data Type Analysis," http://www.omac.org/wgs/MachTool/HMI-API/standards_compare.pdf.

[16.4] http://www.mazak.jp/english/products/cnc/index.html

[16.5] http://www.siemens.de/jobshop

[16.6] http://www.gibbscam.com/

[16.7] ISO 10303-1, 1994, "Part 1: Overview and fundamental principles," *Industrial Automation Systems and Integration – Product Data Representation and Exchange*.

[16.8] ISO 10303-203, 1994, "Part 203: Application protocol: configuration controlled 3D designs of mechanical parts and assemblies," *Industrial Automation Systems and Integration – Product Data Representation and Exchange*.

[16.9] ISO 10303-214, 1994, "Part 214: Application protocol: core data for automotive mechanical design processes," *Industrial Automation Systems and Integration – Product Data Representation and Exchange*.

[16.10] ISO 6983-1, 1982, "Part 1: Data format for positioning, line motion and contouring control systems," *Numerical Control of Machines – Program Format and Definition of Address Words*.

[16.11] ISO 14649-1, 2003, *Data Model for Computerized Numerical Controllers: Part 1 – Overview and Fundamental Principles*.

[16.12] ISO 14649-10, 2003, *Data Model for Computerized Numerical Controllers: Part 10 – General Process Data*.

[16.13] ISO 14649-11, 2003, *Data Model for Computerized Numerical Controllers: Part 11 – Process Data for Milling*.

[16.14] ISO 14649-111, 2001, *Data Model for Computerized Numerical Controllers: Part 111 – Tools for Milling*.

[16.15] ISO/DIS 14649-12, 2003, Data model for Computerized Numerical Controllers: Part 12 – Process data for turning.

[16.16] ISO/DIS 14649-121, 2003, *Data Model for Computerized Numerical Controllers: Part 12 – Tools for Turning*.

[16.17] ISO/DIS 10303-238, 2003, "Part 238: Application protocols: application interpreted model for computerized numerical controllers," *Industrial Automation Systems and Integration – Product Data Representation and Exchange*.

[16.18] Wolf, J., 2003, "Requirements in NC machining and use cases for STEP-NC-Analysis of ISO 14649 (ARM) and AP 238 (AIM)," *White Paper*, ISO T24 STEP-Manufacturing Meeting, San Diego, USA, March 2003.

[16.19] Feeney, A.B., Kramer, T., Proctor, F., Hardwick, M. and Loffredo, D., 2003, "STEP-NC implementation – ARM or AIM?" *White Paper*, ISO T24 STEP-Manufacturing Meeting, San Diego, USA, March 2003.

[16.20] Xu, X., Wang, H., Mao, J., Newman, S.T., Kramer, T.R., Proctor, F.M. and Michaloski, J.L., 2005, "STEP-compliant NC research: the search for intelligent CAD/CAPP/CAM/CNC integration," *International Journal of Production Research*, **43**(17), pp. 3703–3743.

[16.21] Newman, S.T., 2004, "Integrated CAD/CAM/CNC manufacture for the 21st century," *Keynote Speech*, In *Proceedings of the 14th International Conference on Flexible Automation and Intelligent Manufacturing (FAIM2004)*, July 12–14, 2004, Toronto, Canada.

[16.22] Allen, R.D., Newman, S.T., Harding, J.A. and Rosso, Jr. R.S.U., 2003, "The design of a STEP-NC compliant agent based CAD/CAM system," In *Proceedings of the 13th International Conference on Flexible Automation and Intelligent Manufacturing (FAIM2003)*, Tampa, Florida, USA, pp. 530–540.

[16.23] ISO 13030-224, 2001, "Part 224: Application protocol: mechanical product definition for process plans using machining features," *Industrial Automation Systems and Integration – Product Data Representation and Exchange*.

[16.24] Xu, X. and He, Q., 2004, "Striving for a total integration of CAD, CAPP, CAM and CNC," *Robotics and Computer Integrated Manufacturing*, **20**, pp. 101–109.

[16.25] OMAC STEP-NC Working Group, 2002, "The value proposition for STEP-NC," *OMAC Users Group, Draft Version 4*.

[16.26] ISO 10303-11, 1994, "Part 11: Description methods: the EXPRESS language reference manual," *Industrial Automation Systems and Integration – Product Data Representation and Exchange*.

[16.27] ISO 10303-21, 2002, "Part 21: Implementation methods: Clear text encoding of the exchange structure," *Industrial Automation Systems and Integration – Product Data Representation and Exchange*.

[16.28] ISO 10303-22, 1998, "Part 22: Implementation methods: standard data access interface," *Industrial Automation Systems and Integration – Product Data Representation and Exchange*.

[16.29] ISO 10303-23, 2000, "Part 23: C++ language binding to the standard data access interface," *Industrial Automation Systems and Integration – Product Data Representation and Exchange*.

[16.30] ISO 10303-24, 2001, "Part 24: C language binding of standard data access interface," *Industrial Automation Systems and Integration – Product Data Representation and Exchange*.

[16.31] ISO 10303-27, 2000, "Part 27: Java programming language binding to the standard data access interface with Internet/Intranet extensions," *Industrial Automation Systems and Integration – Product Data Representation and Exchange*.

[16.32] ISO/CD TS 10303-28 (Edition 1), 2002, "Product data representation and exchange: Implementation methods: EXPRESS to XML binding," *Draft Technical Specification, ISO TC184/SC4/WG11 N169*, 2002–02–14.

[16.33] ISO/TS 10303-28 (Edition 2), 2004, ISO ISO/WD 10303-28 (Edition 2), "Product data representation and exchange: Implementation methods: XML schema governed representation of EXPRESS schema governed data," *TC184/SC4/WG11 N223*, 2004–02–17.

[16.34] Harold, E.R., 2002, *XML Bible Gold Edition*, Hungry Minds. Inc., New York, NY, USA, ISBN: 0-7645-4819-0.

[16.35] Wosnik, M., Kramer, C., Selig, A. and Klemm, P., 2006, "Enabling feedback of process data by use of STEP-NC," *International Journal of Computer-Integrated Manufacturing,* **19**(6), pp. 559–569.

[16.36] Fichtner, D., Nestler, A., Dang, T.N., Schulze, A., Carlsen, U., Schreiber, S. and Lee, S.W., 2006, "Use of agents and neural networks for acquisition and preparation of distributed NC information to support NC planning," *International Journal of Computer-Integrated Manufacturing,* **19**(6), pp. 581–592.

[16.37] Lee, W. and Bang, Y.B., 2003, "Design and implementation of an ISO14649-compliant CNC milling machine," *International Journal of Production Research,* **41**(3), pp. 3007–3017.

[16.38] Lee, W. and Bang, Y.B., 2003, "Development of STEP-NC milling based on XML," In *Proceedings of the 5th German-Korean Workshop on Manufacturing and control.*

[16.39] Lee, W., Bang, Y.B. and Kwon, W., 2002, "PC-NC STEP-NC milling using STEP-NC in XML form," In *Proceedings of the 3rd Seoul International IMS,* Seoul, Korea, February 2002.

[16.40] Lee, W., Bang, Y.B., Ryou, M.S., Kwon, W.H. and Jee, H.S., 2006, "Development of a PC-based milling machine operated by STEP-NC in XML format," *International Journal of Computer Integrated Manufacturing,* **19**(6), pp. 593–602.

[16.41] Mao, J., 2003, "A STEP-compliant collaborative product development system," *Master of Engineering Thesis,* Department of Mechanical Engineering, School of Engineering, The University of Auckland.

[16.42] Xu, X. and Mao, J., 2004, "A STEP-compliant collaborative product development system", In *Proceedings of the 33rd International Conference on Computers and Industrial Engineering,* Jeju, Korea, 25–27 March, 2004, CIE598.

[16.43] http://www.steptools.com

Index

Printed in the United States
97832LV00002BB/1/A

9 781846 287510